Semi-Markov Models and Applications

Semi-Markov Models and Applications

Edited by

Jacques Janssen
Université Libre de Bruxelles,
Belgium

and

Nikolaos Limnios
Université de Technologie de Compiègne,
France

KLUWER ACADEMIC PUBLISHERS
DORDRECHT / BOSTON / LONDON

A C.I.P. Catalogue record for this book is available from the Library of Congress.

ISBN 0-7923-5963-1

Published by Kluwer Academic Publishers,
P.O. Box 17, 3300 AA Dordrecht, The Netherlands.

Sold and distributed in North, Central and South America
by Kluwer Academic Publishers,
101 Philip Drive, Norwell, MA 02061, U.S.A.

In all other countries, sold and distributed
by Kluwer Academic Publishers,
P.O. Box 322, 3300 AH Dordrecht, The Netherlands.

Printed on acid-free paper

Printed in the Netherlands.

Contents

Part VI Controlled Processes & Maintenance

Part VII Chromatography & Fluid Mechanics

Contributors

Mohamed Abdel-Hameed
Department of Statistics, United Arab Emirates University, P.O. Box 1755, Al-Ain, U.A.E.
E-mail : abdelham@emirates.net.ae

Larisa G. Afanasyeva
Moscow State University, Oktyabrsky prospekt 2, kv 135, Moskovskaya obl., Troitsk, 142092, Russia.
E-mail : simplex@glasnet.ru

Vladimir V. Anisimov
Department of Applied Statistics, Faculty of Cybernetics, Kiev State University, Kiev, 22 252 022, Ukraine, and Bilkent University, Ankara, Turkey.
E-mail : vlanis@bilkent.edu.tr

Søren Asmussen
Department of Mathematical Statistics, Lund Unviversity, Box 118, 221 00 Lund, Sweden.
E-mail : asmus@maths.lth.se

Debasis Bhattacharya
University of California Davis, Division of Statistics, 380 Kerr Hall, Davis, CA 95 616, USA.
E-mail :

Tadashi Dohi
Department of Industrial and Systems Engineering, Hiroshima University, Higashi-Hiroshima 739, Japan.

E-mail : dohi@clalinet.gal.sys.hiroshima-u.ac.jp

M. Luz Gámiz-Pérez
Departamento de Estadística e Investigación Operativa, Universidad de Granada, Facultad de Ciencias, Campus de Fuentenueva, 18071, Granada, Spain.
E-mail : mgamiz@goliat.ugr.es

Ilya Gertsbakh
Ben Gurion University of the Negev, P.O. Box 653, Beer-sheva, 84 105, Israel.
E-mail : eliahu@indigo.cs.bgu.ac.il

Rudy Gideon
Department of Mathematics, University of Montana, Missoula, Montana, USA.
E-mail : gideon@selway.umt.edu

Mats Gyllenberg
Department of Applied Mathematics, University of Turku, Finland.
E-mail : mats.gyllenberg@utu.fi

Boris P. Harlamov
Mechanical Engineering Problem Institute, Russian Academy of Sciences, Av. Bolshoi (V.O.), 61, 199178 St Petersburg, Russia.
E-mail : harlamov@ensure.ipme.ru

Marius Iosifescu
Centre for Mathematical Statistics, Roumanian Academy of Science, Str. Dr. N. Manoloscu 9-11, 76222 Bucharest, Romania.
E-mail : miosifes@valhalla.racai.ro

Jacques Janssen
Université Libre de Bruxelles, Ecole de Commerce SOLVAY- CADEPS, 50 Avenue F.D. Roosevelt CP 194/7, B-1050 Bruxelles, Belgium.
E-mail : janssen@ulb.ac.be

Naoto Kaio
Department of Economic Informatics, Hiroshima Shudo University, 1-1-1, Ozukahigashi, Asaminami-ku, Hiroshima 731-3195, Japan.
E-mail : kaio@mb.infoweb.ne.jp

Vladimir S. Korolyuk
Ukainian Academy of Science, Institut of Mathematics, Repine str., 3, Kiev, 252 601, Ukraine.
E-mail : syta@imath.kiev.ua

Nikolaos Limnios
Division Mathématiques Appliquées, Université de Technologie de Compiègne, B.P. 529, 60205 Compiègne Cedex, France.
E-mail : nikolaos.limnios@utc.fr

Raymondo Manca
Dipartimento di Scienze, Università di Chieti, Viale Pindaro, 42, 65127 Pescara, Italy.
E-mail : Rmanca@sci.unich.it

Sally I. McCLean
Division of Mathematics, School of Information and Software Engineering, University of Ulster, Coleraine BT52 1SA, U.K.
E-mail : si.mcclean@ulst.ac.uk

Shunji Osaki
Department of Industrial and Systems Engineering, Hiroshima University, Higashi-Hiroshima 739, Japan.
E-mail : osaki@gal.sys.hiroshima-u.ac.jp

Brahim Ouhbi
Ecole Nationale Supérieure d'Arts et Métiers, Marjane II, Meknès Ismailia, B. P. 4024 Beni M'Hamed, Meknès, Morocco.
E-mail : ouhbib@hotmail.com

Aleka Papadopoulou
Statistics and Operation Research, Department of Mathematics, Aristotle University of Thessaloniki, 540 06 Thessaloniki, Greece.
E-mail : aleka@cso.auth.gr

Rafael Pérez Ocón
Departamento de Estadística e Investigación Operativa, Universidad de Granada, Facultad de Ciencias, Campus de Fuentenueva, 18071, Granada, Spain.
E-mail : rperezo@goliat.ugr.es

Susan M. Pitts
Statistical Laboratory, 16 Mill Lane, Cambridge CB2 1SB, U.K.
E-mail : S.Pitts@statslab.cam.ac.uk

Ronald Pyke
University of Washington, Department of Mathematics, Box 354350, Seattle, WA 98195-4350, USA.
E-mail : pyke@math.washington.edu

Peter Radchenko
Moscow State University, Oktyabrsky prospekt 2, kv 135, Moskovskaya obl., Troitsk, 142092, Russia.
E-mail : simplex@glasnet.ru

George G. Roussas
University of California Davis, Division of Statistics, 380 Kerr Hall, Davis, CA 95 616, USA.
E-mail : roussas@wald.ucdavis.edu

Juan Eloy Ruiz-Castro
Departamento de Estadística e Investigación Operativa, Universidad de Granada, Facultad de Ciencias, Campus de Fuentenueva, 18071, Granada, Spain.
E-mail : jeloy@goliat.ugr.es

Dmitrii S. Silvestrov
Institute of Mathematical Statistics, Umea University, S-90 187 Umea, Sweden, and Department of Applied Mathematics, University of Turku, FIN-20014 Turku, Finland.
E-mail : dmitrii@matstat.umu.se

I. Spungin
Ben Gurion University of the Negev, P.O. Box 653, Beer-sheva, 84 105, Israel.
E-mail : josef@indigo.cs.bgu.ac.il

Valeri T. Stefanov
The University of Western Australia, Department of Mathematics, Nedlands, WA 6907, Australia.
E-mail : stefanov@maths.uwa.edu.au

Anatoly Swishchuk
International Mathematical Center, Institute of Mathematics, National Academy of Sciences of Ukraine, Rue de Repine, 3, Kiev, 252 601, Ukraine.
E-mail : swan@imath.kiev.ua

Helena Tsakiridou
Statistics and Operation Research, Department of Mathematics, Aristotle University of Thessaloniki, 540 06 Thessaloniki, Greece.
E-mail : heltsaki@auth.gr

George M. Tsaklidis
Statistics and Operation Research, Department of Mathematics, Aristotle University of Thessaloniki, 540 06 Thessaloniki, Greece.
E-mail : tsaklidi@ccf.auth.gr

Panagiotis C. G. Vassiliou
Statistics and Operation Research, Department of Mathematics, Aristotle University of Thessaloniki, 540 06 Thessaloniki, Greece.
E-mail : vasiliou@ccf.auth.gr

Ernesto Volpe
University La Sapienza, Via Castro Laurentano, 9, Roma, Italy.
E-mail : volpe@scec.eco.uniroma.it

Preface

This book presents a selection of papers presented to the Second International Symposium on Semi-Markov Models : Theory and Applications held in Compiègne (France) in December 1998.

This international meeting had the same aim as the first one held in Brussels in 1984 : to make, fourteen years later, the state of the art in the field of semi-Markov processes and their applications, bring together researchers in this field and also to stimulate fruitful discussions.

The set of the subjects of the papers presented in Compiègne has a lot of similarities with the preceding Symposium; this shows that the main fields of semi-Markov processes are now well established particularly for basic applications in Reliability and Maintenance, Biomedicine, Queueing, Control processes and production.

A growing field is the one of insurance and finance but this is not really a surprising fact as the problem of pricing derivative products represents now a crucial problem in economics and finance.

For example, stochastic models can be applied to financial and insurance models as we have to evaluate the uncertainty of the future market behavior in order, firstly, to propose different measures for important risks such as the interest risk, the risk of default or the risk of catastrophe and secondly, to describe how to act in order to optimize the situation in time. Recently, the concept of VaR (Value at Risk) was "discovered" in portfolio theory enlarging so the fundamental model of Markowitz.

It is interesting to note the strong similitudes with some concepts of reliability theory. Indeed, here too, one of the basic concepts is now the concept of random environment which can usefully be modeled under a semi-Markov approach.

It is now clear that semi-Markov processes play a more and more crucial role in Business and Industry; this is partially due to the fact that, with the powerful mathematical softwares now existing, the numerical treatment of basic integral equations is easier and so leads to more concrete applications; moreover the developments in topics like non-homogeneous models and statistical estimation give the possibility to construct more adequate models of real-life problems and to calibrate the basic parameters of the models from real data more accurately.

This book presents a lot of original models that are or could be really useful for applications in real life problems and we strongly hope that it will contribute to stimulate new interactions between the theoretical development and the applications of semi-Markov models.

This volume is divided in seven parts. The first three parts are dedicated to probabilistic and statistical models while the last four ones to applications to important fields such as queuing, traffic, finance, reliability and maintenance, chromatography and fluid mechanics.

Part I, *Extensions of basic models*, includes papers devoted to semi-Markov probabilistic models and methods as solidarity properties, backward and forward equations, quasi-stationary phenomena, random walks, diffusion approximations and approximations of semi-Markov transition function. Part II, *Statistical Estimation*, presents asymptotic results and exponential approximation, exponential families with limit theory, homogeneity test for two samples and product-type estimation of convolutions. Moreover, this part also develops applications in reliability, manpower and lifetime data analysis. Part III, *Non-homogeneous models*, gives limit results and perturbed models, while Part IV, *Queuing systems theory*, discusses queues with heavy tails and Poisson traffic. Part V, *Financial models*, deals with risk processes in random evolution and option pricing with an extension of the Black & Scholes formula. Part VI, *Controlled processes and maintenance*, presents an overview of optimal replacement models and controlled processes in reliability and maintenance and the last part, Part VII, *Chromatography and fluid mechanics*, introduces diffusion semi-Markov processes with applications in chromatography and finally proposes a model of a continuum in non-newtonian fluid mechanics.

So, this book should constitute a basic reference for all researchers and practitioners in the field of stochastic modeling.

This is of course due to the fact that the organizers were lucky enough to put together eminent specialists from all over the world who also accept to contribute to the present book. Many thanks to them for coming in Compiègne and for writing the papers !

We would also thank Professor N. Balakrishnan and Dr J. Martindale for their strong editorial help, C. Papadopoulos for his permanent help for the realization of this volume and also C. Prigent and S. Donovan for their very efficient organizational help.

The Editors,

Jacques Janssen,
Nikolaos Limnios.

I

EXTENSIONS OF BASIC MODELS

Chapter 1

THE SOLIDARITY OF MARKOV RENEWAL PROCESSES

Ronald Pyke
University of Washington, USA

Abstract It is known that for a wide class of irreducible recurrent Markov Chains and Markov Renewal processes, the finiteness of the p-th moment of recurrence times is finite for one state if and only if it is finite for all states. This result also holds for a wide class of functionals of the processes between successive occurrence times. Such results are known as solidarity theorems. In this paper, a solidarity theorem is proved concerning the property of "being in the domain of attraction of a stable law with specified parameters." A basic identity linking additive random functions determined by two distinct states is emphasized, and used to derive solidarity results for moments and for the WLLN.

Keywords: Markov renewal processes, semi-Markov processes, solidarity theorems, moments, Teugel's identity, domains of attraction, stable laws.

1. PROLOGUE

The purpose of this written contribution is to present some remarks and results about the solidarity of states in a recurrent class of a Markov renewal process (MRP). Before doing so, I wish to include some of the preliminary comments from the oral presentation made at the Second International Symposium on Semi-Markov Models of which this volume is the proceedings. The talk was in three parts; (1) Historical reflections, (2) Theoretical remarks and (3) Applied reminders. Here and in what follows, the terms semi-Markov process (S-MP) and (MRP) are used interchangeably when speaking generally of these models. Although their precise definitions may differentiate between them, they are even then essentially equivalent, requiring only a relabeling of states at most to make them equivalent. The terms possibly indicate different emphases with the one following the occupied state and the other following the cumulative counts of visits. However, the literature indicates more often than not that the terms are treated as synonyms.

In any event, historical reflections upon semi-Markov or Markov renewal processes must begin about forty-five years ago in 1954. In this year, three independent papers appeared and introduced this new type of stochastic process. Two of these papers were presented at the International Congress of Mathematicians held in Amsterdam in 1954, namely, Lévy (1954) and Smith (1954). In both of these presentations, the authors described a new class of stochastic processes that generalized both continuous and discrete parameter Markov chains with finite or countable state spaces. Both authors named the processes, semi-Markov processes. The third paper is Takács (1954) in which essentially the same class of processes are introduced. Takács did not name these processes, describing them only as recurrent processes "of a certain type".

Usually, when new classes of stochastic processes are introduced, it is because of important applications that can not be adequately modeled with existing processes. Indeed, the health of a theoretical discipline is closely related to the level of interaction between it and other disciplines. The programme of this Symposium, by virtue of the exceptional breadth and variety of applications reflected therein, is a good indicator of the healthy state of stochastic modelling today. Moreover, most of the applications presented here are of the best kind; ones in which the field of application enters realistically and centrally. In terms of the trichotomy of applied probability given in Pyke (1975), these are the 'hard-hat' applications in contrast with the 'top-hat' or 'bishop mitre' variety described there.

Ironically, when the concepts of an S-MP and MRP burst upon the process scene in 1954, two of the three papers were theoretically motivated. For Lévy, his study was in answer to a question of K. L. Chung about what would happen to the sample paths of Markov chains if the assumption of exponential holding times were dropped. For Smith, the idea of an S-MP was advanced as an illustration of his regenerative processes, though he does describe briefly a 2-state MRP related to a storage model in the full paper, Smith (1955). For Takács, his processes were motivated by problems involving coincidences in particle counting problems, though the emphasis is not on specific applications in his 1954 paper.

The literature on MRPs has grown dramatically since 1954. In my first paper, Pyke (1961a), I gave "a complete list of references" at that time; it numbered 18 and exactly half of these were either abstracts or unpublished preprints. The survey paper by Cinlar (1969), includes 30 papers about MRPs. Five years later, the bibliography Teugels (1974) includes about 540 references. Twelve years later the bibliography Teugels (1986) contains about 1000 references by over 800 authors. A further twelve years brings us to this Symposium which gives evidence that the growth continues. A check of review and citation databases gives further evidence. (As an aside, one needs care as always in choosing words and phrases for optimal database searches. An early search of a physical and chemical science database turned up nearly 200 recent papers with 'MRP' in the title. Most of these, however, were about 'multidrug resistant proteins', some about 'manufacturing resource planning', and only a few involved Markov renewal processes.)

To me, the most important feature of the literature on Markov renewal and semi-Markov processes is the enormous breadth of applications it contains and, consequently, the great diversity of journals in which this research appears.

Without giving complete definitions, the notation for these processes can be summarized as follows. Let

$$(Z,U) = \{(Z_t, U_t) : t \geq 0\} = \{(\text{State, Age}) \text{ at time } t \geq 0\}$$

be a Markov process on $\{0,1,2,\dots\} \times [0,\infty)$ with

$$P[Z_t^+ = j, V_t \leq x | (Z_s, U_s) : 0 \leq s \leq t] = Q_{Z_t,j}(U_t, x) \text{ a.s.}$$

for mass functions $\{Q_{ij}\}$ where

$$V_t = \inf\{v > 0 : Z_{t+v} \neq Z_t\}, \qquad Z_t^+ = Z_{t+V_t}.$$

When explosions (or instantaneous states) are not allowed, as would be the case in most modelling applications, the constructive definition for the resulting regular case would begin with a Markov process $(J,X) \equiv \{(J_n, X_n) : n \geq 0\}$ on $\{0,1,2,\dots\} \times [0,\infty)$ for which $X_0 = 0$, $P[J_0 = k] = a_k$ and

$$P[J_n = j, X_n \leq x | (J_m, X_m) : 0 \leq m < n, J_{n-1} = i] = Q_{ij}(x).$$

The random variables (rv) of interest are then defined as

$$\begin{aligned} S_n &= X_0 + \dots + X_n \\ N(t) &= \#\{n : S_n \leq t\}, & Z(t) &= J_{N(t)} - 1 \\ N_i(t) &= \#\{n : S_n \leq t, J_n = i\} \\ U(t) &= t - S_{N(t)-1}, & V(t) &= S_{N(t)} - t \\ T(t) &= X_{N(t)} = U(t) + V(t). \end{aligned}$$

In terms of these quantities one usually defines a semi-Markov process to be $\{Z(t) : t \geq 0\}$ and a Markov renewal process to be the vector process $\{(N_0(t), N_1(t), \ldots) : t \geq 0\}$ although generically the terms are used interchangeably.

Regardless of the emphasis one requires in a given application, one of the most useful properties of this family of processes is that it is preserved under restrictions upon the states. For example, if $\{i_1, i_2, \ldots\}$ is a subset of the states, then $\{(N_{i_1}(t), N_{i_2}(t), \ldots) : t \geq 0\}$ remains an MRP. (Recall that this preservation under state restrictions does not hold for Markov chains.) In particular, the restriction to a single state is a renewal process (RP) the simplest and most basic of the MRPs.

To make full use of their preservation under restrictions of the states, it is important to emphasize the concept of an MRP with auxiliary paths or excursions. If one views an MRP as a process that moves among a countable set of states with the times required to make these moves depending only upon the pairs of states between which the transitions are being made, it is natural to envisage auxiliary random quantities, possibly quite complex that may be attached to the transitions and be such that their probabilistic behaviors also depend only upon the transition states. It is important to 'think excursions' when working with MRPs. Many basic results for MRPs are consequences of the observation: An MRP is just an RP with excursions. That is, when an MRP starts in state i, it may be viewed as the RP associated with the recurrences of that state together with the inter-occurrence information of the original MRP attached as excursions. The solidarity results presented below in this paper utilize intrinsically an embedded 2-state MRP with the original process' information attached as excursions between the transitions of this restricted process.

At the Symposium, I attempted to emphasize the role of excursions by using the "picnic table" representation given in Figure 1.1. The three separate processes $U(\cdot)$, $V(\cdot)$ and $T(\cdot)$, representing, respectively, age, residual life and total life, are drawn together to form shapes suggestive of picnic tables. The widths of these 'tables' are the inter-occurrence times X_n in the regular case. On the top of these tables, to force the analogy further, are bowls containing the excursions. To represent an MRP itself, let the bowls contain simply the states J_n (or Z_t) as shown. However, if the tables are those of the RP formed by the renewals of a certain state, say i, then the excursions in the bowls would be the full sample paths of the MRP between successive visits to state i. We stress again that excursions may be quite general random quantities including trajectories of stochastic processes. Illustrative examples of this can be derived from Brownian motion by letting the MRP be the embedded sequence of visits of the Brownian motion to a countable set of states (e.g. $\{0, \pm 1, \pm 2, \ldots\}$ or $\{0, \pm 2^n$ for $n = 0, \pm 1, \pm 2, \ldots\}$) and the excursions being the trajectories of conditional Brownian motion between these visits. For the second choice of states, observe that the MRP will have explosions around the zero state. Early references about MRPs with excursions or auxiliary paths are Pyke and Schaufele (1966) and Schäl (1970).

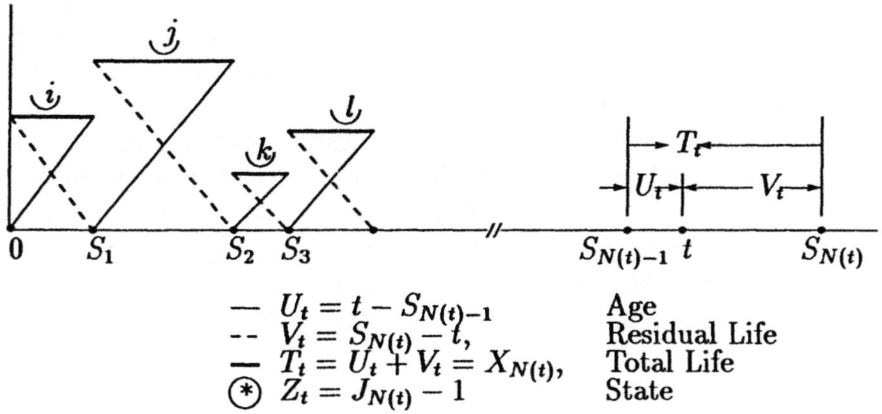

Figure 1.1 Picnic-table processes

2. INTRODUCTION

In the theories of Markov Chains and Markov Renewal processes (MRP's), a solidarity theorem is one that shows that all states in an irreducible class have a certain property if and only if one of the states does. Many of the early key results in the foundations of Markov Chains are in this sense solidarity theorems. For example, one state in a communicating class is recurrent or transient, null or ergodic, periodic or aperiodic if and only if all states are, with the periodicity of a state being constant over a class. To be more specific, let $\{J_n : n \geq 0\}$ be a recurrent irreducible Markov Chain and write m_{ij} for the expected number of transitions to go from state i to state j. Then already in 1936, Kolmogorov proved that $m_{ii} < \infty$ for a fixed i if and only if $m_{jk} < \infty$ for all choices of j and k.

In Chung (1954), it is shown that the *order* of a Markov Chain may be defined in terms of a single state, namely, as the infimum of those

$p > 0$ for which the p-th moment of the recurrence time of this fixed state is infinite. To justify this definition of order, Chung (1954) proves that the p-th moment for one state is finite if and only if it is finite for all states. In the same reference, Chung proves a corresponding result for moments of more general sums over Markov Chains.

Similar solidarity theorems for moments of recurrence times and more general sums are given for Markov Renewal processes in Pyke and Schaufele (1964). Then, starting with Cheong (1967) and Teugels (1968, 1970), results have been obtained regarding the tail behavior of the renewal functions of MRP's; see in particular the comments and references in Anderson and Athreya (1987) and Sgibnev (1996). These solidarity results show that the asymptotic behaviors of the renewal functions $M_{ij}(t)$ are the same for all states in an irreducible MRP. See in particular Section 3 of Teugels (1970).

These results on the tail behaviors of the renewal functions, $M_{ij}(t)$, are closely related to the central limit results of this paper in view of Mohan (1976). In that paper it is shown that for a renewal process with finite mean μ, the regular variation (of order, $0 < \theta < 1$, say) of the difference $M(t) - t/\mu$ is equivalent to the common inter-event distribution function, F, being in the domain of attraction of a Stable law of exponent $\alpha = 2 - \theta$.

The purpose of this paper is to derive solidarity results for the central limit theorem for linear functionals of Markov Chains and MRP's. For the classical central limit theorem in which the limit law is Normal and the sums are normalized by variances, such a solidarity result is an immediate consequence of the previously mentioned solidarity results for moments. Our focus here, therefore, is upon the remaining cases in which the summands are either in the domain of attraction of stable laws of exponents $\alpha \in (0, 2)$ or in the domain of non-normal attraction when $\alpha = 2$.

The general problem and notation is presented in Section 3. The main result on the domain of attraction solidarity is in Section 4. In Section 5, a constructive description is given for the structure underlying a basic identity of Teugels (1968), which is then applied to obtain solidarity results for the WLLN and finiteness of moments.

3. PRELIMINARIES

Solidarity results for an irreducible recurrent Markov Renewal Process (MRP) or Markov Chain (MC) are in fact results about 2-state MRP's. That is, if one is proving that when a property holds for one state i, it also holds for another state j, one uses the inherent coupling between these states embodied in the MRP with states 1 and 2, say, whose transitions into state 1 (2) occur precisely when the original process makes transitions into state i (j). Thus, in the fairly standard notation for MRP's, (see, e.g., Pyke and Schaufele (1964)) this 2-state embedded process has a Q-matrix defined in terms of quantities for the original process by $Q_{11} = {}_jG_{ii}, Q_{12} = {}_iG_{ij}$ and with Q_{22} and Q_{21} obtained by interchanging i and j.

Throughout this paper we assume that the MRP's are regular in the sense that only finitely many transitions to each state occur a.s. in each finite interval. (See e.g. Pyke and Schaufele (1964) for definitions and notation.) Of course, there may be infinitely many transitions in total in such intervals.

We first record an elementary probability calculation concerning the alternation of states. Write ${}_ip_{jk}$ for ${}_iG_{jk}(\infty)$, the taboo probability that state k is visited before i, starting in j. For each state i, let $\{r_n^{(i)}; n \geq 1\}$ be the successive occurrence times of state i and set $r_0^{(i)} = 0$. These times are finite and increase to $+\infty$ with probability one. Set

$$\nu_{ij}(n) = \min\{k \geq 0; r_{k+1}^{(j)} > r_{n+1}^{(i)}\} \text{ for } i \neq j \text{ and } n = 0, 1, \ldots.$$

Thus $\nu_{ij}(n)$ is the number of transitions, into state j before the $(n+1)$-th transition into state i. The distribution function (d.f.) of $\nu_{ij}(n)$ may be derived straightforwardly. In particular, when $J_0 = i$ and hence $r_1^{(i)} = 0$,

$$P[\nu_{ij}(n) = k \mid J_0 = i] = \sum_{m=0}^{\min(n,k)} \binom{n}{m} ({}_jp_{ii})^{n-m} ({}_ip_{ij})^m ({}_jp_{ji})^m \binom{k-1}{m-1} ({}_ip_{jj})^{k-m} \tag{1.1}$$

where $\binom{k-1}{-1} = \delta_{k0}$, the Kronecker delta. The proof of (1.1) is as follows: The differences $D_m = \nu_{ij}(m+1) - \nu_{ij}(m)$, $m \geq 1$, are independent and identically distributed with a common d.f. given by

$$P[D_1 = k] = \begin{cases} {}_jp_{ii} & \text{for } k = 0, \\ {}_ip_{ij}({}_ip_{jj}^{k-1})\,{}_jp_{ji} & \text{for } k = 1, 2, 3, \ldots. \end{cases} \tag{1.2}$$

The event $[\nu_{ij}(n) = k]$ may be interpreted as the occurrence of M_n positive values among $D_1, D_2, \ldots, D_n$, whose sum is k. Clearly M_n is a $B(n, {}_ip_{ij})$ random variable. Each of these M_n positive values of D can be interpreted as the number of "failures" (visiting state j before i) between "successes" (visiting i before j), leading to (1.2). Thus (1.1) follows by compounding the Binomial and the Geometric distributions in the appropriate way. Observe that (1.2) implies that D_1 has the structure of ϵY where ϵ is Bernoulli (π), Y is Geometric (p), and ϵ and Y are independent with $\pi = {}_jp_{ii}$ and $p = {}_jp_{ji}$. Since $\nu_{ij}(n) = D_1 + D_2 + \ldots + D_n$, we record for future reference that

$$E(\nu_{ij}(n) \mid J_0 = i) = nE(D_1) = n({}_ip_{ij}/{}_jp_{ji}) := n\gamma < \infty \tag{1.3}$$

and

$$\mathrm{Var}(\nu_{ij}(n) \mid J_0 = i) = n\mathrm{Var}(D_1) = n\frac{{}_ip_{ij}({}_jp_{ii} + + {}_ip_{jj})}{{}_jp_{ji}^2} := n\sigma^2 < \infty, \tag{1.4}$$

and note further that $\nu_{ij}(n)/n \xrightarrow{\text{a.s.}} \gamma$ and $[\nu_{ij}(n) - n\gamma]/\sqrt{n} \xrightarrow{\text{L}}$ a $N(0, \sigma^2)$ random variable. The notation π, p, γ and σ helps throughout to simplify expressions but their dependence upon i and j should be noted.

To define the inter-occurrence sums studied in this paper, we first introduce independent sequences of i.i.d. r.v.'s $\{V_{kln} : n \geq 1\}$ for k, l in the state space to represent the random summands associated with the MRP's transitions from k to l. Then define for a fixed state i the inter-occurrence sums

$$U_m^{(i)} = \sum_{k,l}\sum\{V_{kln} : n\text{-th occurrence of the transition from } k \text{ to } l \text{ occurs during } (r_m^{(i)}, r_{m+1}^{(i)}]\} \tag{1.5}$$

for $m = 1, 2, \ldots$. This generalizes the type of additive functionals for which limit theorems were obtained in Pyke and Schaufele (1964), in that the random measures associated with each transition may involve auxiliary randomness to that given by a deterministic function of the holding times between transitions. The only restriction upon the distributions of the V's is that the sums in (1.5) converge a.s.

The purpose of this note is to prove that the same limit laws hold for normed sums of $U_n^{(i)}, n \geq 1$, as for sums of $U_n^{(j)}, n \geq 1$, for every pair of states i, j. This extends our knowledge of the solidarity of states in a recurrent class as discussed above. For example, it is well known that for a wide class of r.v.'s $U_n^{(i)}$, a recurrent class possesses solidarity with respect to limit theorems that are characterized by the existence of moments such as the strong law of large numbers and the classical Central Limit Theorem.

For any state i, set $S_n^{(i)} = \sum_{m=1}^n U_m^{(i)}$ for $n \geq 1$ and $S_0^{(i)} = S_{-1}^{(i)} = 0$. The sum of the V-contributions represented by $S_n^{(i)}$ should not differ significantly from the sum of those V-contributions that accrue between the first and last visit to state j following the first and before the n-th visit to i. Let

$$\Delta_{nij} = S_n^{(i)} - S_{\nu_{ij}(n)-1}^{(j)}, \qquad n = 1, 2, \ldots \tag{1.6}$$

denote this difference. We continue to assume that $r_1^{(i)} = 0$ (or $J_0 = i$). Observe first that Δ_{nij} is equal to $S_n^{(i)}$ on the event $[\nu_{ij}(n) \le 1]$, whereas on $[\nu_{ij}(n) > 1]$, the difference is made up of the contributions before the first and after the last visit to j. Specifically,

$$\Delta_{nij} = \left(\sum_{(0, r_1^{(j)}]}^{*} V_{kln} \right) 1_{[\nu_{ij}(n)>0]} + \sum_{(r_{\nu_{ij}(n)+1}^{(j)}, r_{n+1}^{(i)}]}^{*} . \tag{1.7}$$

where Σ_I^* denotes summation of all V's accrued during the interval I.

In this representation we will use the fact that there is a random number (possibly zero) of i-to-i intervals during which j is not visited, before the first visit to j at time $r_1^{(j)}$. Likewise, at the other end of the time scale, the next visit to i following the last visit to j at time $p_{\nu_{ij}(n)}^{(j)}$, will itself be followed by a random number of i-to-i intervals that exclude j. Observe that this may be viewed as a property of a sequence of n Bernoulli-p trials: Let K_n and L_n, respectively, denote the number of the n successes before the first failure, and after the last failure. ($L = 0$ if all n trials are successes.) It is straightforwardly shown that for $k, l = 0, 1, 2, \ldots$

$$P[K_n = k, L_n = l] = \begin{cases} p^{k+l} q^2 & \text{if } k + l \le n - 2, \\ p^{n-1} q & \text{if } k + l = n - 1, \\ p^n & \text{if } k = n, l = 0. \end{cases}$$

We are interested in $R_n := K_n + L_n$, the total number of i-to-i intervals before and after the visits to j. It follows from the above that

$$P[R_n = m] = \begin{cases} (m+1) q^2 p^m & \text{if } m = 0, 1, \ldots, n-2, \\ nqp^{n-1} & \text{if } m = n - 1, \\ p^n & \text{if } m = n. \end{cases} \tag{1.8}$$

Let us for clarity write ${}_jU_m^{(i)}$ for $U_m^{(i)}$ and ${}_jS_m^{(i)}$ for $S_m^{(i)}$ when we need to indicate that we are using the *conditional* distribution for the $U^{(i)}$'s given that no visit to j occurred between the two consecutive visits to i that define them. [This is technically poor notation, since it is not the random variables that are changing, but hopefully its mnemonic value offsets this.] We can then give the following representation for Δ_{nij}.

Lemma 1 $\Delta_{nij} \stackrel{L}{=} X 1_{[V_{ij}(n)>0]} + {}_jS^{(i)}_{R_n}$ *in which the distribution of* X *is the conditional distribution of any* $U^{(i)}_m$ *given there is* exactly *one visit to* j *between the two consecutive visits to* i *defining this* $U^{(i)}_m$, *and* $X, \{{}_jS^{(i)}_m : m \geq 1, \nu_{ij}(n)\}$ *and* R_n *are assumed to be independent with distributions consistent with their symbols.*

Proof. The representation has been described above, but the diagram in Figure 1.2 may help to clarify the terms, in which one can see that the difference between $S^{(i)}_n$ and $S^{(j)}_{\nu_{ij}(n)-1}$ is made up of the outer portions that comprise the contributions from the R_n "i-to-i-without-j" intervals together with that of a single "i-to-j-to-i" interval for X. The Markov property is of course central here. Figure 1.2 also stresses the basic 2-state nature of the problem. □

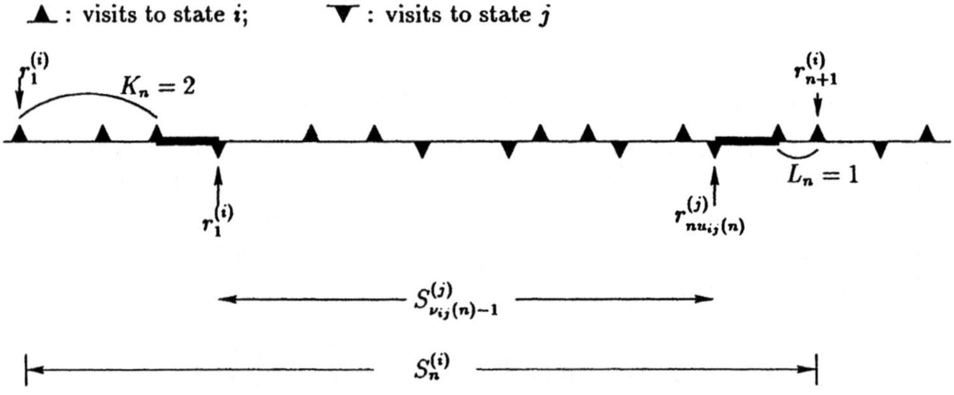

Figure 1.2 X is the sum of the V's over the two thick portions. $R_n = 3$ in this illustration since $K_n = 2$ and $L_n = 1$.

Since X does not depend on n and ${}_jS^{(i)}_{R_n}$ is $O_p(1)$ uniformly in n, we obtain

Corollary 1 $\Delta_{nij} = O_p(1)$ *uniformly in* n.

4. STABLE LAWS AND SOLIDARITY

In this section we prove that if $U^{(i)}_1$ satisfies a Central Limit Theorem then so also does $U^{(j)}_1$. In order to utilize the above results to obtain this solidarity result, it is necessary to relate $S^{(j)}_{\nu_{ij}(n)-1}$, a sum of a random number of $U_m^{(j)}$'s, to a corresponding sum of a fixed number of terms. To this end, let $k_n = [n\gamma]$, the greatest integer not exceeding $E(\nu_{ij}(n)|J_0 = i)$; see (1.3). For $0 < \alpha \le 2$ and $|\beta| \le 1$, write "$U^{(j)} \in \mathcal{D}_{\alpha,\beta}$" for the statement that $U^{(j)}$ is in the domain of attraction of a stable distribution with primary parameters α and β. We write $\mathcal{S}_{\alpha,\beta}$ to denote such a distribution. For information about the $\mathcal{S}_{\alpha,\beta}$ distributions and their domains $\mathcal{D}_{\alpha,\beta}$, the reader is referred to Chapter 7 of Gnedenko and Kolmogorov (1954) and Chapter 17 of Feller (1966). Recall that the designation $\mathcal{S}_{\alpha,\beta}$ actually refers to a type, in that it is specified only up to translation and scalar constants.

Assume that $U^{(j)} \in \mathcal{D}_{\alpha,\beta}$ for some $0 < \alpha \le 2$ and $|\beta| \le 1$. By definition, this means that there exist constants $a_n^{(j)} \in \mathbb{R}^1$ and $b_n^{(j)} > 0$ such that

$$\frac{S_n^{(j)} - a_n^{(j)}}{b_n^{(j)}} \xrightarrow{L} Z_{\alpha,\beta}, \tag{1.9}$$

where $Z_{\alpha,\beta}$ is an $\mathcal{S}_{\alpha,\beta}$ random variable. By Kimbleton (1970), this implies that

$$\left\{S^{(j)}_{\nu_{ij}(n)-1} - \frac{\nu_{ij}(n)-1}{k_n} a_{k_n}\right\} / b_{k_n}^{(j)} \xrightarrow{L} Z_{\alpha,\beta},$$

since the random sample size $\nu_{ij}(n) - 1$ satisfies the required assumption that $\{\nu_{ij}(n) - 1\}/n \xrightarrow{P} \gamma$ as noted following (1.4). From (1.6) and Corollary 1, this implies that

$$\left\{S_n^{(i)} - \frac{\nu_{ij}(n)-1}{k_n} a_{k_n}^{(j)}\right\} / b_{k_n}^{(j)} \xrightarrow{L} Z_{\alpha,\beta},$$

or, equivalently,

$$\frac{S_n^{(i)} - a_{k_n}^{(j)}}{b_{k_n}^{(j)}} - \left\{\frac{\nu_{ij}(n)-1}{k_n} - 1\right\} \frac{a_{k_n}^{(j)}}{b_{k_n}^{(j)}} \xrightarrow{L} Z_{\alpha,\beta}. \tag{1.10}$$

From XVII, (5.16) of Feller (1966) we know that we may assume that $a_n^{(j)}/\sqrt{n} b_n^{(j)} \to 0$. Since $\nu_{ij}(n) - n\gamma = O_p(\sqrt{n})$ by the remark following (1.4), this means that the second term in (1.10) is $o_p(1)$, which proves the following result.

Theorem 1 *For any two states i, j of an irreducible recurrent Markov Chain, if $U_1^{(j)} \in \mathcal{D}_{\alpha,\beta}$ for $0 < \alpha \le 2$ and $|\beta| \le 1$, then $U_1^{(i)} \in \mathcal{D}_{\alpha,\beta}$ as* *well. If for $U_1^{(j)}$ the normalizing constants of (1.9) are $a_n^{(j)}$ and $b_n^{(j)}$, then for $U_1^{(i)}$ one may use $a_n^{(i)} = a_{k_n}^{(j)}$ and $b_n^{(i)} = b_{k_n}^{(j)}$ in which $k_n = [n\gamma]$ and $\gamma = \gamma_{ij} \equiv {}_ip_{ij}/{}_jp_{ji}$.*

Since appropriate normalizing constants for sums of i.i.d. random variables in the domain of attraction of a stable law are known, it is worthwhile to compare these known values with those implied by the relationships of Theorem 1. First of all, the scalar constants are known to satisfy $b_n^{(j)} = n^{1/\alpha}L_j(n)$ for some slowly varying sequence L_j. It may be shown that then $L_j([n\gamma]) = L_i(n)$ say, is slowly varying, which implies that $b_n^{(i)} = b_{k_n}^{(j)} = [n\gamma]^{1/\alpha}L_j([n\gamma])$ is of the required form, $b_n^{(i)} = \gamma^{1/\alpha}n^{1/\alpha}L_i(n)$. The constant $\gamma^{1/\alpha}$ could have been included in L_i but its separation is helpful in relating the scales of the two distributions.

For the centering constants, $a_n^{(j)}$, recall that for the case of $1 < \alpha \leq 2$, they may be taken to be the means, $nE(U_1^{(j)})$. Thus $a_{k_n}^{(j)}$ should be asymptotically equivalent to $nE(U_1^{(i)})$. To see this, consider $a_{k_n}^{(j)} = [n\gamma]E(U_1^{(j)}) \approx n\{\gamma E(U_1^{(j)})\}$. Then, an application of Corollary 4.4 below shows that $\gamma E(U_1^{(j)}) = E(U_1^{(i)})$. It may be worthwhile as an aside to compare the simplicity of this approach with one making use of moment calculations such as those for Markov Chains in Chung ((1967), theorem I.14.6) and for MRP's in Pyke and Schaufele ((1964), Lemma 4.1). For example, the latter would say that for a restricted (essentially deterministic) linear functional of an MRP and under added summability conditions, the mean inter-occurrence sum $E(U_1^{(i)})$ has the form

$$E(U_1^{(i)}) = \sum_k {}_iM_{ik}(\infty)\zeta_k$$

where $\zeta_k = E(V_k|J_0 = k)$ and ${}_iM_{ik}(\infty)$ is the indicated taboo renewal function, namely, the expected number of visits to state k before a visit to state i given $J_0 = i$. Since ${}_iM_{ik}(\infty) = {}_jM_{jk}(\infty){}_iM_{ij}(\infty)$, it follows that

$$E(U_1^{(i)}) = {}_iM_{ij}(\infty)\sum_k {}_jM_{jk}(\infty)\zeta_k = {}_iM_{ij}(\infty)E(U_1^{(j)}).$$

To see the desired proportionality in this more restrictive situation, one would then observe that ${}_iM_{ij}(\infty) = {}_iG_{ij}(\infty)/\{1 - {}_iG_{jj}(\infty)\} = {}_ip_{ij}/{}_jp_{ji} = \gamma$, with ${}_jG_{kl}$ being the usual taboo first passage distributions.

If $0 < \alpha < 1$, the centering constants, $a_n^{(j)}$, may be taken to be zero, so that then all of the other $a_n^{(i)}$'s may be zero as well. In the singular

Cauchy case of $\alpha = 1$, one may center r.v.'s in $\mathcal{D}_{\alpha,\beta}$ by truncated means

$$a_n^{(j)} = E(U_1^{(j)} : |U_1^{(j)}| \leq \tau b_n^{(j)})$$

as in Theorem 25.4 of Gnedenko and Kolmogorov (1954). Alternatively, one may use

$$a_n^{(j)} = nb_n^{(j)} E(\sin(U_1^{(j)}/b_n^{(j)})); \tag{1.11}$$

cf. Theorem XVII.5, 2 of Feller (1966). In view of Theorem 2 below and the result of Pitman (1956) that relates the limiting average of truncated means to the differentiability of characteristic functions, one has the limiting proportionality of the truncated means of $U_1^{(i)}$ and $U_1^{(j)}$, thereby being consistent with Theorem 1 in this case.

5. A BASIC IDENTITY FOR COMPARING INTER-OCCURRENCE SUMS

The structure of an MRP leads to simplifying relationships involving first passage distributions, renewal functions and transition quantities. In particular, if G_{ij} and ${}_kG_{ij}$ denote absolute and taboo first passage distribution functions, with M_{ij} and ${}_kM_{ij}$ being their corresponding renewal functions, identities (3.1) and (3.3) of Pyke and Schaufele (1964) imply that

$$G_{ii} = {}_jG_{ii} + {}_iG_{ij} * G_{ji} = {}_jG_{ii} + {}_iG_{ij} * \{{}_iM_{jj} * {}_jG_{ji}\}$$

for any $i \neq j$. Convolution over $[0, \infty)$ of both sides with $1 - {}_iG_{jj}$, the inverse of ${}_iM_{jj}$, then gives

$$(1 - {}_iG_{jj}) * (1 - G_{ii}) = (1 - {}_iG_{jj}) * (1 - {}_jG_{ii}) - {}_iG_{ij} * {}_jG_{ji}.$$

Since the right hand side is symmetric in i and j, it follows that

$$(1 - {}_iG_{jj}) * (1 - G_{ii}) = (1 - {}_jG_{ii}) * (1 - G_{jj}). \tag{1.12}$$

This identity was first set down by J. Teugels (see Teugels (1968), equation (18)) where it is used in his study of solidarity results regarding the order of exponential decay of the renewal functions M_{ii}. More recently it was used in Sgibnev (1996) to establish the solidarity of MRP's with respect to the asymptotic behavior of the renewal functions in the case of infinite sequences. This identity however, holds in a much broader context, depending as it does only upon additive random excursions between successive occurrences of states in a two state Markov Chain. I believe this underlying structure is revealed best using random variables through the equations (1.16) below. The motivation in our MRP context is as follows.

Fix distinct states i and j, and let $\{{}_jU_r^{(i)} : r \geq 1\}$ and $\{U_r^{(i|j)} : r \geq 1\}$ denote independent sequences of i.i.d. r.v.'s whose common distributions are respectively equal to the conditional distribution of $U_r^{(i)}$ given that state j is not visited, is visited between the successive visits to state i that determine $U_r^{(i)}$. If then we let A_r denote the event that state j *is* visited between the r-th and $(r+1)$-th visit to i, then

$$U_1^{(i)} \stackrel{L}{=} U_1^{(i|j)} 1_{A_1} + {}_jU_1^{(i)} 1_{A_1^c}. \tag{1.13}$$

Since on A_1 state j is visited, a further decomposition is possible. Suppose one defines $U_m^{(ij)}$ to be the sum of the V's for all transitions that occur between the m-th visit to i and the subsequent visit to j. (Thus the $U_m^{(i)}$ of (1.5) would be the same as $U_m^{(ii)}$.) Then let $\{U_r^{(i\to j)} : r \geq 1\}$ be a sequence of i.i.d. r.v.'s whose common distribution is the conditional distribution of $U_1^{(ij)}$ given that the subsequent visit to j occurs before the next visit to i. Also, let ${}_iN_j$ be a Geometric r.v. on $\{0, 1, 2, \dots\}$ with parameter ${}_jp_{ii}$. Then one may write

$$U_1^{(i|j)} \stackrel{L}{=} U_1^{(i\to j)} + \sum_{r=1}^{{}_iN_j} {}_iU_r^{(j)} + U_1^{(j\to i)} \tag{1.14}$$

where all r.v.'s on the right hand side are assumed independent with distributions as specified above. Substitution of this into (1.13) leads to the key representation,

$$U_1^{(i)} = \left\{ U_1^{(i\to j)} + U_1^{(j\to i)} + \sum_{r=1}^{{}_iN_j} {}_iU_r^{(j)} \right\} 1_{A_m} + {}_jU_1^{(i)} 1_{A_m^c}, \tag{1.15}$$

again, with all r.v.'s on the right hand side being independent. A similar expression for $U_1^{(j)}$ follows by interchanging i and j in (1.15).

It is the structure in (1.15) that is important and this structure can be seen more clearly by simplifying the notation way from our MRP context, even though the above notation was chosen to be visually suggestive of the terms involved.

To this end, introduce three r.v.'s, X, Y, Z, with independent copies of them indicated by subscripts. Let M, N be Geometric r.v.s on $\{0, 1, \dots\}$ with parameters p, π, respectively. Finally, let I (J) be the indicator r.v. of an event A (B) of probability $1 - \pi$ ($1 - p$). Assume all r.v.'s are independent of each other and that $0 < p, \pi < 1$. Define

$$\begin{aligned} U &= XI + \left(Z + \sum_{m=1}^{M} Y_m\right)(1 - I), \\ V &= YJ + \left(Z + \sum_{n=1}^{N} X_n\right)(1 - J). \end{aligned} \tag{1.16}$$

It is clear that (4.4) and its counterpart for $U_1^{(j)}$ are of this form: Note that Z plays the role of $U_1^{(i\to j)} + U_1^{(j\to i)}$. In terms of a 2-state MRP, U and V represent the inter-recurrence sums for the two states when the random contributions for transitions $1 \to 1, 2 \to 2, 1 \to 2, 2 \to 1$ are distributed, respectively, as X, Y, Z_1 and Z_2 with $Z_1 + Z_2 = Z$.

Let f, g, h, φ, ψ, denote the characteristic functions of X, Y, Z, U, V, respectively. Then for real u,

$$\begin{aligned} \varphi(u) &= (1-\pi)f(u) + \pi h(u) E\left(e^{iu\sum_{m=1}^{M} Y_m}\right) \\ &= (1-\pi)f(u) + \pi p \frac{h(u)}{1-(1-p)g(u)} \end{aligned} \tag{1.17}$$

and, similarly,

$$\psi = (1-p)g + \pi p \frac{h}{1-(1-\pi)f}, \tag{1.18}$$

with the quotients well defined since both p and π are assumed to be positive. Consequently,

$$\frac{\varphi - 1}{(1-\pi)f - 1} = \frac{\psi - 1}{(1-p)g - 1}. \tag{1.19}$$

To return to the general MRP case, set

$$\varphi_k(u) = E(e^{iuU_1^{(k)}}), \; {}_jf_k(u) = E(e^{iu\, {}_jU_1^{(k)}}). \tag{1.20}$$

A comparison of (1.15) and (1.16) shows that in this context, $p = {}_jp_{ji}$ and $\pi = {}_ip_{ij}$, and the resulting identity implied by (1.19) leads to

Theorem 2 *For distinct states k and j of an irreducible recurrent MRP, the inter-occurrence sums $U_1^{(k)}$ and $U_1^{(j)}$ defined by (1.5) have characteristic functions that satisfy*

$$\varphi_j(u) - 1 = \rho_{ji}(u)\{\varphi_i(u) - 1\} \tag{1.21}$$

for all real u, where

$$\rho_{ji}(u) := \frac{1-(1-{}_ip_{ij})_jf_i(u)}{1-(1-{}_jp_{ji})_if_j(u)} \rightarrow \frac{{}_ip_{ij}}{{}_jp_{ji}} = \frac{\pi}{p} = \gamma \tag{1.22}$$

as $u \rightarrow 0$.

Corollary 2 (WLLN Solidarity) *For any two states $i \neq j$ of an irreducible recurrent MRP, $S_n^{(i)}$ satisfies the WLLN (i.e., there exist constants A_{ni} such that $(S_n^{(i)} - A_{ni})/n \xrightarrow{P} 0)$ if and only if $S_n^{(j)}$ satisfies the WLLN.*

Proof. This follows immediately from (1.21) since it is known that the WLLN holds if and only if the common characteristic function of the summands is differentiable at the origin. (Cf. Fisz (1963), Problem 6.38. Also, Feller (1966), Thm. VII, 7.1 together with Pitman (1956).) □

An important consequence of the representation (1.15) is the fact that moment solidarity results may be proved simply without requiring conditions about absolute convergence of series,k such as are assumed in the results of Chung (1967; Sect. I.14) and Pyke and Schaufele (1964; Lemma 4.1). Here is the result.

Theorem 3 (Moment Solidarity) *For any two distinct states, i and j, of an irreducible recurrent MRP and any $r > 0$, $E|U_1^{(i)}|^r < \infty$ if and only if $E|U_1^{(j)}|^r < \infty$.*

Proof. This is a direct consequence of the representation (1.15). To simplify notation we work with the generic representations of (1.16). Clearly $E|U|^r < \infty$ implies $E|X|^r < \infty$ and $E|Z + \sum_{m=1}^{M} Y_m|^r < \infty$ by independence since π is different from 0 and 1. Again, by the independence of Z and $\{M, Y_1, Y_2, \ldots\}$, this in turn implies (and of course is implied by) the finiteness of $E|Z|^r$ and $E|\sum_{m=1}^{M}, Y_m|^r$. However, the latter's finiteness implies that of $E|Y_1|^r$ since $0 < p < 1$. In summary then, the finiteness of $E|U|^r$ implies through (1.16) the finiteness of $E|X|^r$, $E|Z|^r$ and $E|Y|^r$. Now, using this in the second equation of (1.16) one obtains the finiteness of $E|V|^r$ as desired. □

Corollary 3 *For any two distinct states, i and j, of an irreducible recurrent MRP, $E|U_1^{(i)}| < \infty$ implies $E|U_1^{(j)|}| < \infty$ and $E(U_1^{(i)}) = \gamma E(U_1^{(j)})$.*

Proof. In view of Theorem 3, the result follows by dividing both sides of (1.21) by u and taking limits as $u \to 0$. Alternatively, it is an interesting calculation using (1.16) to obtain, under the assumed integrability of U, that

$$E(U) = (1-\pi)E(X) + \pi\{E(Z) + (1/p-1)E(Y)\}$$

and

$$E(V) = (1-p)E(Y) + p\{E(Z) + (1/\pi-1)E(X)\},$$

proving thereby that $pE(U) = \pi E(V)$. Since in our application of (1.16), $p = {}_jp_{ji}$ and $\pi = {}_ip_{ij}$, the proportionality of the means follows. □

It would be of interest to know whether the solidarity of the domains of attraction given in Theorem 1 is provable directly from the identity of Theorem 2. It appears that the main challenge would be the case when $\alpha = 1$. When $\alpha < 1$, the centering constants may be taken to be zero, so that (1.21) is directly applicable to the condition given in Lemma 1, Section XVII.5 of Feller (1966). When $\alpha > 1$, the centering constants may be taken to be means. Substitution of (1.21) into Feller's characterization requires the use of Theorem 3 to handle the order of the difference $\rho_{ji}(u) - \rho_{ji}(0)$ that arises, but otherwise this case seems to work straightforwardly. The author was not able, however, to handle this difference satisfactorily when $\alpha = 1$.

6. REMARK

The additive functionals implicit in (1.5) are fairly general and represent natural situations in which real-valued random measurements (the V's) are recorded at each transition of an MRP. The distribution of each measurement depends only upon the two states between which the transition is being made. Moreover, the measurements are mutually conditionally independent given the MRP. Nevertheless, all that entered into the arguments of this paper, as has hopefully been adequately stressed, were the global measurements such as ${}_jU_r^{(i)}$ that represent cumulative measurements assigned to the embedded transitions among the two specified states. We remark further that the basic structure embodied in (1.16) is applicable to more general situations, such as when the V's are not real, but possibly random processes themselves. For example, V_{kln}, might be a Brownian Motion whose mean, variance and lifetime depend only upon k and l, and the additivity in (1.5) might be that of attaching these "excursions" end to end.

A conceptually simple illustration of this would be the reconstruction of Brownian Motion from itself as follows. Let $\{Z(t); t \geq 0\}$ be standard Brownian Motion starting at $Z(0) = 1$, say. Consider the MRP embedded in Z that is defined as the one that is said to be in state s at time $t > 0$ if s is the last value among $\{\pm 2^n : n = 0, \pm 1, \pm 2, \ldots\}$ taken on by Z during $(0, t]$. The Q_{ij}'s for this MRP are well known first passage probabilities for Brownian Motion. In particular, $p_{rs} = 1/3$ if $s = 2r$, $= 2/3$ if $s = r/2$ and $= 0$, otherwise. If for each transition r to $2r$, the associated random summand $V_{r,2r,m}$ is an independent segment of Brownian Motion/Bridge of the relevant length that is conditioned to go from 0 to r and to have remained within $(-r/2, r)$, the knitting together of the pieces will yield a Brownian Motion again (though in this case, one might prefer never to have cut up the original one in the first place to form the embedding!) Notice that this embedded MRP is an example of one with many "explosions to infinity", manifested here by the zeros of the underlying Brownian Motion. It also satisfies the continuity-at-infinity condition, thereby satisfying the hypothesis A regularity assumed in Pyke and Schaufele (1964).

References

[1] Anderson, K. K. and Athreya, K. B. (1987). A renewal theorem in the infinite mean case, *Ann. Prob.* **15**, 388–393.

[2] Cheong, C. K. (1967). Geometric convergence of semi-Markov transition probabilities, *Z. Wahrshein.* **7**, 122–130.

[3] Chung, K. L. (1954). Contributions to the theory of Markov chains, II. *Trans. A.M.S.* **76**, 397–419.

[4] Chung, K. L. (1967). *Markov Chains with Stationary Transition Probabilities*, Second Edition. Springer-Verlag, New York.

[5] Cinlar, E. (1959). Markov renewal theory, *Adv. Appl. Prob.* **1**, 123–187.

[6] Feller, W. (1966). *An Introduction to Probability Theory and its Applications*, Vol. II. J. Wiley and Sons, New York.

[7] Fisz, M. (1963). *Probability Theory and Mathematical Statistics*, Third edition. J. Wiley and Sons, New York.

[8] Gnedenko, B. V. and Kolmogorov, A. N. (1954). *Limit Distributions for Sums of Independent Random Variables*, Addison-Wesley, Reading, MA.

[9] Kimbleton, S. R. (1970). A simple proof of a random stable limit theorem, *J. Appl. Prob.* **7**, 502–504.

[10] Kolmogorov, A. N. (1936). Anfangsgrunde der Theorie der Markoffschen Ketten mit unendlich vielen möglichen Zustanden, *Mat. Sb.* **1**, 607–610.

[11] Lévy, P. (1954). Processus Semi-Markoviens, *Proc. Int. Congr. Math.* **3**, 416–426.

[12] Mohan, N. R. (1976). Teugel's renewal theorem and stable laws, *Ann. Prob.* **4**, 863–868.

[13] Pitman, E. J. G. (1956). On the derivatives of a characteristic function at the origin, *Ann. Math. Statist.* **27**, 1156–1160.

[14] Pyke, R. (1961a). Markov renewal processes: Definitions and preliminary properties, *Ann. Maht. Statist.* **32**, 1231–1242.

[15] Pyke, R. (1961b). Markov renewal processes with finitely many states, *Ann. Math. Statist.* **32**, 1243–1259.

[16] Pyke, R. and Schaufele, R. (1964). Limit theorems for Markov renewal processes, *Ann. Math. Statist.* **35**, 1746–1764.

[17] Pyke, R. and Schaufele, R. (1966). Stationary measures for Markov Renewal processes, *Ann. Math. Statist.* **35**, 1746–1764.

[18] Pyke, R. (1975). Applied probability: An editor's dilemma, *Suppl. Adv. Appl. Prob.* **7**, 17–37.

[19] Schäl, M. (1970). Markov renewal processes with auxiliary paths, *Ann. Math. Statist.* **41**, 1604–1623.

[20] Sgibnev, M. S. (1996). An infinite variance solidarity theorem for Markov renewal functions, *J. Appl. Prob.* **33**, 434–438.

[21] Smith, W. L. (1954). Regenerative stochastic processes, *Proc. Int. Congr. Math.* **2**, 304–305.

[22] Smith, W. L. (1955). Regenerative stochastic processes, *Proc. Roy. Soc. A*, **232**, 6–31.

[23] Takács, L. (1954). Some investigations concerning recurrent stochastic processes of a certain type, *Magyar Tud. Akad. Mat. Kutató Int. Kŏzl.* **3**, 115–128.

[24] Teugels, J. L. (1968). Exponential ergodicity in Markov renewal processes, *J. Appl. Prob.* **5**, 387–400.

[25] Teugels, J. L. (1970). Regular variation of Markov renewal processes, *J. London Math. Soc.* **2**, 179–190.

[26] Teugels, J. L. (1976). A bibliography on semi-Markov processes, *J. Comput. Appl. Math.* **2**, 125–144.

[27] Teugels, J. L. (1986). A second bibliography on semi-Markov processes, *Semi-Markov Models*, 507–584. Plenum, New York-London.

Chapter 2

A GENERALIZATION OF SEMI-MARKOV PROCESSES

Marius Iosifescu
Romanian Academy, Romania

Abstract We consider a slight generalization of semi-Markov processes obtained by allowing a more flexible transition mechanism. W. Feller's 1964 treatment of semi-Markov processes with a denumerable state space can be easily adapted to the present case.

Keywords: Backward equations, minimal solution, forward equations.

1. INTRODUCTION AND SUMMARY

This paper deals with a slight generalization of semi-Markov processes obtained by allowing a more flexible transition mechanism. The latter is assumed to work time-invariantly between any two consecutive terms of a given increasing sequence of stopping times, which does not coincide with the sequence of jump times of the process considered. The restriction to a denumerable state space and to the time-stationarity of the transition mechanism is introduced for the sake of clarity. The theory should go through for the corresponding analogue of semi-Markov processes with an arbitrary state space. Such a generalization has been already described in a recent paper by Hou, Liu and Zou (1997). It should be nevertheless noted that the results we obtain in our restricted framework are more precise and complete.

We follow W. Feller's 1964 treatment, which can be easily adapted to the present case. His paper [Feller (1964)], apparently little known among those interested in semi-Markov processes, is still useful and inspiring.

We conclude by indicating two open problems for semi-Markov processes.

2. DEFINITION AND BACKWARD EQUATIONS

Let $\mathbb{N}_+ = \{1, 2, \cdots\}, \mathbb{N} = \mathbb{N}_+ \cup \{0\}$, and $\mathbb{R}_+ = \{x : x \geq 0\}$. Consider a complete probability space $(\Omega, \mathcal{K}, P)$ and a random variable $\tau : \Omega \to \mathbb{R}_+$ on it. For any $i, j \in \mathbb{N}_+$ consider a non-decreasing mapping $f(\cdot, i, j) : \mathbb{R}_+ \to \mathbb{R}_+$ and a measurable mapping $g(\cdot, i, j) : \mathbb{R}_+ \to \mathbb{R}_+$ such that :

(i) $f(0, i, j) = 0, i, j \in \mathbb{N}_+$;

(ii) $\lim_{t\to\infty} S_i(t) = 1$, where $S_i(t) = \sum_{j\in\mathbb{N}_+} f(t, i, j)$, $t \in \mathbb{R}_+$, $i \in \mathbb{N}_+$;

(iii) for any $i \in \mathbb{N}_+$ there exists $t_i > 0$ such that $S_i(t) < 1$ for all $t \in [0, t_i]$;

(iv) $g(0, i, j) = \delta_{ij}$ (Kronecker's delta), $i, j \in \mathbb{N}_+$;

(v) $\lim_{t\to\infty} g(t, i, j) = 0, i, j \in \mathbb{N}_+$;

(vi) $\sum_{j\in\mathbb{N}_+} g(t, i, j) = 1 - S_i(t), t \in \mathbb{R}_+, i \in \mathbb{N}_+$.

Note that it follows from (ii) that putting :

$$p_{ij} = \lim_{t\to\infty} f(t, i, j), i, j \in \mathbb{N}_+,$$

the matrix $\mathbf{P} = (p_{ij})_{i,j\in\mathbb{N}_+}$ is a stochastic one.

Definition 1 *Let* $\mathbf{f} = (f(\cdot, i, j))_{i,j\in\mathbb{N}_+}$ *and* $\mathbf{g} = (g(\cdot, i, j))_{i,j\in\mathbb{N}_+}$ *be matrix functions satisfying properties (i)-(vi) above. A right-continuous* $\mathbb{N}_+$*-valued stochastic process* $X = (X(t, \omega) : 0 \leq t \leq \tau(\omega), \omega \in \Omega)$ *is said to be an* $(\mathbf{f}, \mathbf{g})$*-process iff there exists a sequence* $(\tau_n)_{n\in\mathbb{N}}$ *of stopping times for* X *satisfying* $0 = \tau_0 < \tau_1 < \cdots$ *and* $\lim_{n\to\infty} \tau_n = \tau P$*-a.s., for which the equations :*

$$P(X(\tau_{n+1}) = j, \tau_{n+1} - \tau_n \leq t | X(\tau_m), \tau_m, 0 \leq m \leq n)$$

$$= P(X(\tau_{n+1}) = j, \tau_{n+1} - \tau_n \leq t | X(\tau_n)) = f(t, X(\tau_n), j),$$

and

$$P(X(\tau_n + t) = j, \tau_{n+1} - \tau_n > t | X(\tau_m), \tau_m, 0 \leq m \leq n)$$

$$= P(X(\tau_n + t) = j, \tau_{n+1} - \tau_n > t | X(\tau_n)) = g(t, X(\tau_n), j),$$

hold P-a.s. for all $t \in \mathbb{R}_+$ and $n, j \in \mathbb{N}_+$.

(Clearly, as usual, $X(\tau_{n+1})$ is an abbreviation for $X(\tau_{n+1}(\omega), \omega)$ etc. Note also that the two equations above are consistent with property (vi)). It follows from Definition 1 that $(X(\tau_n))_{n \in \mathbb{N}_+}$ is an $\mathbb{N}_+$-valued Markov chain with transition matrix $\mathbf{P}$. Put :

$$p(t, i, j) = P(X(t) = j | X(0) = i)$$

for all $t \in \mathbb{R}_+$ and $i, j \in \mathbb{N}_+$. On account of Definition 1, $p(t, i, j)$ is the conditional probability of the random event $(X(t + u) = j)$ given that a jump to i occurs at time $u \in \mathbb{R}_+$.

Considering the first jump following u, it is seen that :

$$p(t, i, j) = \sum_{k \in \mathbb{N}_+} \int_0^t f(\mathrm{d}u, i, k) p(t - u, k, j) + g(t, i, j) \qquad (2.1)$$

for all $t \in \mathbb{R}_+$ and $i, j \in \mathbb{N}_+$.

It is easy to check that in the special case where X is the minimal Markov process associated with a conservative infinitesimal transition matrix and $(\tau_n)_{n \in \mathbb{N}_+}$ is the sequence of the jump times of the process, equations (2.1) are equivalent to Kolmogorov's system of backward differential equations [cf. Chung (1967, Part II, § 17)].

3. THE MINIMAL SOLUTION

Passing to Laplace transforms and matrix notation, for $\lambda > 0$ and $i, j \in \mathbb{N}_+$, let us put :

$$\pi_{ij}(\lambda) = \int_{\mathbb{R}_+} e^{-\lambda t} p(t, i, j) \mathrm{d}t,$$

$$\varphi_{ij}(\lambda) = \int_{\mathbb{R}_+} e^{-\lambda t} f(\mathrm{d}t, i, j),$$

$$\gamma_{ij}(\lambda) = \int_{\mathbb{R}_+} e^{-\lambda t} g(t, i, j) \mathrm{d}t,$$

and

$$\Pi = (\pi_{ij})_{i,j \in \mathbb{N}_+}, \Phi = (\varphi_{ij})_{i,j \in \mathbb{N}_+}, \Gamma = (\gamma_{ij})_{i,j \in \mathbb{N}_+}.$$

(The dependence on λ of Π, Φ, Γ and other matrices denoted by Greek letters should be borne in mind). Clearly, in terms of Laplace transforms equations (2.1) read :

$$\Pi = \Phi \Pi + \Gamma. \qquad (2.2)$$

Let $\mathbf{0}$ and $\mathbf{1}$ denote the (column) vectors $(1-\delta_{ii})_{i\in \mathbb{N}_+}$ and $(\delta_{ii})_{i\in \mathbb{N}_+}$, respectively. It is easy to check using assumption (iii) above that :

$$\Phi\mathbf{1} < \mathbf{1}. \tag{2.3}$$

Hence, denoting by $\mathbf{I}$ the unit matrix $(\delta_{ij})_{i,j\in \mathbb{N}_+}$, the matrix series $\mathbf{I}+\Phi+\Phi^2+\cdots=\sum_{n\in \mathbb{N}}\Phi^n$ is convergent and represents the unique minimum non-negative inverse $(\mathbf{I}-\Phi)^{-1}$ of the matrix $\mathbf{I}-\Phi$. See Proposition 5.11 in Kemeny, Snell and Knapp (1976).

Theorem 1 *Equations (2.1) possess a minimal solution whose matrix Laplace transform* Π_{min} *is given by :*

$$\Pi_{\text{min}} = (\mathbf{I}-\Phi)^{-1}\Gamma.$$

The minimal solution is either stochastic or sub-stochastic. It is stochastic for all $t \in \mathbb{R}_+$ *iff for some* $\lambda > 0$ *the only solution of the equation* $\Phi\xi = \xi$, *where* $\mathbf{0} \le \xi \le \mathbf{1}$ *(component-wise), is* $\xi = \mathbf{0}$.

Proof. It follows from property (vi) that :

$$\lambda \sum_{j\in \mathbb{N}_+} \gamma_{ij}(\lambda) = 1 - \sum_{j\in \mathbb{N}_+} \varphi_{ij}(\lambda), \lambda > 0, i \in \mathbb{N}_+,$$

which implies that :

$$\lambda\, \Gamma\, \mathbf{1} = \mathbf{1} - \Phi\, \mathbf{1},\ \lambda > 0. \tag{2.4}$$

Putting :

$$\Pi_n = (\mathbf{I}+\Phi+\cdots+\Phi^n)\Gamma,\ n \in \mathbb{N}_+,$$

by (2.4) we obtain :

$$\lambda\, \Pi_n\, \mathbf{1} = \mathbf{1} - \Phi^{n+1}\, \mathbf{1},\ \lambda > 0,\ n \in \mathbb{N}_+. \tag{2.5}$$

By (2.3), the sequence of vectors $(\Phi^n\, \mathbf{1})_{n\in \mathbb{N}_+}$ is monotone and converges to a vector $\xi = \xi(\lambda)$ such that :

$$\Phi\xi = \xi, \mathbf{0} \le \xi \le \mathbf{1} \tag{2.6}$$

Now, the matrix Γ is non-negative, so that :

$$\Pi_1 \le \Pi_2 \le \cdots \tag{2.7}$$

Since (2.2) implies $\Pi = \Phi^{n+1}\Pi + \Pi_n\Gamma, n \in \mathbb{N}_+$, every non-negative solution to (2.2) is greater than the (formal) limit :

$$\Pi_{\text{min}} = (\mathbf{I}+\Phi+\Phi^2+\cdots)\Gamma = (\mathbf{I}-\Phi)^{-1}\Gamma \tag{2.8}$$

of Π_n as $n \to \infty$. Actually, (2.8) is the genuine limit of Π_n as $n \to \infty$ by the Beppo Levi monotone convergence theorem as applied to matrices [see Proposition 1.53 in Kemeny, Snell and Knapp (1976)]. By the same theorem, it follows from (2.5) and (2.7) that

$$\lambda\, \Pi_{\min}\, \mathbf{1} = \mathbf{1} - \xi, \ \lambda > 0. \tag{2.9}$$

Hence all entries of $(\mathbf{I} - \Phi)^{-1}\Gamma$ are finite.

Since all entries of the matrix $(\mathbf{I} - \Phi)^{-1} = \mathbf{I} + \Phi + \Phi^2 + \cdots$ are limits of completely monotone functions [see, e.g., Feller (1966, XIII.4)], they are completely monotone functions, too. As the entries of Γ are Laplace transforms (with respect to Lebesgue measure), for any $i, j \in \mathbb{N}_+$ the ij-entry of $\Pi_{\min}$ is the Laplace transform of a point function, say $p_{\min}(t, i, j)$. Equation (2.9) then shows that the row sums $\sum_{j \in \mathbb{N}_+} p_{\min}(t, i, j)$, $i \in \mathbb{N}_+$, are ≤ 1, and identically equal to 1 iff $\xi = \mathbf{0}$. Since the components of ξ are completely monotone functions, we have either $\xi = \mathbf{0}$ for all $\lambda > 0$ or $\xi \neq \mathbf{0}$ for all $\lambda > 0$.

The proof is complete.

Corollary 1 *The solution to equations (2.1) is unique iff the minimal solution is strictly stochastic.*

Proof. It is clear that the minimal solution by its very minimal character is unique whenever it is strictly stochastic. If not, then proceeding as in the Markovian case [see, e.g., Chung (1967, Part II, § 19)], it is possible to construct infinitely many solutions to equations (2.1).

Theorem 2 (Forward equation)
Assume that there exists a non-negative matrix $\widehat{\Phi}$ such that $\Gamma\widehat{\Phi} = \Phi\Gamma$. Then $\Pi_{\min}$ is the minimal solution to the equation :

$$\Pi = \Gamma + \Pi\widehat{\Phi}. \tag{2.10}$$

In particular, if Γ has a right inverse Γ_r^{-1} (i.e., $\Gamma\Gamma_r^{-1} = \mathbf{I}$), then we can take $\widehat{\Phi} = \Gamma_r^{-1}\Phi\Gamma$ whenever $\Gamma_r^{-1}\Phi\Gamma$ is a non-negative matrix.

Proof. It is easy to check that we also have $\Gamma\widehat{\Phi}^n = \Phi^n\Gamma$ for all $n \in \mathbb{N}_+$. Then we can write

$$\Pi_{\min} = (\mathbf{I} + \Phi + \Phi^2 + \cdots)\Gamma = \Gamma(\mathbf{I} + \widehat{\Phi} + \widehat{\Phi}^2 + \cdots).$$

Now, equation (2.10) is clearly satisfied by $\Pi_{\min}$ and, as in the case of equation (2.2), every non-negative solution to (2.10) is greater than $\Pi_{\min}$. In the special case $\widehat{\Phi} = \Gamma_r^{-1}\Phi\Gamma$, equation (2.10) reads :

$$\Pi = \Gamma + \Pi\Gamma_r^{-1}\Phi\Gamma. \tag{2.11}$$

We shall discuss equation (2.11) in Section 6 in the special case of semi-Markov processes.

4. CRITERIA FOR NON-UNIQUENESS

Since

$$\lim_{\lambda \downarrow 0} \Phi(\lambda) = \mathbf{P},$$

it is possible to relate the vector ξ occurring in the statement of Theorem 2 to $\mathbf{P}$. Write :

$$\varphi_{ij}(\lambda) = p_{ij} - \lambda \psi_{ij}(\lambda),$$

where

$$\psi_{ij}(\lambda) = \lambda^{-1} \int_{\mathbb{R}_+} (1 - e^{-\lambda t}) f(dt, i, j), \lambda > 0, \; i, j \in \mathbb{N}_+.$$

Equation (2.6) becomes :

$$\mathbf{P}\xi = \xi + \lambda \Psi \xi, \lambda > 0,$$

where $\Psi = (\psi_{ij})_{i,j \in \mathbb{N}_+}$. Hence

$$\mathbf{P}^n \xi = \xi + \lambda \left(\sum_{k=0}^{n-1} \mathbf{P}^k \right) \Psi \xi, \; \lambda > 0, \; n \in \mathbb{N}_+. \tag{2.12}$$

Theorem 3 *If all states of the Markov chain with transition matrix* $\mathbf{P}$ *are recurrent, then the minimal solution is strictly stochastic (and hence unique).*

Proof. The result follows from Theorem 2 and equation (2.12) since under the assumption made any entry of the matrix $\sum_{k=1}^{n-1} \mathbf{P}^k$ corresponding to two states in the same class, converges to ∞ as $n \to \infty$. [cf., e.g., Kemeny, Snell and Knapp (1976, Chapter 6.)].

The above theorem applies to indecomposable subsets of states, so that in studying non-uniqueness we may assume that all states of the Markov chain with transition matrix $\mathbf{P}$ are transient.

Let A be a subset of $\mathbb{N}_+$ and denote by ${}_A\mathbf{P}$ the sub-stochastic matrix $(p_{ij})_{i,j \in A}$. The matrix ${}_A\mathbf{P}$ governs the restriction to A of the Markov chain with transition matrix $\mathbf{P}$, i.e., the chain obtained by stopping the latter as soon as it leaves A. The row sums ${}_A\mathbf{P}^n\mathbf{1}$ of ${}_A\mathbf{P}^n$ are precisely the probabilities of staying n steps within A. So

$$\lim_{n \to \infty} {}_A\mathbf{P}^n \mathbf{1} = (y_i)_{i \in A},$$

where y_i is the probability that, starting from $i \in A$, the restricted process never terminates. According to Feller (1956) a set $A \subset \mathbb{N}_+$ is said to be a *sojourn set* for the Markov chain with transition matrix **P** iff $(y_i)_{i\in A} \neq 0$.

Consider the truncated moments

$$m_{ij} = \int_{\mathbb{R}_+} \frac{t}{t+1} f(\mathrm{d}t, i, j), i, j \in \mathbb{N}_+.$$

Put $\mathbf{M} = (m_{ij})_{i,j\in\mathbb{N}_+}$ and

$$\mathbf{U} = \left(\sum_{n\geq 0} \mathbf{P}^n\right) \mathbf{M} = (\mathbf{I} - \mathbf{P})^{-1}\mathbf{M} = (u_{ij})_{i,j\in\mathbb{N}_+}.$$

Theorem 4 *Assume all states of the Markov chain with transition matrix* **P** *are transient. The minimal solution to equations (1) is substochastic iff there exists a sojourn set* $A \subset \mathbb{N}_+$ *such that all row sums of the matrix* ${}_A\mathbf{U} = (u_{ij})_{i,j\in A}$ *are finite.*

The *proof* is entirely similar to that working in the case of a semi-Markov process [see Feller (1964, Theorem 4)] and is therefore omitted.

We also give [see Feller (1964)] a sufficient condition for non-uniqueness. Assume that

$$m_i = \int_{\mathbb{R}_+} tS_i(\mathrm{d}t) = \int_{\mathbb{R}_+} (1 - S_i(t))\mathrm{d}t < \infty, \ i \in \mathbb{N}_+. \tag{2.13}$$

Note that by (iii) we have $m_i > 0, i \in \mathbb{N}_+$.

Proposition 1 *Assume all states of the Markov chain with transition matrix* **P** *are transient. If there exists a sojourn set* A *such that the components of the vector* $(\mathbf{I} - {}_A\mathbf{P})^{-1}(m_i)_{i\in A}$ *are finite, then the minimal solution is not strictly stochastic.*

5. ASYMPTOTIC BEHAVIOR

We can describe the asymptotic behavior of the probabilities $p(t, i, j)$, $i, j \in \mathbb{N}_+$, as $t \to \infty$ under suitable (undoubtedly too strong) assumptions on **f** and **g**.

Let us first remember that a *strongly* recurrent Markov chain is a positive recurrent Markov chain for which the second moments of all return times are finite. The transition matrix of a strongly recurrent Markov chain is itself called strongly recurrent. For details see Kemeny, Snell and Knapp (1976, p.274).

Theorem 5 *Assume that the matrix* **P** *is indecomposable and strongly recurrent and let* $(p_i)_{i\in\mathbb{N}_+}$ *be the (unique) strictly positive stationary* *distribution for* **P**. *Assume that*

$$\sup_{i \in \mathbb{N}_+} m_i < \infty,$$

where the $m_i, i \in \mathbb{N}_+$, *are defined in (2.13). Assume also that there exists a non-negative matrix* $\mathbf{G} = (g_{ij})_{i,j \in \mathbb{N}_+}$ *with* $\mathbf{G1} = (m_i)_{i \in \mathbb{N}_+}$ *such that*

$$\lim_{\lambda \downarrow 0} \sup_{i \in \mathbb{N}_+} \sum_j |\gamma_{ij}(\lambda) - g_{ij}| = 0. \tag{2.14}$$

Under these assumptions

$$\lim_{t \to \infty} p(t,i,j) = \frac{\sum_{k \in \mathbb{N}_+} p_k g_{kj}}{\sum_{k \in \mathbb{N}_+} p_k m_k}, \quad i,j \in \mathbb{N}_+. \tag{2.15}$$

The *proof* is similar to that of Theorem 5.9 in Koroliuk and Turbin (1976) and is therefore omitted.

6. THE SPECIAL CASE OF SEMI-MARKOV PROCESSES

A semi-Markov process is an $(\mathbf{f}, \mathbf{g})$-process where :

$$g(t,i,j) = \delta_{ij}(1 - \sum_{j \in \mathbb{N}_+} f(t,i,j)), t \in \mathbb{R}_+, i,j \in \mathbb{N}_+,$$

while $(\tau_n)_{n \in \mathbb{N}_+}$ is the very sequence of jump times of the latter.

In this case we thus have

$$\Gamma(\lambda) = \lambda^{-1}(\mathbf{I} - \Sigma(\lambda)), \lambda > 0,$$

where

$$\Sigma = (\delta_{ij}\sigma_i)_{i,j \in \mathbb{N}_+}$$

with $\sigma_i = \sum_{j \in \mathbb{N}_+} \varphi_{ij}$ being the Laplace transform of $S_i, i \in \mathbb{N}_+$.

6.1 THE FORWARD EQUATION

Since Γ is a diagonal matrix the inverse

$$\Gamma^{-1} = \lambda(\delta_{ij}(1 - \sigma_i)^{-1})_{i,j \in \mathbb{N}_+}$$

exists, is non-negative, and the forward equation (2.11) reads

$$\Pi = \lambda^{-1}(\mathbf{I} - \Sigma) + \Pi(\mathbf{I} - \Sigma)^{-1}\Phi(\mathbf{I} - \Sigma), \lambda > 0. \tag{2.16}$$

Note that the diagonal matrix $(\mathbf{I}-\Sigma)^{-1}$ contains the Laplace transforms of the renewal functions ${}_{n\in\mathbb{N}}S_i^{n*}, i \in \mathbb{N}_+$. (Here $*$ denotes convolution.) Hence (2.16) leads to the equations

$$p(t,i,j) = \delta_{ij}(1 - S_i(t)) \tag{2.17}$$

$$+ \sum_{k\in\mathbb{N}_+}\sum_{n\in\mathbb{N}}\int_0^t p(t-u,i,k)S_k^{n*} * f(\cdot,k,j) * T_j(u)du, t \in \mathbb{R}_+, i,j \in \mathbb{N}_+,$$

where $T_j = 1 - S_j, j \in \mathbb{N}_+$.

In the special case of the minimal Markov process associated with a conservative infinitesimal transition matrix, equations (2.17) are equivalent to Kolmogorov's system of forward differential equations [cf. Chung (1967, Part II, § 17)]. As Feller (1964) notes, in that case the forward equations are obtained formally by considering the *last* jump preceding t (although they may hold even when *no* last jump exists). It is difficult to see how equations (2.17) might be interpreted in such terms.

6.2 ASYMPTOTIC BEHAVIOR

In the present case condition (2.14) is satisfied with

$$g_{ij} = \delta_{ij}m_i, i,j \in \mathbb{N}_+,$$

so that Theorem 7 reduces to Theorem 5.9 in Koroliuk and Turbin (1976).

In Feller (1964), the existence of the limits $\lim_{t\to\infty} p(t,i,j), i,j \in \mathbb{N}_+$, is claimed under much weaker assumptions. Namely, $\mathbf{P}$ is assumed to be indecomposable and (just) recurrent so that [cf.Derman (1954)] there exists a unique strictly positive invariant vector $\mathbf{r} = (1, r_2, r_3, \cdots)$ for $\mathbf{P}$, that is, $\mathbf{rP} = \mathbf{r}$; and it is also assumed that

$$0 < m_i < \infty, \ i \in \mathbb{N}_+.$$

Under these only assumptions it is stated that for the (unique) solution to the backward equations we have

$$\lim_{t\to\infty} p(t,i,j) = \frac{r_j m_j}{\sum_{k\in\mathbb{N}_+} r_k m_k}, \ i,j \in \mathbb{N}_+, \tag{2.18}$$

if $\sum_{k\in\mathbb{N}_+} r_k m_k < \infty$, and

$$\lim_{t\to\infty} p(t,i,j) = 0, i,j \in \mathbb{N}_+, \tag{2.19}$$

if $\sum_{k\in\mathbb{N}_+} r_k m_k = \infty$.

Feller's proof of (2.18) and (2.19) is a heuristic one, and we have been unable to translate it into a rigorous proof.

6.3 OPEN PROBLEMS

The remarks in the two preceding subsections point out to the following open problems.

1. What is the probabilistic meaning of the forward equations (2.6) ?

2. Is it possible to prove (2.18) and (2.19) under Feller's assumptions ?

3. We think both problems are important and difficult.

References

[1] K.L. Chung (1967), *Markov Chains with Stationary Transition Probabilities*, 2nd Edition. Springer, Berlin.

[2] C. Derman (1954), A solution to a set of fundamental equations in Markov chains. *Proc. Amer. Math. Soc.* **5**, 332-334.

[3] W.Feller (1956), Boundaries induced by non-negative matrices. *Trans. Amer. Math. Soc.* **83**, 19-54.

[4] W. Feller (1964), On semi-Markov processes. *Proc. Nat. Acad. Sci. USA*, **51**, 653-659

[5] W. Feller (1966), *An Introduction to Probability Theory and Its Applications,* Vol.II. Wiley, New York.

[6] Z. Hou, Z. Liu and J. Zou (1997), QNQL processes– (H, Q)-processes and their applications. *Chinese Sci.Bull.* **42**, no.11, 881-886.

[7] J.G. Kemeny, J.L. Snell and A.W. Knapp (1976), *Denumerable Markov Chains,* 2nd Edition. Springer, New York.

[8] V.S. Koroliuk and A.F. Turbin (1976), *Semi-Markov Processes and Their Applications.* Naukova Dumka, Kiev. (Russian)

Chapter 3

QUASI-STATIONARY PHENOMENA FOR SEMI-MARKOV PROCESSES

Mats Gyllenberg
University of Turku, Finland

Dmitrii S. Silvestrov
Umeå University, Sweden and
University of Turku, Finland

Abstract New types of *nonlinear asymptotical expansions* are obtained for distribution of first hitting times and quasi-stationary distributions for *nonlinearly perturbed* semi-Markov processes.

Keywords: Semi-Markov process, large deviation theorem, ergodic theorem, renewal equation, nonlinear perturbation, nonlinear asymptotical expansion, absorption, quasi-stationary distribution.

1. INTRODUCTION

The aim of this paper is to present a new type of asymptotical expansions for distributions of lifetime functionals and quasi-stationary distributions for nonlinearly perturbed semi-Markov processes.

We consider a semi-Markov process $\eta^{(\varepsilon)}(t), t \geq 0$ with phase space $X = \{0, 1, \dots, N\}$ and transition probabilities $Q_{ij}^{(\varepsilon)}(u)$. The state 0 in an absorption state, that is, the transition probabilities $Q_{0j}^{(\varepsilon)}(\infty) = 0$ for all $j \neq 0$.

Our studies are concerned with the random functional $\mu_0^{(\varepsilon)}$ which is the first hitting time of the process $\eta^{(\varepsilon)}(t), t \geq 0$ into the absorption state 0. In applications the absorption times $\mu_0^{(\varepsilon)}$ are often interpreted as transition times for different stochastic systems described by semi-

Markov processes (occupation times or waiting times in queuing systems, lifetimes in reliability models, extinction times in population dynamic models, etc.).

The semi-Markov process $\eta^{(\varepsilon)}(t), t \geq 0$ is assumed to depend on a small perturbation parameter $\varepsilon \geq 0$. The processes $\eta^{(\varepsilon)}(t), t \geq 0$ for $\varepsilon > 0$ are considered as perturbations of the process $\eta^{(0)}(t), t \geq 0$ and therefore we assume some weak continuity conditions for the transition quantities, namely the moment functionals of the transition probabilities, as functions of ε at point $\varepsilon = 0$.

The object of our study is the joint distributions $P_i\{\mu_0^{(\varepsilon)} > t, \eta^{(\varepsilon)}(t) = j\}$ and their asymptotic behaviour as time $t \to \infty$ and the parameter $\varepsilon \to 0$.

One of the main new elements in the present study is that we consider the model with *nonlinear* perturbations. This means that the moment functionals $\varphi_{ij}^{(\varepsilon)}(\rho, r) = \int_0^\infty u^r e^{\rho u} Q_{ij}^{(\varepsilon)}(du), r = 1, \dots, k$ are nonlinear functions of ε. However, we restrict our attention to the case of a smooth nonlinear perturbation and assume that these moment functionals have k derivatives at $\varepsilon = 0$, i.e., these functions can be expanded in a power series with respect to ε up to and including the order k.

It turns out that the relation between the speeds at which ε tends to zero and time t tends to infinity has a delicate influence upon the results. Without loss of generality we assume that time $t = t^{(\varepsilon)}$ is a function of the parameter ε. The balance between the rate of perturbation and the rate of growth of time is characterized by the following condition which is assumed to hold for some $1 \leq r \leq k$:

$$\varepsilon^r t^{(\varepsilon)} \to \lambda_r < \infty \quad \text{as} \quad \varepsilon \to 0. \tag{3.1}$$

Under the assumptions mentioned above and some natural additional conditions of Cramér type for transition probabilities $Q_{ij}^{(\varepsilon)}(u)$ we obtain the asymptotical relation

$$\frac{P_i\{\mu_0^{(\varepsilon)} > t^{(\varepsilon)}, \eta^{(\varepsilon)}(t^{(\varepsilon)}) = j\}}{\exp\{-(\rho^{(0)} + a_1\varepsilon + \dots + a_{r-1}\varepsilon^{r-1})t^{(\varepsilon)}\}} \to \tag{3.2}$$

$$e^{-\lambda_r a_r} \pi_{ij}^{(0)} \quad \text{as} \quad \varepsilon \to 0.$$

We derive an equation to find the coefficient $\rho^{(0)}$, and systems of linear equation to find $\pi_{ij}^{(0)}, i, j \neq 0$ and finally an explicit recurrent algorithm for calculation of the coefficients and $a_1, \dots, a_k$ as rational functions of the coefficients in the expansions for the moments of the transition probabilities.

The asymptotical behaviour of $P_i\{\mu_0^{(\varepsilon)} > t, \eta^{(\varepsilon)}(t) = j\}$ depends on which of the following two alternatives hold: (1) $\sum_{i \neq 0} Q_{i0}^{(0)}(\infty) > 0$ or (2) $\sum_{i \neq 0} Q_{i0}^{(0)}(\infty) = 0$.

The first alternative means that absorption in 0 is possible for the limiting process $\eta^{(0)}(t), t \geq 0$ and that the coefficient $\rho^{(0)} > 0$. In this case the absorption times $\mu_0^{(\varepsilon)}$ are stochastically bounded as $\varepsilon \to 0$.

The second alternative means that absorption is impossible for the limiting process $\eta^{(0)}(t), t \geq 0$ and that the coefficient $\rho^{(0)} = 0$. In this case the absorption times $\mu_0^{(\varepsilon)} \xrightarrow{P} \infty$ as $\varepsilon \to 0$.

In the literature the asymptotics related to the first model is known as *quasi-stationary* asymptotics. To distinguish between the asymptotics (3.2) for the first and the second cases, Gyllenberg and Silvestrov (1997) coined the term *pseudo-stationary* for the second one.

The origin of research in the case of quasi-stationary asymptotics stems from Yaglom (1947) and intensive studies began in the sixties by Vere-Jones (1962), Kingman (1963), Darroch and Seneta (1965), Seneta and Vere-Jones (1966). The studies of pseudo-stationary asymptotics were began by Simon and Ando (1961), Schweitzer (1968) and Korolyuk (1969). The annotated bibliography by Pollett (1996), the report by Gyllenberg and Silvestrov (1997) and the book by Kijima (1997) contain extensive bibliographies of works in the area.

The results known in the literature and related to asymptotics given in (3.2) cover mainly the case $k = 1$ which corresponds to the model of linearly perturbed processes. In this case the condition (3.1) reduces to $\varepsilon t^{(\varepsilon)} \to \lambda_1$ and the asymptotical relation (3.2) can be rewritten in the form

$$\frac{P_i\{\mu_0^{(\varepsilon)} > t^{(\varepsilon)}, \eta^{(\varepsilon)}(t^{(\varepsilon)}) = j\}}{\exp\{-\rho^{(0)} t^{(\varepsilon)}\}} \to e^{-\lambda_1 a_1} \pi_{ij}^{(0)} \quad \text{as} \quad \varepsilon \to 0. \tag{3.3}$$

The asymptotical relation (3.3) does not give satisfactory information about the asymptotics of the probabilities $P_i\{\mu_0^{(\varepsilon)} > t, \eta^{(\varepsilon)}(t) = j\}$ in zones of large deviations when $\varepsilon t^{(\varepsilon)} \to \infty$.

The relation (3.2) for $k > 1$ provides a means to describe this asymptotics. This case corresponds to the model of nonlinearly perturbed processes. This is one of the main new element of our results. Another new element is the high order asymptotical expansions which are new even in the case of linearly perturbed processes.

In the case $k = 2$ (3.1) reduces to $\varepsilon^2 t^{(\varepsilon)} \to \lambda_2$ and the asymptotical relation (3.2) can be rewritten in the form

$$\frac{P_i\{\mu_0^{(\varepsilon)} > t^{(\varepsilon)}, \eta^{(\varepsilon)}(t^{(\varepsilon)}) = j\}}{\exp\{-(\rho^{(0)} + a_1\varepsilon)t^{(\varepsilon)}\}} \to e^{-\lambda_2 a_2}\pi_{ij}^{(0)} \quad \text{as} \quad \varepsilon \to 0. \tag{3.4}$$

The limit on the right hand side is the limit of the relative error of the approximation (3.2) in the large deviation zone of the orders $\varepsilon t^{(\varepsilon)} = o(\varepsilon^{-1})$ or $\varepsilon t^{(\varepsilon)} = O(\varepsilon^{-1})$.

If $\lambda_2 = 0$ then $\varepsilon t^{(\varepsilon)} = o(\varepsilon^{-1})$ and the asymptotic relative error is 1. Note that this case covers also the situation with bounded $\varepsilon t^{(\varepsilon)}$, which corresponds to the asymptotical relation (3.2) in the case $k = 1$. This is already an extension of that asymptotic result since it is possible that $\varepsilon t^{(\varepsilon)} \to \infty$.

If $\lambda_2 > 0$ then $\varepsilon t^{(\varepsilon)} = O(\varepsilon^{-1})$ and the asymptotic relative error is $e^{-\lambda_2 a_2}$. It differs from 1. Therefore $o(\varepsilon^{-1})$ is the asymptotical bound for the large deviation zone with the asymptotic relative error 1.

Using relation (3.2) for $k > 2$ one can also describe the asymptotics of relative errors in large deviation zones of higher orders $\varepsilon t^{(\varepsilon)} = o(\varepsilon^{-k+1})$ or $\varepsilon t^{(\varepsilon)} = O(\varepsilon^{-k+1})$.

The relation (3.2) can be interpreted as a new kind of mixed large deviation and ergodic theorems for nonlinearly perturbed semi-Markov processes.

Under some natural conditions of nonarithmeticity of the transition probabilities $Q_{ij}^{(\varepsilon)}(u)$, there exists a so-called quasi-stationary distribution for the semi-Markov process $\eta^{(\varepsilon)}(t), t \geq 0$ which is given by the formula

$$\kappa_j^{(\varepsilon)} = \lim_{t\to\infty} P_i\{\eta^{(\varepsilon)}(t) = j \mid \mu_0^{(\varepsilon)} > t\}, j \neq 0. \tag{3.5}$$

Using the asymptotics (3.2) we also obtain nonlinear asymptotical expansions for the quasi-stationary distributions of the nonlinearly perturbed semi-Markov processes

$$\kappa_j^{(\varepsilon)} = \kappa_j^{(0)} + g_j[1]\varepsilon + \cdots + g_j[k]\varepsilon^k + o(\varepsilon^k). \tag{3.6}$$

We give an explicit recurrent algorithm for calculation of the coefficients and $g_j[1], \ldots, g_j[k]$ as rational functions of the coefficients in the expansions for the moment functionals of transition probabilities.

The results known in the literature relate to the model of linearly perturbed Markov chains and the first order asymptotics. These studies were initiated by Schweitzer (1968). The related references can be found in the papers by Seneta (1988) and Stewart (1991).

As in the case of the asymptotic relation (3.2) the new elements of our studies are the nonlinear perturbations and nonlinear high order asymptotical expansions of type (3.6) for quasi-stationary and usual stationary distributions of the perturbed semi-Markov processes.

Another new element is the unified treatment of different types of asymptotics: Both the quasi- and pseudo-stationary cases are covered by the same analytical approach. This approach is based on results by Silvestrov (1976, 1978, 1979) concerning an extension of renewal theory to a model of perturbed renewal equations and the recent improvements of these results achieved in works by Silvestrov (1995) and Gyllenberg and Silvestrov (1997, 1998a, 1998b) concerning exponential asymptotics for perturbed renewal equations. These papers contain also examples of applications of exponential asymptotical expansions for nonlinearly perturbed semi-Markov processes to the analysis of quasi- and pseudo-stationary phenomena in stochastic systems, in particular metapopulation models and highly reliable queuing systems.

We finally comment on the structure of the paper. In Section 2. we present the results concerning exponential asymptotics for perturbed renewal equation, which we think are interesting in their own right. Section 3. contains results concerning asymptotical expansions for moments of first hitting times for perturbed semi-Markov processes. We postpone the presentation of our main results concerning mixed large deviation and ergodic asymptotics of the type (3.2) to Sections 4. in order to make the presentation of these results more readable. In Sections 5. we present asymptotical expansions of the type (3.6) for quasi-stationary distributions.

2. EXPONENTIAL ASYMPTOTICS FOR PERTURBED RENEWAL EQUATION

Consider the family of renewal equations

$$x^{(\varepsilon)}(t) = q^{(\varepsilon)}(t) + \int_0^t x^{(\varepsilon)}(t-s)F^{(\varepsilon)}(ds), t \geq 0, \tag{3.7}$$

where for every $\varepsilon \geq 0$, (a) $q^{(\varepsilon)}(t)$ is a measurable and locally bounded (that is, bounded on every finite interval) real-valued function on $[0, \infty)$ and (b) $F^{(\varepsilon)}(s)$ is a distribution function on $[0, \infty)$ which is not concentrated at 0 and can be improper, i.e., its total variation $F^{(\varepsilon)}(\infty)$ can be equal to or less than 1. As is well known, there is a unique measurable and locally bounded solution $x^{(\varepsilon)}(t)$ of equation (3.7).

We denote the expectation of $F^{(\varepsilon)}(s)$ by $m^{(\varepsilon)}$:

$$m^{(\varepsilon)} = \int_0^\infty sF^{(\varepsilon)}(ds).$$

We assume that the functions $q^{(\varepsilon)}(t)$ and distributions $F^{(\varepsilon)}(s)$ satisfy the following continuity conditions at $\varepsilon = 0$:

A (a) $F^{(\varepsilon)}(\cdot) \Rightarrow F^{(0)}(\cdot)$ as $\varepsilon \to 0$ (converge weakly), where $F^{(0)}(s)$ is a non-arithmetic distribution function which is not concentrated at the origin;

(b) $m^{(\varepsilon)} \to m^{(0)} < \infty$ as $\varepsilon \to 0$.

B (a) $\lim_{u\to 0} \overline{\lim}_{0\le\varepsilon\to 0} \sup_{|v|\le u} |q^{(\varepsilon)}(t+v) - q^{(0)}(t)| = 0$ almost everywhere with respect to Lebesgue measure on $[0,\infty)$;

(b) $\overline{\lim}_{0\le\varepsilon\to 0} \sup_{0\le t\le T} \mid q^{(\varepsilon)}(t) \mid < \infty$ for every $T \ge 0$;

(c) $\lim_{T\to\infty} \overline{\lim}_{0\le\varepsilon\to 0} h \sum_{r\ge T/h} \sup_{rh\le t\le(r+1)h} \mid q^{(\varepsilon)}(t) \mid = 0$ for some $h > 0$.

Here and henceforth we use the symbol $\varepsilon \to 0$ to denote that ε tends to 0 being positive and the symbol $0 \le \varepsilon \to 0$ if zero values of ε are also admitted.

Note that condition **A** implies that $F^{(\varepsilon)}(\infty) \to F^{(0)}(\infty)$ as $\varepsilon \to 0$, but that the limiting total variation $F^{(0)}(\infty)$ can be equal to or less than 1.

The conditions **A** and **B** reduce to the conditions of the classical renewal theorem (Feller, 1966) if $F^{(\varepsilon)}(s) \equiv F^{(0)}(s)$ and $q^{(\varepsilon)}(t) \equiv q^{(0)}(t)$ do not depend on ε and $F^{(0)}(s)$ is a proper distribution function, i.e., if $F^{(0)}(\infty) = 1$. In particular, condition **A** reduces to the assumption that $F^{(0)}(s)$ is a non-arithmetic distribution function with a finite expectation $m^{(0)}$ and **B** to the assumption that the function $q^{(0)}(t)$ is directly Riemann integrable on $[0,\infty)$.

The starting point for our investigation is the following lemma (Silvestrov, 1976, 1978, 1979) generalizing the classical renewal theorem to the model of perturbed renewal equations.

Lemma 1 *Let $F^{(\varepsilon)}(s)$ be proper distributions for every $\varepsilon \ge 0$ and conditions* **A** *and* **B** *hold. Assume $t^{(\varepsilon)} \to \infty$ as $\varepsilon \to 0$. Then:*

$$x^{(\varepsilon)}(t^{(\varepsilon)}) \to x^{(0)}(\infty) = \frac{\int_0^\infty q^{(0)}(s)ds}{m^{(0)}} \quad \text{as} \quad \varepsilon \to 0. \tag{3.8}$$

We introduce the moment generating functions

$$\varphi^{(\varepsilon)}(\rho) = \int_0^\infty e^{\rho s} F^{(\varepsilon)}(du), \quad \rho \ge 0$$

and assume that the distributions $F^{(\varepsilon)}(s)$ have finite exponential moments and that the functions $q^{(\varepsilon)}(t)$ are exponentially integrable, that is, we assume that the following conditions are fulfilled:

C There exists a $\delta > 0$ such, that:

(a) $\overline{\lim}_{0\leq\varepsilon\to 0}\varphi^{(\varepsilon)}(\delta) < \infty$;

(b) $\varphi^{(0)}(\delta) \in (1,\infty)$.

D The functions $e^{\delta t}q^{(\varepsilon)}(t)$ satisfy condition **B**.

As is well-known, the asymptotic behaviour of the solution of the renewal equation depends on the root of the characteristic equation

$$\varphi^{(\varepsilon)}(\rho) = \int_0^\infty e^{\rho s}F^{(\varepsilon)}(ds) = 1. \tag{3.9}$$

Lemma 2 *If conditions* **A** *and* **C** *hold, then for all ε small enough: (a) there exists a unique nonnegative root $\rho^{(\varepsilon)}$ of equation (3.9), (b) $\rho^{(\varepsilon)} < \delta$; (c) $\rho^{(\varepsilon)} \to \rho^{(0)}$ as $\varepsilon \to 0$.*

The root $\rho^{(0)}$ of equation (3.9) for $\varepsilon = 0$ plays a special role. Note first of all that $\rho^{(0)} = 0$ if $F^{(0)}(\infty) = 1$ or $\rho^{(0)} > 0$ if $F^{(0)}(\infty) < 1$.

We define for $r = 0, 1, \ldots$ the moment functionals

$$\varphi^{(\varepsilon)}(\rho, r) = \int_0^\infty s^r e^{\rho s}F^{(\varepsilon)}(ds), \rho \geq 0.$$

Note that by definition $\varphi^{(\varepsilon)}(\rho, 0) = \varphi^{(\varepsilon)}(\rho)$.

It follows from conditions **A** and **C** that for any $\delta' \in [0,\delta)$ moment functionals $\varphi^{(\varepsilon)}(\delta', r) < \infty, r = 0, 1, \ldots$ for all ε small enough.

The following perturbation condition plays a crucial role in the subsequent analysis.

E$_k$ $\varphi^{(\varepsilon)}(\rho^{(0)}, r) = \varphi^{(0)}(\rho^{(0)}, r) + b_{1r}\varepsilon + \cdots + b_{kr}\varepsilon^k + o(\varepsilon^k)$ for $r = 0, \ldots, k$, where $|b_{nr}| < \infty, n = 1, \ldots, k,\ r = 0, \ldots, k$.

It is convenient to define $b_{0r} = \varphi^{(0)}(\rho^{(0)}, r), r = 0, 1, \ldots$. By the definition of $\rho^{(0)}$ it is clear that $b_{00} = \varphi^{(0)}(\rho^{(0)}, 0) = 1$.

We denote

$$\tilde{x}^{(\varepsilon)}(\infty) = \frac{\int_0^\infty e^{\rho^{(\varepsilon)}s}q^{(\varepsilon)}(s)ds}{\int_0^\infty se^{\rho^{(\varepsilon)}s}F^{(\varepsilon)}(ds)}.$$

Lemma 3 *Let conditions* **A**–**D** *hold. Then: $\tilde{x}^{(\varepsilon)}(\infty) \to \tilde{x}^{(0)}(\infty)$ as $\varepsilon \to 0$.*

The following theorem is the main result concerning exponential asymptotics for perturbed renewal equations.

Theorem 1 *Let conditions* **A**–$\mathbf{E}_k$ *be satisfied. Then:*

(i) *The root* $\rho^{(\varepsilon)}$ *of equation (3.9) has the asymptotic expansion*

$$\rho^{(\varepsilon)} = \rho^{(0)} + a_1\varepsilon + \cdots + a_k\varepsilon^k + o(\varepsilon^k), \tag{3.10}$$

where the coefficients a_n *are given by the recursion formulas* $a_1 = -b_{01}^{-1}b_{10}$ *and in general for* $n = 1, \ldots, k$*:*

$$a_n = b_{01}^{-1}(-b_{n0} - \sum_{q=1}^{n-1} b_{n-q1}a_q - \tag{3.11}$$

$$\sum_{2\leq m\leq n}\sum_{q=2}^{n} b_{n-qm} \cdot \sum_{n_1,\ldots,n_{q-1}\in D_{mq}} \prod_{p=1}^{q-1} a_p^{n_p}/n_p!),$$

where D_{mq} *is the set of all nonnegative, integer solutions of the system* $n_1 + \cdots + n_{q-1} = m,\ n_1 + \cdots + (q-1)n_{q-1} = q$.

(ii) *If* $t^{(\varepsilon)} \to \infty$*, the following asymptotic relation holds*

$$\frac{x^{(\varepsilon)}(t^{(\varepsilon)})}{\exp\{-\rho^{(\varepsilon)}t^{(\varepsilon)}\}} \to \tilde{x}^{(0)}(\infty) \text{ as } \varepsilon \to 0. \tag{3.12}$$

(iii) *If* $t^{(\varepsilon)} \to \infty$ *such that* $\varepsilon^r t^{(\varepsilon)} \to \tau_r \in [0,\infty)$ *for some* $1 \leq r \leq k$*, the following asymptotic relation holds:*

$$\frac{x^{(\varepsilon)}(t^{(\varepsilon)})}{\exp\{-(\rho^{(0)} + a_1\varepsilon + \cdots + a_{r-1}\varepsilon^{r-1})t^{(\varepsilon)}\}} \to \tag{3.13}$$

$$e^{-\tau_r a_r}\tilde{x}^{(0)}(\infty) \text{ as } \varepsilon \to 0.$$

Proof. By Lemma 2

$$\Delta^{(\varepsilon)} = \rho^{(\varepsilon)} - \rho^{(0)} \to 0 \text{ as } \varepsilon \to 0. \tag{3.14}$$

The Taylor expansion for the function e^s yields

$$e^{s\rho^{(\varepsilon)}} = e^{s\rho^{(0)}}(1 + s\Delta^{(\varepsilon)}/1! + \cdots + s^k(\Delta^{(\varepsilon)})^k/k! + \tag{3.15}$$

$$s^{k+1}(\Delta^{(\varepsilon)})^{k+1}e^{s|\Delta^{(\varepsilon)}|}\theta_{ks}^{(\varepsilon)}/(k+1)!,$$

where $0 \leq \theta_{ks}^{(\varepsilon)} \leq 1, 0 \leq s < \infty$.

Recall that $\rho^{(0)} < \delta$ and $\Delta^{(\varepsilon)} \to 0$. Therefore there exist $\delta_0 < \delta$ such that $\rho^{(0)} + | \Delta^{(\varepsilon)} | < \delta_0$ for ε small enough, say $\varepsilon \leq \varepsilon_0$. Integrating (3.15) and taking condition **C** into account one obtains for $\varepsilon \leq \varepsilon_0$

$$1 = \int_0^\infty e^{s\rho^{(\varepsilon)}} F^{(\varepsilon)}(ds) = \varphi^{(\varepsilon)}(\rho^{(0)}, 0) + \varphi^{(\varepsilon)}(\rho^{(0)}, 1)\Delta^{(\varepsilon)}/1! + \cdots + \tag{3.16}$$

$$\varphi^{(\varepsilon)}(\rho^{(0)}, k)(\Delta^{(\varepsilon)})^k/k! + (\Delta^{(\varepsilon)})^{k+1} M_k \theta_k^{(\varepsilon)}.$$

Here

$$M_k = ((k+1)!)^{-1} \sup_{\varepsilon \leq \varepsilon_0} \int_0^\infty s^{k+1} e^{s\delta_0} F^{(\varepsilon)}(ds) < \infty, \ 0 \leq \theta_k^{(\varepsilon)} \leq 1.$$

Formula (3.16) can be rewritten in the form

$$\varphi^{(\varepsilon)}(\rho^{(0)}, 1)\Delta^{(\varepsilon)}/1! + \varphi^{(\varepsilon)}(\rho^{(0)}, 2)(\Delta^{(\varepsilon)})^2/2! + \cdots + \tag{3.17}$$

$$\varphi^{(\varepsilon)}(\rho^{(0)}, k)(\Delta^{(\varepsilon)})^k/k! + (\Delta^{(\varepsilon)})^{k+1} M_k \theta_k^{(\varepsilon)} = 1 - \varphi^{(\varepsilon)}(\rho^{(0)}, 0).$$

The difference $\Delta^{(\varepsilon)} \to 0$ and the sum of all terms on the left side in (3.17) beginning from the second one is $o(\Delta^{(\varepsilon)})$. Recall also that $b_{00} = 1$ and therefore the expression on the right hand side is of order ε. Dividing both sides in (3.17) by $\varphi^{(\varepsilon)}(\rho^{(0)}, 1)\varepsilon$ and evaluating the corresponding limits we obtain using $\mathbf{E}_k$ that $\Delta^{(\varepsilon)}/\varepsilon \to -b_{10}/b_{01}$. This means that $\Delta^{(\varepsilon)}$ can be represented in the form

$$\Delta^{(\varepsilon)} = a_1\varepsilon + \Delta_1^{(\varepsilon)}, \tag{3.18}$$

where $a_1 = -b_{10}/b_{01}$ and $\Delta_1^{(\varepsilon)} = o(\varepsilon)$. The relation (3.18) reduces to (3.10) in the case $k = 1$.

Substituting the expansions given in condition **E** and (3.18) into (3.17) one obtains

$$\Delta_1^{(\varepsilon)} = a_2\varepsilon^2 + \Delta_2^{(\varepsilon)}, \tag{3.19}$$

where $a_2 = b_{01}^{-1}(-b_{20} - b_{11}a_1 - \frac{1}{2}b_{02}a_1^2)$ and $\Delta_2^{(\varepsilon)} = o(\varepsilon^2)$. The relations (3.18) and (3.19) yield the relation (3.10) for the case $k = 2$. The expression for a_2 given above is exactly formula (3.11) with $k = 2$.

Repeating the above argument we obtain the expansion (3.10) and the formula (3.11) for $k > 2$. However, the formula (3.11) can be obtained in a simpler way when the asymptotic expansion (3.10) has already been proven. From (3.17) we get the following formal equation

$$(b_{01} + b_{11}\varepsilon + \cdots)(a_1\varepsilon + a_2\varepsilon^2 + \cdots)/1! + \tag{3.20}$$

$$(b_{02} + b_{12}\varepsilon + \cdots)(a_1\varepsilon + a_2\varepsilon^2 + \cdots)^2/2! + \cdots = -(b_{10}\varepsilon + b_{20}\varepsilon^2 + \cdots).$$

Equalizing the coefficients of ε^n for $n \geq 1$ on the right and left hand sides of (3.20) we obtain the formula (3.11) for calculating the coefficients $a_1, \ldots, a_k$.

The first summand in (3.11) reflects the contribution of the sum on the right side in (3.20). The second sum term in (3.11) reflects the contribution of the first product of sums on the left side in (3.20). The sum with index $2 \leq m \leq n$ in (3.11) reflects the contribution of the product of sums $(b_{0m}+\cdots)(a_1\varepsilon+\cdots)^m/m!$ on the right side in (3.20). In this case we collect all terms in this product such that the sum $(b_{0m}+\cdots)$ contribute with ε^{n-q} and $(a_1\varepsilon + \cdots)^m/m!$ with the ε^q for $2 \leq q \leq n$. Note that the case $q = 1$ is impossible since the minimal power degree of ε which can contribute to this product is 2. The coefficient for ε^q coming from $(a_1\varepsilon + \cdots)^m/m!$ is the sum of the products $a_1^{n_1} \cdots a_q^{n_q}$ where nonnegative integer coefficients $n_1, \ldots, n_q$ must satisfy to obvious relations $n_1+\cdots+n_q = m$ and $n_1+\cdots+qn_q = q$. Only the values 0 and 1 are possible for the coefficient n_q. However, if $n_q = 1$ then all other coefficients $n_1, \ldots, n_{q-1}$ must equal 0. In this case $n_1 + \cdots + n_q = 1$, which contradicts the restriction $2 \leq m \leq n$. Therefore the coefficient n_q must equal 0. Finally, the combinatorial counting of the coefficient for the product $a_1^{n_1} \cdots a_q^{n_q}$ yields the number $m!/n_1! \cdots n_q!$. The coefficient $m!$ cancels since it also appears in as the denominator of $(a_1\varepsilon+\cdots)^m/m!$. This completes the proof of (i).

To prove statement (ii) we multiply the renewal equation (3.7) by $e^{t\rho^{(\varepsilon)}}$ and transform it to the equivalent form

$$\tilde{x}^{(\varepsilon)}(t) = \tilde{q}^{(\varepsilon)}(t) + \int_0^t \tilde{x}^{(\varepsilon)}(t-s)\tilde{F}^{(\varepsilon)}(ds), \ t \geq 0, \tag{3.21}$$

where

$$\tilde{x}^{(\varepsilon)}(t) = e^{t\rho^{(\varepsilon)}}x^{(\varepsilon)}(t), \ \tilde{q}^{(\varepsilon)}(t) = e^{t\rho^{(\varepsilon)}}q^{(\varepsilon)}(t), \ \tilde{F}^{(\varepsilon)}(t) = \int_0^t e^{s\rho^{(\varepsilon)}}F^{(\varepsilon)}(ds).$$

Because $\rho^{(\varepsilon)} \to \rho^{(0)}$ and $\rho^{(\varepsilon)} \leq \delta_0 < \delta$ for all ε small enough it is obvious that condition **A** is satisfied for the distribution functions $\tilde{F}^{(\varepsilon)}(s)$ if conditions **A** and **C** are satisfied for the distribution functions $F^{(\varepsilon)}(s)$. Note also that the distribution functions $\tilde{F}^{(\varepsilon)}(s)$ are proper for all $\varepsilon \geq 0$. Also, the mean of the distribution $\tilde{F}^{(\varepsilon)}(s)$ coincides with the mean with exponential weight $\varphi^{(\varepsilon)}(\rho^{(\varepsilon)}, 1)$ for the distribution $F^{(\varepsilon)}(s)$. Similarly condition **B** is satisfied for the functions $\tilde{q}^{(\varepsilon)}(t) = e^{t\rho^{(\varepsilon)}}q^{(\varepsilon)}(t)$ if this condition is satisfied for the functions $e^{\delta t}q^{(\varepsilon)}(t)$.

Now we can apply Lemma 1 to equation (3.21). Conditions **A** and **B** are satisfied. According to the remark made above the function $e^{t\rho^{(0)}}q^{(0)}(t)$ and the distribution function $\tilde{F}^{(0)}(s)$ must be used instead of $q^{(0)}(t)$ and $F^{(0)}(s)$, respectively, when calculating the constant $x^{(0)}(\infty)$, i.e., this constant must be replaced by the constant $\tilde{x}^{(0)}(\infty)$.

The limit (3.8), written for the functions $\tilde{x}^{(\varepsilon)}(t)$, takes the form

$$\tilde{x}^{(\varepsilon)}(t^{(\varepsilon)}) = \frac{x^{(\varepsilon)}(t^{(\varepsilon)})}{e^{-\rho^{(\varepsilon)}t^{(\varepsilon)}}} \to \tilde{x}^{(0)}(\infty) \text{ as } \varepsilon \to 0. \tag{3.22}$$

Under the balancing condition $\varepsilon^r t^{(\varepsilon)} \to \tau_r$ the asymptotical relation (3.22) can be rewritten in the following equivalent form:

$$\tilde{x}^{(\varepsilon)}(t^{(\varepsilon)}) = \frac{x^{(\varepsilon)}(t^{(\varepsilon)})}{e^{-(\rho^{(0)}+a_1\varepsilon+\cdots+a_r\varepsilon^r+o(\varepsilon^r))t^{(\varepsilon)}}} \sim \tag{3.23}$$

$$\frac{x^{(\varepsilon)}(t^{(\varepsilon)})}{e^{-(\rho^{(0)}+a_1\varepsilon+\cdots+a_r\varepsilon^r)t^{(\varepsilon)}}} \to \tilde{x}^{(0)}(\infty) \text{ as } \varepsilon \to 0.$$

The latter asymptotical relation is equivalent to (3.13). □

We define for $r = 0, 1, \ldots$ the moment functionals

$$\omega^{(\varepsilon)}(\rho, r) = \int_0^\infty s^r e^{\rho s} q^{(\varepsilon)}(s) ds, \rho \geq 0.$$

It follows from conditions **B**, **D** that for any $0 \leq \delta' < \delta$ one has $\omega^{(\varepsilon)}(\delta', r) < \infty, r = 0, 1, \ldots$ for all ε small enough.

We now formulate the perturbation condition $\mathbf{E}_k$ for functions $q^{(\varepsilon)}(t)$:

F $\omega^{(\varepsilon)}(\rho^{(0)}, r) = \omega^{(0)}(\rho^{(0)}, r) + c_{1r}\varepsilon + \cdots + c_{kr}\varepsilon^k + o(\varepsilon^k)$ for $r = 0, \ldots, k$, where $|c_{nr}| < \infty, n = 1, \ldots, k, r = 0, \ldots, k$.

It is convenient to define $c_{0r} = \omega^{(0)}(\rho^{(0)}, r), r = 0, 1, \ldots$.

The following theorem supplements the result of Theorem 1.

Theorem 2 *Let conditions* **A–D**, $\mathbf{E}_{k+1}$ *and* **F** *be satisfied. Then the functional* $\tilde{x}^{(\varepsilon)}(\infty)$ *has the asymptotic expansions:*

$$\tilde{x}^{(\varepsilon)}(\infty) = \frac{\omega^{(0)}(\rho^{(0)}, 0) + f'_1\varepsilon + \cdots + f'_k\varepsilon^k + o(\varepsilon^k)}{\varphi^{(0)}(\rho^{(0)}, 1) + f''_1\varepsilon + \cdots + f''_k\varepsilon^k + o(\varepsilon^k)} = \tag{3.24}$$

$$\tilde{x}^{(0)}(\infty) + f_1\varepsilon + \cdots + f_k\varepsilon^k + o(\varepsilon^k),$$

where the coefficients f'_n, f''_n *are given by the formulas:* $f'_1 = c_{00}, f'_1 = c_{10} + c_{01}a_1$, $f''_{01} = b_{01}, f''_1 = b_{11} + b_{02}a_1$, *and in general for* $n = 0, \ldots, k$:

$$f'_n = c_{n0} + \sum_{q=1}^{n} c_{n-q1}a_q + \sum_{2\leq m\leq n}\sum_{q=2}^{n} c_{n-qm} \cdot \sum_{n_1,\ldots,n_{q-1}\in D_{mq}} \prod_{p=1}^{q-1} a_p^{n_p}/n_p!$$

and

$$f_n'' = b_{n1} + \sum_{q=1}^{n} b_{n-q2} a_q + \sum_{2 \le m \le n} \sum_{q=2}^{n} b_{n-qm+1} \cdot \sum_{n_1, \dots, n_{q-1} \in D_{mq}} \prod_{p=1}^{q-1} a_p^{n_p} / n_p!,$$

where the coefficients f_n *are given by the recursion formulas:* $f_0 = \tilde{x}^{(0)}(\infty) = f_0'/f_0''$ *and in general for* $n = 0 \dots, k$*:*

$$f_n = (f_n' - \sum_{q=0}^{n-1} f_{n-q}'' f_q)/f_0''.$$

Proof. Note first of all that

$$\tilde{x}^{(\varepsilon)}(\infty) = \frac{\omega^{(\varepsilon)}(\rho^{(\varepsilon)}, 0)}{\varphi^{(\varepsilon)}(\rho^{(\varepsilon)}, 1)}. \tag{3.25}$$

We can write down the asymptotic expansion of the numerator and the denominator in the quotient in (3.25) in powers of the difference $\Delta^{(\varepsilon)} = \rho^{(\varepsilon)} - \rho^{(0)}$. Recall that $\rho^{(0)} < \delta$, where δ is defined in **C**, and $\Delta^{(\varepsilon)} \to 0$. Hence there exists $\delta_0 < \delta$ such that $\rho^{(0)} + \mid \Delta^{(\varepsilon)} \mid \le \delta_0$ for ε small enough, say $\varepsilon \le \varepsilon_0$. Substituting the Taylor expansion (3.15) for the function $e^{\rho^{(\varepsilon)} s}$ into the integrals defining $\omega^{(\varepsilon)}(\rho^{(\varepsilon)}, 0)$ and $\varphi^{(\varepsilon)}(\rho^{(\varepsilon)}, 1)$ one obtains for $\varepsilon \le \varepsilon_0$

$$\omega^{(\varepsilon)}(\rho^{(\varepsilon)}, 0) = \omega^{(\varepsilon)}(\rho^{(0)}, 0) + \Delta^{(\varepsilon)} \omega^{(\varepsilon)}(\rho^{(0)}, 1) + \cdots + \tag{3.26}$$

$$\frac{(\Delta^{(\varepsilon)})^k}{k!} \omega^{(\varepsilon)}(\rho^{(0)}, k) + (\Delta^{(\varepsilon)})^{(k+1)} M_k' \theta_k'^{(\varepsilon)},$$

where

$$M_k' = ((k+1)!)^{-1} \sup_{\varepsilon \le \varepsilon_0} \int_0^\infty s^{k+1} e^{s\delta_0} g^{(\varepsilon)}(s) ds < \infty, \ 0 \le \theta_k'^{(\varepsilon)} \le 1$$

and

$$\varphi^{(\varepsilon)}(\rho^{(\varepsilon)}, 0) = \varphi^{(\varepsilon)}(\rho^{(0)}, 1) + \Delta^{(\varepsilon)} \varphi^{(\varepsilon)}(\rho^{(0)}, 2) + \cdots + \tag{3.27}$$

$$\frac{(\Delta^{(\varepsilon)})^k}{k!} \varphi^{(\varepsilon)}(\rho^{(0)}, k+1), + (\Delta^{(\varepsilon)})^{(k+1)} M_k'' \theta_k''^{(\varepsilon)},$$

where

$$M_k'' = ((k+1)!)^{-1} \sup_{\varepsilon \le \varepsilon_0} \int_0^\infty s^{k+2} e^{s\delta_0} F^{(\varepsilon)}(ds) < \infty, \ 0 \le \theta_k''^{(\varepsilon)} \le 1.$$

It was proved in Theorem 1 that the functions $\Delta^{(\varepsilon)}$ can be expanded in asymptotical series. The functions $\omega^{(\varepsilon)}(\rho^{(0)}, n)$ and $\varphi^{(\varepsilon)}(\rho^{(0)}, n)$ can also be expanded in asymptotical series due to conditions **F** and $\mathbf{E}_{k+1}$, respectively. Substituting these expansions into (3.26) and (3.27) we obtain

$$\omega^{(\varepsilon)}(\rho^{(\varepsilon)}, 0) = (c_{00} + \cdots c_{k0}\varepsilon^k + o(\varepsilon^k)) + \tag{3.28}$$

$$(c_{01} + \cdots c_{k1}\varepsilon^k + o(\varepsilon^k))(a_1\varepsilon + \cdots + a_k\varepsilon^k + o(\varepsilon^k))/1! + \cdots +$$

$$(c_{0k} + \cdots c_{kk}\varepsilon^k + o(\varepsilon^k))(a_1\varepsilon + \cdots + a_k\varepsilon^k + o(\varepsilon^k))^k/k! + o(\varepsilon^k),$$

and

$$\varphi^{(\varepsilon)}(\rho^{(0)}, 1) = (b_{01} + \cdots b_{k1}\varepsilon^k + o(\varepsilon^k)) + \tag{3.29}$$

$$(b_{02} + \cdots b_{k2}\varepsilon^k + o(\varepsilon^k))(a_1\varepsilon + \cdots + a_k\varepsilon^k + o(\varepsilon^k))/1! + \cdots +$$

$$(b_{0k+1} + \cdots b_{kk+1}\varepsilon^k + o(\varepsilon^k))(a_1\varepsilon + \cdots + a_k\varepsilon^k + o(\varepsilon^k))^k/k! + o(\varepsilon^k).$$

By grouping the coefficients for powers of ε in these expansions one can transform them to the form $\omega^{(\varepsilon)}(\rho^{(\varepsilon)}, 0) = f_0' + \cdots f_k'\varepsilon^k + o(\varepsilon^k)$ and $\varphi^{(\varepsilon)}(\rho^{(\varepsilon)}, 1) = f_0'' + \cdots f_k''\varepsilon^k + o(\varepsilon^k)$ described in (3.24).

The transition of the expansion (3.24) to the polynomial form is obvious. □

3. ASYMPTOTICAL EXPANSIONS FOR MOMENTS OF HITTING TIMES

In this section we describe the recursive algorithm for asymptotic analysis of exponential type moments of hitting times for nonlinearly perturbed semi-Markov processes. The algorithm is important in its own right and form an essential part of the proof of our main results concerning semi-Markov processes.

Let for every $\varepsilon \geq 0$, $\eta^{(\varepsilon)}(t), t \geq 0$ be a semi-Markov process, continuous from the right, with phase space $X = \{0, 1, \ldots, N\}$, and with transition probabilities $Q_{ij}^{(\varepsilon)}(u) = p_{ij}^{(\varepsilon)} F_{ij}^{(\varepsilon)}(u)$. Here $p_{ij}^{(\varepsilon)} = Q_{ij}^{(\varepsilon)}(\infty)$ are the transition probabilities of the embedded Markov chain and $F_{ij}^{(\varepsilon)}(u)$ are the distribution functions of transition times. We exclude instant states by assuming that $F_{ij}^{(\varepsilon)}(0) = 0, i, j \in X$.

We consider a model in which 0 is an absorbing state, that is, the following condition holds:

G $p_{0j}^{(\varepsilon)} = 0, j \neq 0$ for all $\varepsilon \geq 0$.

We denote the moments of time at which the jumps of the semi-Markov process $\eta^{(\varepsilon)}(t)$ occur by $\tau^{(\varepsilon)}(n)$ and we let $\eta_n^{(\varepsilon)} = \eta^{(\varepsilon)}(\tau^{(\varepsilon)}(n))$ be the positions of this process at these moments. The sequence $\eta_n^{(\varepsilon)}$ is an embedded Markov chain for semi-Markov process $\eta^{(\varepsilon)}(t)$.

We also define $\nu_j^{(\varepsilon)} = \min(n \geq 1 : \eta_n^{(\varepsilon)} = j)$ and $\mu_j^{(\varepsilon)} = \tau^{(\varepsilon)}(\nu_j^{(\varepsilon)})$. By definition $\nu_j^{(\varepsilon)}$ and $\mu_j^{(\varepsilon)}$ are the first hitting times to the state j for the embedded Markov chain $\eta_n^{(\varepsilon)}$ and the semi-Markov process $\eta^{(\varepsilon)}(t)$, respectively. The first hitting times $\nu_0^{(\varepsilon)}$ and $\mu_0^{(\varepsilon)}$ are called the absorption times for the embedded Markov chain $\eta_n^{(\varepsilon)}$ and the semi-Markov process $\eta^{(\varepsilon)}(t)$, respectively.

We assume that the following perturbation conditions hold:

H $p_{ij}^{(\varepsilon)} \to p_{ij}^{(0)}$ as $\varepsilon \to 0,\ i \neq 0, j \in X$.

I $F_{ij}^{(\varepsilon)}(\cdot) \Rightarrow F_{ij}^{(0)}(\cdot)$ as $\varepsilon \to 0,\ i \neq 0, j \in X$, where $F_{ij}^{(0)}(s)$ is a proper non-arithmetic distribution function for $i \neq 0, j \in X$.

The object of our interest is the asymptotic behaviour of the transition probabilities $P_{ij}^{(\varepsilon)}(t) = P_i\{\eta^{(\varepsilon)}(t) = j, \mu_0^{(\varepsilon)} > t\}$ as $t \to \infty$ and $\varepsilon \to 0$.

We formulate results for the basic case where the following communication condition holds:

J The set $X_1 = \{1, \ldots, N\}$ is a communicative class of states for the limiting semi-Markov process $\eta^{(0)}(t), t \geq 0$.

We define for $i \neq 0, j \in X$ the moment generating functions

$$\psi_{ij}^{(\varepsilon)}(\rho) = \int_0^\infty e^{\rho s} F_{ij}^{(\varepsilon)}(ds),\ \rho \geq 0.$$

We assume that the following Cramér type condition holds:

K There exists $\delta > 0$ such that $\overline{\lim}_{0 \leq \varepsilon \to 0} \psi_{ij}^{(\varepsilon)}(\delta) < \infty,\ i \neq 0, j \in X$.

It follows from the condition **K** that for any $\delta' \in [0, \delta)$ one has $\psi_{ij}^{(\varepsilon)}(\delta') < \infty, i \neq 0, j \in X$ for all ε small enough.

We introduce for $i, j \neq 0$ the moment generating functions of the first hitting times

$$\varphi_{ij}^{(\varepsilon)}(\rho) = \int_0^\infty e^{\rho s} G_{ij}^{(\varepsilon)}(ds),\ \rho \geq 0, \text{where } G_{ij}^{(\varepsilon)}(s) = P_i\{\mu_j^{(\varepsilon)} \leq s\}.$$

Lemma 4 *Let conditions* **G–K** *hold. Then there exist* $\beta \in (0, \delta]$ *such that* $\varphi_{ij}^{(\varepsilon)}(\beta) < \infty, i, j \neq 0$ *for all* ε *small enough; (b)* $\varphi_{ij}^{(\varepsilon)}(\rho) \to \varphi_{ij}^{(0)}(\rho)$

as $\varepsilon \to 0$ *for* $\rho \in [0, \beta]$ *and every* $i, j \neq 0$; *(c)* $G_{ij}^{(\varepsilon)}(\cdot) \Rightarrow G_{ij}^{(0)}(\cdot)$ *as* $\varepsilon \to 0$ *for every* $i, j \neq 0$.

The root $\rho^{(\varepsilon)}$ of the characteristic equation

$$\varphi_{jj}^{(\varepsilon)}(\rho) = \int_0^\infty e^{\rho s} G_{jj}^{(\varepsilon)}(ds) = 1 \tag{3.30}$$

plays an important role for the asymptotics of the probabilities $P_{ij}^{(\varepsilon)}(t)$ as $t \to \infty$ and $\varepsilon \to 0$. Lemma 4 guarantees that under conditions **G**–**K** there exists a $\beta \in (0, \delta]$, such that $\varphi_{ij}^{(\varepsilon)}(\beta) < \infty, i, j \neq 0$ for all ε small enough. However, these conditions do not guarantee the existence of a root of equation (3.30). We therefore assume that the following condition holds for some $j \neq 0$:

L $\varphi_{jj}^{(0)}(\beta_j) \in (1, \infty)$ for some $0 < \beta_j \leq \delta$.

Lemma 5 *Let conditions* **J**–**K** *hold. If condition* **L** *holds for some* $j \neq 0$, *then it holds for all* $j \neq 0$.

A priori the root $\rho^{(\varepsilon)}$ depends on j. However, it is possible to show using the results by Teugels (1968) that it is independent of j under an additional condition which guarantees that equation (3.30) has a root.

Lemma 6 *Let conditions* **G**–**L** *hold. Then there exists* $\beta \in (0, \delta]$ *as in Lemma 4 and such that for all* ε *small enough: (a)* $\varphi_{jj}^{(\varepsilon)}(\beta) \in (1, \infty), j \neq 0$; *(b)* $\varphi_{ij}^{(\varepsilon)}(\beta) < \infty, i, j \neq 0$; *(c) for every* $j \neq 0$ *there exists a unique nonnegative root* $\rho^{(\varepsilon)}$ *of equation (3.30); (d) this root is independent of* $j \neq 0$; *(e)* $\rho^{(\varepsilon)} < \beta$; *(f)* $\rho^{(\varepsilon)} \to \rho^{(0)}$ *as* $\varepsilon \to 0$.

We introduce the matrix $V_j^{(\varepsilon)}(\rho) = \| p_{il}^{(\varepsilon)} \psi_{il}^{(\varepsilon)}(\rho)(1 - \delta(l, j)) \|_{i,l=1}^N$. By definition, the n-the power of this matrix is given by

$$[V_j^{(\varepsilon)}(\rho)]^n = \| E_i e^{\rho\tau^{(\varepsilon)}(n)} \chi(\nu_j^{(\varepsilon)} > n, \eta_n^{(\varepsilon)} = l) \|_{i,l=1}^N$$

Lemma 7 *Let conditions* **G**–**L** *hold. Then for* $i, l, j \neq 0$ *and* β *defined in Lemma 6 for all* ε *small enough* $E_i e^{\beta\tau^{(\varepsilon)}(n)} \chi(\nu_j^{(\varepsilon)} > n, \eta_n^{(\varepsilon)} = l) \to 0$ *as* $n \to \infty$ *and therefore for* $\rho \in [0, \beta]$ *for all* ε *small enough there exists the inverse matrix* $U_j^{(\varepsilon)}(\rho) = [I - V_j^{(\varepsilon)}(\rho)]^{-1} = I + V_j^{(\varepsilon)}(\rho) + [V_j^{(\varepsilon)}(\rho)]^2 + \cdots$.

Lemma 8 *Let conditions* **G**–**L** *hold. Then for every* $j \neq 0$ *and* $\rho \in [0, \beta]$ *for all* ε *small enough* $\varphi_{ij}^{(\varepsilon)}(\rho), i \neq 0$ *is the unique solution of the following system of linear equations:*

$$\varphi_{ij}^{(\varepsilon)}(\rho) = p_{ij}^{(\varepsilon)} \psi_{ij}^{(\varepsilon)}(\rho) + \sum_{l \neq 0, j} p_{il}^{(\varepsilon)} \psi_{il}^{(\varepsilon)}(\rho) \varphi_{lj}^{(\varepsilon)}(\rho), \quad i \neq 0. \tag{3.31}$$

We also introduce for $i \neq 0, j \in X, r = 0, 1, \ldots$ the moment functionals

$$\psi_{ij}^{(\varepsilon)}(\rho, r) = \int_0^\infty s^r e^{\rho s} F_{ij}^{(\varepsilon)}(ds), \ \rho \geq 0.$$

Notice that $\psi_{ij}^{(\varepsilon)}(\rho, 0) = \psi_{ij}^{(\varepsilon)}(\rho)$.

It follows from condition **K** that for every $\beta' \in [0, \beta)$ one has $\psi_{ij}^{(\varepsilon)}(\beta', r) < \infty, i \neq 0, j \in X, r = 0, 1, \ldots$ for all ε small enough.

Obviously $\psi_{ij}^{(\varepsilon)}(\rho, r)$ is the derivative of order r of the moment generating function $\psi_{ij}^{(\varepsilon)}(\rho)$.

We define for $i, j \neq 0, r = 0, 1, \ldots$ the moment functionals

$$\varphi_{ij}^{(\varepsilon)}(\rho, r) = \int_0^\infty s^r e^{\rho s} G_{ij}^{(\varepsilon)}(ds), \ \rho \geq 0.$$

Notice that $\varphi_{ij}^{(\varepsilon)}(\rho, 0) = \varphi_{ij}^{(\varepsilon)}(\rho)$.

Lemma 9 *Let conditions* **G–L** *hold. Then for every $\beta' \in [0, \beta)$ one has $\varphi_{ij}^{(\varepsilon)}(\beta', r) < \infty, i, j \neq 0, \ r = 1, 2, \ldots$ for all ε small enough.*

Let us define for $i, j \neq 0, r = 0, 1, \ldots$

$$\lambda_{ij}^{(\varepsilon)}(\rho, r) = p_{ij}^{(\varepsilon)} \psi_{ij}^{(\varepsilon)}(\rho, r) + \sum_{l \neq 0, j} \sum_{n=1}^{r} C_r^n p_{il}^{(\varepsilon)} \psi_{il}^{(\varepsilon)}(\rho, r - n) \varphi_{lj}^{(\varepsilon)}(\rho, n). \tag{3.32}$$

Obviously $\varphi_{ij}^{(\varepsilon)}(\rho, r)$ is the derivative of the order r of the moment generating function $\varphi_{ij}^{(\varepsilon)}(\rho)$. Differentiation in (3.32) yields a system of linear equations for the functional $\varphi_{ij}^{(\varepsilon)}(\rho, r)$.

Lemma 10 *Let conditions* **G–L** *hold. Then for every $r = 1, 2, \ldots, j \neq 0$ and $\rho \in [0, \beta']$ for all ε small enough $\varphi_{ij}^{(\varepsilon)}(\rho, r), i \neq 0$ is the unique solution of the following system of linear equations:*

$$\varphi_{ij}^{(\varepsilon)}(\rho, r) = \lambda_{ij}^{(\varepsilon)}(\rho, r) + \sum_{l \neq 0, j} p_{il}^{(\varepsilon)} \psi_{il}^{(\varepsilon)}(\rho) \varphi_{lj}^{(\varepsilon)}(\rho, r), \quad i \neq 0. \tag{3.33}$$

Note that the systems (3.31) and (3.33) have the same matrix of coefficients but different inhomogeneous terms. These systems can be solved recursively since the expressions for $\lambda_{ij}^{(\varepsilon)}(\rho, r), i \neq 0$ given in (3.32) include the solutions $\varphi_{ij}^{(\varepsilon)}(\rho, n), i \neq 0$ of systems (3.31) and (3.33) for $n = 0, \ldots, r - 1$.

The following nonlinear perturbation conditions play a crucial role in what follows:

M $p_{ij}^{(\varepsilon)} = p_{ij}^{(0)} + \varepsilon b_{ij}[1] + \cdots + \varepsilon^k b_{ij}[k] + o(\varepsilon^k)$ for $i \neq 0, j \in X$, where $|b_{ij}[n]| < \infty, n = 1, \ldots, k,\ i \neq 0, j \in X$.

N$_k$ $\psi_{ij}^{(\varepsilon)}(\rho^{(0)}, r) = \psi_{ij}^{(0)}(\rho^{(0)}, r) + \varepsilon c_{ij}[1, r] + \cdots + \varepsilon^k c_{ij}[k, r] + o(\varepsilon^k)$ for $r = 0, \ldots, k,\ i \neq 0, j \in X$, where $|c_{ij}[n, r]| < \infty, n = 1, \ldots k,\ r = 0, \ldots, k,\ i \neq 0, j \in X$.

It is convenient to define $b_{ij}[0] = p_{ij}^{(0)}$ and $c_{ij}[0, r] = \psi_{ij}^{(0)}(\rho^{(0)}, r)$ for $i \neq 0, j \in X, r = 0, \ldots, k$.

Consider the following system of linear equations

$$x_{i,j}^{(\varepsilon)} = h_{i,j}^{(\varepsilon)} + \sum_{l \neq 0, j} d_{il}^{(\varepsilon)} x_{l,j}^{(\varepsilon)}, i \neq 0, \tag{3.34}$$

Let $W_j^{(\varepsilon)} = \| d_{il}^{(\varepsilon)}(1 - \delta(l, j)) \|_{i,l=1}^N$ and $[I - W_j^{(\varepsilon)}]^{-1} = \| w_{il,j}^{(\varepsilon)} \|_{i,l=1}^N$ if the inverse matrix exists.

We assume that the following conditions hold:

O $d_{ij}^{(\varepsilon)} = d_{ij}[0] + \varepsilon d_{ij}[1] + \cdots + \varepsilon^k d_{ij}[k] + o(\varepsilon^k)$ for $i, j \neq 0$, where $|d_{ij}[n]| < \infty, n = 0, \ldots, k,\ i, j \neq 0$.

P $h_{i,j}^{(\varepsilon)} = h_{i,j}[0] + \varepsilon h_{i,j}[1] + \cdots + \varepsilon^k h_{i,j}[k] + o(\varepsilon^k)$ for $i, j \neq 0$, where $|h_{i,j}[n]| < \infty,\ n = 0, \ldots, k,\ i, j \neq 0$.

Lemma 11 *Let conditions* **O** *and* **P** *hold and assume that* $\det(I - W_j^{(0)}) \neq 0$ *for* $j \neq 0$. *Then for all* ε *small enough there exists a unique solution of the system (3.34) for every* $j \neq 0$ *and the following asymptotic expansions hold for* $i, j \neq 0$:

$$x_{i,j}^{(\varepsilon)} = x_{i,j}^{(0)} + \varepsilon x_{i,j}[1] + \cdots + \varepsilon^k x_{i,j}[k] + o(\varepsilon^k), \tag{3.35}$$

where $x_{i,j}[0] = x_{i,j}^{(0)}$ *and the coefficients* $x_{i,j}[n]$ *are given for* $n = 0, \ldots, k$, $i, j \neq 0$ *by the recursion formulas:*

$$x_{i,j}[n] = \sum_{l \neq 0} w_{il,j}^{(0)} H_{l,j}[n],$$

where

$$H_{i,j}[n] = h_{i,j}[n] + \sum_{l \neq 0, j} \sum_{q=1}^{n} d_{il}[q] x_{l,j}[n - q].$$

Proof. Condition **O** implies that the matrix $I - W_j^{(\varepsilon)}$ has nonzero determinant for all ε small enough. We can therefore give a rational formula for the corresponding solution of system (3.34). Both the numerator and the denominator in this formula can be represented as expansions with respect to powers of ε under the conditions of the lemma. Because the limiting determinant $\det(I - W_j^{(0)}) \neq 0$, the zero-order coefficient in the expansion for the denominator is not equal to zero. Therefore we can use the quotient rule for asymptotic expansions and expand the solutions of systems (3.34) with respect to powers of ε. So, the expansion (3.35) exists.

The coefficients of the expansion are obtained in the following way. The system (3.34) can be represented formally as follows:

$$x_{i,j}[0] + \varepsilon x_{i,j}[1] + \cdots = h_{i,j}[0] + \varepsilon h_{i,j}[1] + \cdots + \tag{3.36}$$

$$\sum_{l \neq 0,j} (d_{il}[0] + \varepsilon d_{il}[1] + \cdots)(x_{l,j}[0] + \varepsilon x_{l,j}[1] + \cdots), \ i \neq 0.$$

Equalizing the corresponding coefficients of powers of ε on the left and right hand sides of (3.36) one gets for $n = 0, \ldots, k$ the system of linear equations

$$x_{i,j}[n] = H_{i,j}[n] + \sum_{l \neq 0,j} d_{il}^{(0)} x_{l,j}[n], \ i \neq 0. \tag{3.37}$$

This system yields in an obvious way the expansions (3.35). □

Applying Lemma 11 to the system (3.31) and (3.33) in the case where $\rho = \rho^{(0)}$ we obtain the desired asymptotic expansions for $\varphi_{ij}^{(\varepsilon)}(\rho^{(0)}, r)$.

The condition **O** holds for these systems due to the following simple lemma.

Lemma 12 *Let conditions* **G**– $\mathbf{N}_k$ *hold. Then functionals* $p_{ij}^{(\varepsilon)} \psi_{ij}^{(\varepsilon)}(\rho^{(0)})$ *have for every* $i \neq 0, j \in X$ *the asymptotic expansion:*

$$p_{ij}^{(\varepsilon)} \psi_{ij}^{(\varepsilon)}(\rho^{(0)}) = d_{ij}[0] + \varepsilon d_{ij}[1] + \cdots + \varepsilon^k d_{ij}[k] + o(\varepsilon^k), \tag{3.38}$$

where the coefficients $d_{ij}[n]$ *are given for* $n = 0, \ldots, k, \ i \neq 0, j \in X$ *by the formulas:*

$$d_{ij}[n] = \sum_{q=0}^{n} b_{ij}[q] c_{ij}[n-q, 0].$$

In the case of system (3.31) the condition **P** is also provided by Lemma 12. A direct application of Lemma 11 to this system yields the desired expansion.

Lemma 13 *Let conditions* **G–N**$_k$ *hold. Then the following asymptotic expansions hold for* $i, j \neq 0$*:*

$$\varphi_{ij}^{(\varepsilon)}(\rho^{(0)}) = e_{ij}[0,0] + \varepsilon e_{ij}[1,0] + \cdots + \varepsilon^k e_{ij}[k,0] + o(\varepsilon^k), \qquad (3.39)$$

where coefficient $e_{ij}[n,0]$ *coincides for every* $n = 0, \dots, k$ *with the coefficient* $x_{i,j}[n]$ *defined in Lemma 11 for the case where* $d_{ij}^{(\varepsilon)} = p_{ij}^{(\varepsilon)} \psi_{ij}^{(\varepsilon)}(\rho^{(0)})$, $i, j \neq 0$ *and* $h_{i,j}^{(\varepsilon)} = p_{ij}^{(\varepsilon)} \psi_{ij}^{(\varepsilon)}(\rho^{(0)}), i, j \neq 0$. *The corresponding expansions for these coefficients used in conditions* **O** *and* **P** *are given in Lemma 12.*

As far as the system (3.33) is concerned the recursive character of this system and the formula (3.32) for the free terms of systems (3.33) must be taken into account when drawing the conclusion that condition **p** holds for all these systems and when writing down expansions for the inhomogeneous terms of these systems for $r = 1, \dots$. The role of the inhomogeneous terms $h_{i,j}^{(\varepsilon)}$ is now played by coefficients $\lambda_{ij}^{(\varepsilon)}(\rho^{(0)}, r)$ defined in (3.32). The corresponding formula for given r contains the functionals $\varphi_{ij}^{(\varepsilon)}(\rho^{(0)}, n)$ for $i \neq 0, n = 0, \dots, r-1$. Under the assumption that these functionals have already been expanded one can also expand the functionals $\lambda_{ij}^{(\varepsilon)}(\rho^{(0)}, r), i \neq 0$ using the standard sum and product rules for asymptotical power series in (3.32). Then the functionals $\varphi_{ij}^{(\varepsilon)}(\rho^{(0)}, r), i \neq 0$ can be expanded using Lemma 11. The two steps described above have to be repeated for $r = 1, \dots, k$. The possibility to start this recursive procedure is provided by Lemma 13 which gives the corresponding expansions for $\varphi_{ij}^{(\varepsilon)}(\rho^{(0)}, 0), i \neq 0$.

Lemma 14 *Let conditions* **G–N**$_k$ *hold. Then the following asymptotic expansions hold for* $i, j \neq 0,\ r = 1, \dots, k$*:*

$$\varphi_{ij}^{(\varepsilon)}(\rho^{(0)}, r) = e_{ij}[0,r] + \varepsilon e_{ij}[1,r] + \cdots + \varepsilon^k e_{ij}[k,r] + o(\varepsilon^k), \qquad (3.40)$$

where for every $r = 1, \dots, k$ *the coefficient* $e_{ij}[n,r]$ *coincides for every* $n = 0, \dots, k$ *with the coefficient* $x_{i,j}[n]$, *defined in Lemma 11, for the case where* $d_{ij}^{(\varepsilon)} = p_{ij}^{(\varepsilon)} \psi_{il}^{(\varepsilon)}(\rho^{(0)}), i, j \neq 0$ *and* $h_{i,j}^{(\varepsilon)} = \lambda_{ij}^{(\varepsilon)}(\rho^{(0)}, r), i, j \neq 0$. *The coefficients* $d_{ij}[n], n = 0, \dots, k,\ i, j \neq 0$ *appearing in* **O** *are given in this case by Lemma 12. The coefficients* $h_{i,j}[n], n = 0, \dots, k,\ i, j \neq 0$ *appearing in* **P** *are given in this case by the following recursive formulas* *where they are denoted as* $h_{i,j}[n,r]$ *for every* $r = 1, \dots, k$*:*

$$
\begin{aligned}
h_{i,j}[n,r] &= \sum_{q=0}^{n} b_{ij}[q] c_{ij}[n-q,r] \\
&+ \sum_{l \neq 0,j} \sum_{m=1}^{r} C_r^m \sum_{q=0}^{n} \sum_{p=0}^{n-q} b_{il}[q] c_{il}[p,m] e_{lj}[n-q-p, r-m].
\end{aligned}
$$

4. NONLINEARLY PERTURBED SEMI-MARKOV PROCESSES

Our approach is based on the use of Theorems 1 and 2 and the fact that the transition probabilities $P_{jj}^{(\varepsilon)}(t)$ satisfy the following renewal equation

$$
P_{jj}^{(\varepsilon)}(t) = 1 - F_j^{(\varepsilon)}(t) + \int_0^t P_{jj}^{(\varepsilon)}(t-s) G_{jj}^{(\varepsilon)}(ds), \; t \geq 0, \tag{3.41}
$$

where

$$
F_j^{(\varepsilon)}(s) = \sum_{l=0}^{N} p_{jl}^{(\varepsilon)} F_{jl}^{(\varepsilon)}(s), \; j \neq 0.
$$

Let

$$
\pi_j^{(\varepsilon)} = \frac{\int_0^\infty e^{\rho^{(\varepsilon)} s} [1 - F_j^{(\varepsilon)}(s)] ds}{\int_0^\infty s e^{\rho^{(\varepsilon)} s} G_{jj}^{(\varepsilon)}(ds)}, \; \pi_{ij}^{(\varepsilon)} = \psi_{ij}^{(\varepsilon)}(\rho^{(\varepsilon)}) \pi_j^{(\varepsilon)}, \; i,j \neq 0.
$$

Note that (3.30) implies that $\pi_{jj}^{(0)} = \pi_j^{(0)}$ for all $j \neq 0$. Also, if $\rho^{(0)} = 0$ then $\pi_{ij}^{(0)} = \pi_j^{(0)}$ for all $i,j \neq 0$ since in this case $\psi_{ij}^{(0)}(0) = 1$ for all $i,j \neq 0$.

One of the main results of the paper is the following theorem.

Theorem 3 *Let conditions* **G**–$\mathbf{N}_k$ *be satisfied. Then:*

(i) *The root* $\rho^{(\varepsilon)}$ *of the characteristic equation (3.30) can be expanded as*

$$
\rho^{(\varepsilon)} = \rho^{(0)} + a_1 \varepsilon + \cdots + a_k \varepsilon^k + o(\varepsilon^k), \tag{3.42}
$$

where the coefficients $a_1, \dots, a_k$ *can be explicitly calculated using a recursion formulas given in Theorem 1 and applied to the functions* $q^{(\varepsilon)}(t) = 1 - F_j^{(\varepsilon)}(s)$ *and the distribution functions* $F^{(\varepsilon)}(s) = G_{jj}^{(\varepsilon)}(s)$*. The corresponding expansions used in condition* $\mathbf{E}_k$ *are given in Lemmas 13 and 14 for every* $j \neq 0$.

(ii) *If $t^{(\varepsilon)} \to \infty$ as $0 < \varepsilon \to 0$, then for $i, j \neq 0$:*

$$\frac{P_{ij}^{(\varepsilon)}(t^{(\varepsilon)})}{\exp\{-\rho^{(\varepsilon)}t^{(\varepsilon)}\}} \to \pi_{ij}^{(0)} \text{ as } \varepsilon \to 0. \tag{3.43}$$

(iii) *If $k \geq 1$ and $t^{(\varepsilon)} \to \infty$ such that $\varepsilon^r t^{(\varepsilon)} \to \lambda_r \in [0, \infty)$ as $\varepsilon \to 0$ for some $1 \leq r \leq k$, then for $i, j \neq 0$:*

$$\frac{P_{ij}^{(\varepsilon)}(t^{(\varepsilon)})}{\exp\{-(\rho^{(0)} + a_1\varepsilon + \cdots + a_{r-1}\varepsilon^{r-1})t^{(\varepsilon)}\}} \to \tag{3.44}$$

$$e^{-a_r\lambda_r}\pi_{ij}^{(0)} \text{ as } \varepsilon \to 0.$$

Proof. The proof is based on the application of Theorem 1 to the renewal equation (3.41). The conditions **G**–$\mathbf{N}_k$ imply the conditions **A**–$\mathbf{E}_k$ for the functions $1 - F_j^{(\varepsilon)}(t)$ and distribution functions $G_{jj}^{(\varepsilon)}(s)$.

Obviously **H** and **I** imply **B**. It follows from Lemma 4 that conditions **G**–**K** imply **A** and **D**. Moreover, conditions **G**–**L** imply **C**.

The most difficult step of the proof, related to the nonlinear perturbation condition $\mathbf{E}_k$, is based on Lemmas 13 and 14. It follows from these lemmas that **G**-$\mathbf{N}_k$ imply $\mathbf{E}_k$. The corresponding expansions used in $\mathbf{E}_k$ are also given in these lemmas.

After that, Theorem 1 can be applied. The expansion (3.42) is a variant of expansion (3.10). We can also write down for every $j \neq 0$ the following variant of relation (3.43):

$$\frac{P_{jj}^{(\varepsilon)}(t^{(\varepsilon)})}{e^{-\rho^{(\varepsilon)}t^{(\varepsilon)}}} \to \pi_j^{(0)} \text{ as } \varepsilon \to 0. \tag{3.45}$$

Relation (3.45) coincides with (3.43) for the case $i = j$. To prove (3.43) for the case $i \neq j$ we employ the fact that in this case the transition probabilities $P_{ij}^{(\varepsilon)}(t)$ satisfy the following renewal type relation:

$$P_{ij}^{(\varepsilon)}(t) = \int_0^t P_{jj}^{(\varepsilon)}(s)G_{ij}^{(\varepsilon)}(ds), t \geq 0. \tag{3.46}$$

This relation can be rewritten in the form

$$e^{\rho^{(\varepsilon)}t^{(\varepsilon)}}P_{ij}^{(\varepsilon)}(t^{(\varepsilon)}) = \tag{3.47}$$

$$\int_0^{t^{(\varepsilon)}} e^{\rho^{(\varepsilon)}(t^{(\varepsilon)}-s)}P_{jj}^{(\varepsilon)}(t^{(\varepsilon)} - s)e^{\rho^{(\varepsilon)}s}G_{ij}^{(\varepsilon)}(ds), t \geq 0.$$

The function $e^{\rho^{(\varepsilon)}(t^{(\varepsilon)}-s)}P_{jj}^{(\varepsilon)}(t^{(\varepsilon)}-s)$ is bounded for $0 \le s \le t^{(\varepsilon)}$ uniformly in $\varepsilon \to 0$, that is,

$$\overline{\lim}_{\varepsilon\to 0} \sup_{0\le s\le t^{(\varepsilon)}} e^{\rho^{(\varepsilon)}(t^{(\varepsilon)}-s)}P_{jj}^{(\varepsilon)}(t^{(\varepsilon)}-s) \le R_j < \infty. \tag{3.48}$$

To see this, suppose on the contrary that there exists an $s^{(\varepsilon)}$ such that $e^{\rho^{(\varepsilon)}(t^{(\varepsilon)}-s^{(\varepsilon)})}P_{jj}^{(\varepsilon)}(t^{(\varepsilon)}-s^{(\varepsilon)}) \to \infty$. We can also suppose that $t^{(\varepsilon)}-s^{(\varepsilon)} \to w$ where $0 \le w \le \infty$ (if not, we can go to subsequences). But w cannot be finite since in this case $e^{\rho^{(\varepsilon)}(t^{(\varepsilon)}-s^{(\varepsilon)})} \to e^{\rho^{(0)}w}$ and so $e^{\rho^{(\varepsilon)}(t^{(\varepsilon)}-s^{(\varepsilon)})}P_{jj}^{(\varepsilon)}(t^{(\varepsilon)}-s^{(\varepsilon)})$ is bounded. Neither can w be ∞ since then $u^{(\varepsilon)} = t^{(\varepsilon)}-s^{(\varepsilon)}$ would converge to ∞ but from (3.45) it follows that in this case $e^{\rho^{(\varepsilon)}u^{(\varepsilon)}}P_{jj}^{(\varepsilon)}(u^{(\varepsilon)}) \to \pi_j^{(0)}$. From (3.45) it also follows that $e^{\rho^{(\varepsilon)}(t^{(\varepsilon)}-s)}P_{jj}^{(\varepsilon)}(t^{(\varepsilon)}-s) \to \pi_j^{(0)}$ for fixed $s \ge 0$. In addition, $e^{\rho^{(\varepsilon)}s} \to e^{\rho^{(0)}s}$. Using these facts, (3.48), and Lemma 4 we have for any $0 < T < \infty$ which is a point of continuity for the distribution $G_{ij}^{(0)}(s)$ that

$$\lim_{\varepsilon\to 0}\int_0^T e^{\rho^{(\varepsilon)}(t^{(\varepsilon)}-s)}P_{jj}^{(\varepsilon)}(t^{(\varepsilon)}-s)e^{\rho^{(\varepsilon)}s}G_{ij}^{(\varepsilon)}(ds) = \tag{3.49}$$

$$\pi_j^{(0)}\int_0^T e^{\rho^{(0)}s}G_{ij}^{(0)}(ds).$$

Let us choose $\rho \in (\rho^{(0)}, \beta)$ where β was defined in Lemma 6. Then from Lemmas 4 and 6 it follows also that

$$\lim_{T\to\infty}\overline{\lim}_{\varepsilon\to 0}\int_T^\infty e^{\rho^{(\varepsilon)}s}G_{ij}^{(\varepsilon)}(ds) \le \tag{3.50}$$

$$\lim_{T\to\infty}\overline{\lim}_{\varepsilon\to 0}\int_T^\infty e^{\rho s}G_{ij}^{(\varepsilon)}(ds) = \lim_{T\to\infty}\overline{\lim}_{\varepsilon\to 0}(\varphi_{ij}^{(\varepsilon)}(\rho) - \int_0^T e^{\rho s}G_{ij}^{(\varepsilon)}(ds)) =$$

$$\lim_{T\to\infty}(\varphi_{ij}^{(0)}(\rho) - \int_0^T e^{\rho s}G_{ij}^{(0)}(ds)) = 0.$$

Using (3.48) and (3.50) we have

$$\lim_{T\to\infty}\overline{\lim}_{\varepsilon\to 0}\int_T^{t^{(\varepsilon)}} e^{\rho^{(\varepsilon)}(t^{(\varepsilon)}-s)}P_{jj}^{(\varepsilon)}(t^{(\varepsilon)}-s)e^{\rho^{(\varepsilon)}s}G_{ij}^{(\varepsilon)}(ds) \le \tag{3.51}$$

$$\le \lim_{T\to\infty}\overline{\lim}_{\varepsilon\to 0}\int_T^\infty 2R_j e^{\rho^{(\varepsilon)}s}G_{ij}^{(\varepsilon)}(ds) = 0.$$

Finally, from (3.46), (3.49), and (3.51) we get

$$\lim_{\varepsilon \to 0} e^{\rho^{(\varepsilon)} t^{(\varepsilon)}} P_{ij}^{(\varepsilon)}(t^{(\varepsilon)}) = \tag{3.52}$$

$$\lim_{T\to\infty} \lim_{\varepsilon\to 0} \int_0^T e^{\rho^{(\varepsilon)}(t^{(\varepsilon)}-s)} P_{jj}^{(\varepsilon)}(t^{(\varepsilon)} - s) e^{\rho^{(\varepsilon)} s} G_{ij}^{(\varepsilon)}(ds) =$$

$$\lim_{T\to\infty} \pi_j^{(0)} \int_0^T e^{\rho^{(0)} s} G_{ij}^{(0)}(ds) = \pi_j^{(0)} \varphi_{ij}^{(0)}(\rho^{(0)}).$$

This completes the proof of statement (ii). The statement (iii) follows immediately from (ii). □

5. QUASI-STATIONARY DISTRIBUTIONS FOR PERTURBED SEMI-MARKOV PROCESSES

We introduce the quasi-stationary distribution of the semi-Markov process $\eta^{(\varepsilon)}(t), t \geq 0$. To avoid non-arithmetic effects we assume that the following condition holds:

R $F_{ij}^{(\varepsilon)}(\cdot), i \neq 0, j \in X$ are non-arithmetic distribution functions for all ε small enough.

Under conditions **G**–**K** and **R** the quasi-stationary distribution exists for all ε small enough and it is defined in the following way (as usual, quasi-stationary probabilities do not depend on the choice of initial state $i \neq 0$):

$$\kappa_j^{(\varepsilon)} = \lim_{t\to\infty} P_i\{\eta^{(\varepsilon)}(t) = j \mid \mu_0^{(\varepsilon)} > t\}, j \neq 0. \tag{3.53}$$

Theorem 1 (applied to the case where the parameters of the semi-Markov processes do not depend on ε) allows us represent the quasi-stationary probabilities in the following form:

$$\kappa_j^{(\varepsilon)} = \frac{\pi_j^{(\varepsilon)}}{\sum_{l\neq 0} \varphi_{jl}^{(\varepsilon)}(\rho^{(\varepsilon)}) \pi_l^{(\varepsilon)}}, j \neq 0. \tag{3.54}$$

This formula prompts that the quasi-stationary probabilities $\kappa_j^{(\varepsilon)}$ can be expanded in asymptotical series via the corresponding expansions of the functionals $\pi_j^{(\varepsilon)}$ and $\varphi_{ij}^{(\varepsilon)}(\rho^{(\varepsilon)})$ taken in points $\rho^{(\varepsilon)}$ instead of $\rho^{(0)}$.

Lemma 15 *Let conditions* $\mathbf{G}$*–*$\mathbf{N}_{k+s}$ *hold for some integer* $s \geq 0$. *Then the following asymptotical expansions hold for* $r = 0, \dots, s$, $i, j \neq 0$:

$$\varphi_{ij}^{(\varepsilon)}(\rho^{(\varepsilon)}, r) = y_{ij}[0, r] + \varepsilon y_{ij}[1, r] + \dots + \varepsilon^k y_{ij}[k, r] + o(\varepsilon^k), \qquad (3.55)$$

where the coefficients $y_{ij}[n, r]$ *are given for* $n = 0, \dots, k$, $r = 0, \dots, s$ *by the formulas:*

$$y_{ij}[n, r] = e_{ij}[n, r] + \sum_{q=1}^{n} e_{ij}[n - q, r + 1]a_q +$$

$$\sum_{2 \leq m \leq n} \sum_{q=2}^{n} e_{ij}[n - q, r + m] \cdot \sum_{n_1, \dots, n_{p-1} \in D_{mq}} \prod_{p=1}^{q-1} a_p^{n_p}/n_p!.$$

Proof. The proof follows the method used the proof of Theorem 3. We can write down the asymptotic expansion of the $\varphi_{ij}^{(\varepsilon)}(\rho^{(\varepsilon)}, r)$ in powers of the difference $\Delta^{(\varepsilon)} = \rho^{(\varepsilon)} - \rho^{(0)}$. Recall that $\rho^{(0)} < \beta$, where β is defined in Lemma 6, and $\Delta^{(\varepsilon)} \to 0$. Hence there exists $\beta_0 < \beta$ such that $\rho^{(0)} + \mid \Delta^{(\varepsilon)} \mid \leq \beta_0$ for ε small enough, say $\varepsilon \leq \varepsilon_0$. Substituting the Taylor expansion (3.15) for the function $e^{\rho^{(\varepsilon)} s}$ into the integral defining $\varphi_{ij}^{(\varepsilon)}(\rho^{(\varepsilon)}, r)$ one obtains for $\varepsilon \leq \varepsilon_0$

$$\varphi_{ij}^{(\varepsilon)}(\rho^{(\varepsilon)}, r) = \varphi_{ij}^{(\varepsilon)}(\rho^{(0)}, r) + \Delta^{(\varepsilon)} \varphi_{ij}^{(\varepsilon)}(\rho^{(0)}, r + 1) + \dots + \qquad (3.56)$$

$$\frac{(\Delta^{(\varepsilon)})^k}{k!} \varphi_{ij}^{(\varepsilon)}(\rho^{(0)}, r + k) + (\Delta^{(\varepsilon)})^{(k+1)} M_{ij,r+k} \theta_{ij,r+k}^{(\varepsilon)},$$

where

$$M_{ij,r+k} = ((k+1)!)^{-1} \sup_{\varepsilon \leq \varepsilon_0} \int_0^{\infty} s^{r+k+1} e^{s\beta_0} G_{ij}^{(\varepsilon)}(ds)) < \infty, \; 0 \leq \theta_{ij,r+k}^{(\varepsilon)} \leq 1.$$

It was proved in Theorem 3 that functions $\Delta^{(\varepsilon)}$ can be expanded in asymptotical series. Functions $\varphi_{ij}^{(\varepsilon)}(\rho^{(0)}, r + n)$ can also be expanded in asymptotical series due to condition $\mathbf{N}_{k+s}$. Substituting these expansions in (3.56) we can represent this relation in the following form

$$\varphi_{ij}^{(\varepsilon)}(\rho^{(\varepsilon)}, r) = (e[0, r] + \cdots e[k, r]\varepsilon^k + o(\varepsilon^k)) + \qquad (3.57)$$

$$(e[0, r + 1] + \cdots e[k, r + 1]\varepsilon^k + o(\varepsilon^k))(a_1\varepsilon + \dots + a_k\varepsilon^k + o(\varepsilon^k))/1! + \dots +$$

$$(e[0, r + k] + \cdots e[k, r + k]\varepsilon^k + o(\varepsilon^k))(a_1\varepsilon + \dots + a_k\varepsilon^k + o(\varepsilon^k))^k/k! + o(\varepsilon^k).$$

By grouping coefficients for powers of ε in these expansions one can transform them to the form given in Lemma 15. □

Define for $i \neq 0, r = 0, 1, \dots$ the moment functionals

$$\vartheta_i^{(\varepsilon)}(\rho, r) = \int_0^\infty s^r e^{\rho s}(1 - F_i^{(\varepsilon)}(s))ds, \ \rho \geq 0$$

and

$$\theta_i^{(\varepsilon)}(\rho, r) = \int_0^\infty s^r e^{\rho s} F_i^{(\varepsilon)}(ds) = \sum_{l=0}^{N} p_{il}^{(\varepsilon)} \psi_{il}^{(\varepsilon)}(\rho, r), \ \rho \geq 0.$$

It follows from condition **K** that for every $\beta' \in [0, \beta)$ one has $\vartheta_i^{(\varepsilon)}(\beta', r) < \infty$ and $\theta_i^{(\varepsilon)}(\beta', r) < \infty$ for all $i \neq 0, r = 0, 1, \dots$ for all ε small enough.

Lemma 16 *Let conditions* **G**–$\mathbf{N}_{k+1}$ *hold. Then the following asymptotical expansions hold for* $r = 0, \dots, k, \ i \neq 0$:

$$\vartheta_i^{(\varepsilon)}(\rho^{(0)}, r) = z_i[0, r] + \varepsilon z_i[1, r] + \cdots + \varepsilon^k z_i[k, r] + o(\varepsilon^k), \tag{3.58}$$

where the coefficients $z_i[n, r]$ *are given for* $n, r = 0, \dots, k, \ i \neq 0, j \in X$ *by the recursion formulas:*

$$z_i[n, r] = \begin{cases} \frac{d_i[n,r+1]}{r+1} & \text{if } \rho^{(0)} = 0, r = 0, \dots, k, \\ \frac{d_i[n,0]-1}{\rho^{(0)}} & \text{if } \rho^{(0)} > 0, r = 0, \\ \frac{d_i[n,r]-1-rz_i[n,r-1]}{\rho^{(0)}} & \text{if } \rho^{(0)} > 0, r = 1, \dots, k, \end{cases}$$

where coefficients $d_i[n, r]$ *are given for* $n = 0, \dots k, \ r = 0, \dots k+1, \ i \neq 0$ *by the formulas:*

$$d_i[n, r] = \sum_{l=0}^{N} \sum_{q=0}^{n} d_{il}[q] c_{il}[n - q, r].$$

Proof. The proof follows directly from the following recursion formulas obtained by partial integration in the formula which defines the moment functionals $\vartheta_i^{(\varepsilon)}(\rho^{(0)}, r)$:

$$\vartheta_i^{(\varepsilon)}(\rho^{(0)}, r) = \begin{cases} \frac{\theta_i^{(\varepsilon)}(\rho^{(0)},r+1)}{r+1} & \text{if } \rho^{(0)} = 0, r = 0, \dots, k, \\ \frac{\theta_i^{(\varepsilon)}(\rho^{(0)},0)-1}{\rho^{(0)}} & \text{if } \rho^{(0)} > 0, r = 0, \\ \frac{\theta_i^{(\varepsilon)}(\rho^{(0)},r)-1-r\vartheta_i^{(\varepsilon)}(\rho^{(0)},r-1)}{\rho^{(0)}} & \text{if } \rho^{(0)} > 0, r = 1, \dots, k. \end{cases}$$

□

To expand the functionals $\pi_j^{(\varepsilon)}$ we can apply Theorem 2 to the renewal equation (3.41) since $\pi_j^{(\varepsilon)}$ coincides with the functional $\tilde{x}^{(\varepsilon)}(\infty)$ for this equation.

Theorem 4 *Let conditions* **G**–**N**$_{k+1}$ *be satisfied. Then the functionals* $\pi_j^{(\varepsilon)}, j \neq 0$ *have the asymptotic expansions:*

$$\pi_j^{(\varepsilon)} = \frac{f'_j[0] + f'_j[1]\varepsilon + \cdots + f'_j[k]\varepsilon^k + o(\varepsilon^k)}{f''_j[0] + f''_j[1]\varepsilon + \cdots + f''_j[k]\varepsilon^k + o(\varepsilon^k)} = \tag{3.59}$$

$$\pi_j^{(0)} + f_j[1]\varepsilon + \cdots + f_j[k]\varepsilon^k + o(\varepsilon^k),$$

where the coefficients $f'_j[n]$, $f''_j[n], n = 0, \ldots, k$ *and* $f_j[n], n = 1, \ldots k$ *can be explicitly calculated using the formulas given in Theorem 2 and applied to the function* $q^{(\varepsilon)}(t) = 1 - F_j^{(\varepsilon)}(t)$ *and the distribution functions* $F^{(\varepsilon)}(s) = G_{jj}^{(\varepsilon)}(s)$. *The corresponding expansions used in condition* **E**$_k$ *and* **F** *are given respectively in Lemmas 13, 14 and 15 for every* $j \neq 0$.

The concluding theorem, which yields the asymptotical expansions for quasi-stationary distributions $\kappa_j^{(\varepsilon)}$, follows in an obvious way from Lemma 15, Theorem 4 and formula (3.54).

Theorem 5 *Let conditions* **G**–**N**$_{k+1}$ *and* **R** *be satisfied. Then the quasi-stationary probabilities* $\kappa_j^{(\varepsilon)}, j \neq 0$ *have the asymptotic expansions:*

$$\kappa_j^{(\varepsilon)} = \frac{g'_j[0] + g'_j[1]\varepsilon + \cdots + g'_j[k]\varepsilon^k + o(\varepsilon^k)}{g''_j[0] + g''_j[1]\varepsilon + \cdots + g''_j[k]\varepsilon^k + o(\varepsilon^k)} = \tag{3.60}$$

$$\kappa_j^{(0)} + g_j[1]\varepsilon + \cdots + g_j[k]\varepsilon^k + o(\varepsilon^k),$$

where coefficients $g'_j[l]$, $g''_j[l], n = 0, \ldots, k,\ j \neq 0$ *are given by the formulas:*

$$g'_j[n] = f_j[n],\ g''_j[n] = \sum_{l \neq 0} \sum_{m=0}^{n} y_{jl}[n-m, 0] f_l[m],$$

and the coefficients $g_j[n], n = 0, \ldots, k,\ j \neq 0$ *are given by the formulas* $g_j[0] = \kappa_j^{(0)}$ *and in general for* $n = 0, \ldots, k$:

$$g_j[n] = (g'_j[n] - \sum_{m=0}^{n-1} g''_j[n-m] g_j[m]) / g''_j[0].$$

Note that if the local absorption probabilities $p_{i0}^{(\varepsilon)} = 0, i \neq 0$ for all $\varepsilon \geq 0$, then Theorem 6 yields nonlinear expansions for ordinary stationary distributions of nonlinearly perturbed semi-Markov processes.

References

[1] Darroch, J. and Seneta, E. (1965). On quasi-stationary distributions in absorbing discrete-time finite Markov chains. *Journal of Applied Probability*, **2**, 88-100.

[2] Feller, W. (1966). *An Introduction to Probability Theory and its Applications*, Vol. 2. Wiley & Sons, New York.

[3] Gyllenberg, M. and Silvestrov, D.S. (1994). Quasi-stationary distributions of a stochastic meta-population model. *Journal of Mathematical Biology*, **33**, 35-70.

[4] Gyllenberg, M. and Silvestrov, D.S. (1997). *Exponential asymptotics for perturbed renewal equations and pseudo-stationary phenomena for stochastic systems.* Research Report 3, Department of Mathematical Statistics, Umeå University.

[5] Gyllenberg, M. and Silvestrov, D.S. (1998a). *Nonlinearly perturbed regenerative processes and pseudo-stationary phenomena for stochastic systems.* Research Reports A22, Department of Applied Mathematics, University of Turku.

[6] Gyllenberg, M. and Silvestrov, D.S. (1998b). *Cramér-Lundberg approximation for nonlinearly perturbed risk processes.* Research Reports A26, Department of Applied Mathematics, University of Turku.

[7] Kijima, M. (1997). *Markov Processes for Stochastic Modelling.* Chapman & Hall, London.

[8] Kingman, J.F. (1963). The exponential decay of Markovian transition probabilities. *Proceedings of London Mathematical Society*, **13**, 337-358.

[9] Korolyuk, V.S. (1969). Asymptotical behaviour of sojourn time of semi-Markov processes in a subset of states. *Ukrainian Mathematical Journal*, **21**, 842-845.

[10] Pollett, P.K. (1996). *Quasistationary distributions bibliography.* `http://www.maths.uq.oz.au/~pkp/papers/qsds.html`

[11] Schweitzer, P.J. (1968). Perturbation theory and finite Markov chains. *Journal of Applied Probability*, **5**, 401-413.

[12] Seneta, E. (1991). Sensitivity Analysis, Ergodicity Coefficients, and rank-one updates for finite Markov chains. In *Numerical Solution of Markov Chains.* Edited by W.J. Stewart, 121-129, Marcel Dekker, New York.

[13] Seneta, E. and Vere-Jones, D. (1966). On quasi-stationary distributions in discrete-time Markov chains with a denumerable infinity of states. *Journal of Applied Probability*, **3**, 403-434.

[14] Silvestrov, D.S. (1976). On one generalization of the renewal theorem. *Dokladi Akademii Nauk USSR*, **Ser. A, 11**, 978-982.

[15] Silvestrov, D.S. (1978, 1979). The renewal theorem in a series scheme. *Theory Probability and Mathematical Statistics*, Part 1: **18**, 144-161, Part 2: **20**, 97-116.

[16] Silvestrov, D.S. (1995). Exponential asymptotic for perturbed renewal equations. *Theory Probability and Mathematical Statistics*, **52**, 143-153.

[17] Simon, H.A. and Ando, A. (1961). Aggregation of variables in dynamic systems. *Econometrica*, **29**, 11-138.

[18] Stewart, G.W. (1991). On the sensitivity of nearly uncoupled Markov chains. In *Numerical Solution of Markov Chains.* Edited by W.J. Stewart, 105-119, Marcel Dekker, New York.

[19] Teugels, J.L. (1968). Exponential ergodicity in Markov renewal processes. *Journal of Applied Probability*, **33**, 434-438.

[20] Vere-Jones, D. (1962). Geometric ergodicity in denumerable Markov chains. *Quarterly Journal of Mathematics*, **13**, 7-28.

[21] Yaglom, A. M. (1947). Certain limit theorems of the theory of branching processes. *Dokladi Akademii Nauk SSSR*, **56**, 795-798.

Chapter 4

SEMI-MARKOV RANDOM WALKS

Vladimir S. Korolyuk
National Academy of Sciences, Ukraine

Abstract The boundary problems for random walks have various interpretations in applied fields of probability theory, first of all, in mathematical risk theory, connected with insurance problems, in analysis of queueing systems and storage processes and others. The boundary problems for semi-Markov random walks are considered on a bounded interval of the real line with a delaying screen.

Keywords: Semi-Markov random walk, boundary problem, superposition of renewal processes, queueing system, diffusion approximation.

1. THE SEMI-MARKOV RANDOM WALK

The Semi-Markov Random Walk (SMRW) is determined on superposition of two independent renewal processes :

$$\nu_{\pm}(t) = \max\{n : \sum_{k=1}^{n} \alpha_k^{\pm} \leq t\}, t \geq 0. \tag{4.1}$$

The time intervals $\alpha_k^{\pm}, k \geq 1$, are jointly independent and are given by the distribution functions :

$$P_{\pm}(t) = \mathcal{P}\{\alpha_k^{\pm} \leq t\}, t \geq 0.$$

The superposition of two renewal processes (4.1) is given by :

$$\nu(t) = \nu_+(t) + \nu_-(t), t \geq 0. \tag{4.2}$$

The SMRW is defined by the following sums :

$$\xi(t) = \xi_0 + \sum_{r=1}^{\nu_+(t)} \eta_r^+ - \sum_{r=1}^{\nu_-(t)} \eta_r^- \tag{4.3}$$

The jumps $\eta_r^{\pm}, r \geq 1$, are jointly independent and are given by the distribution functions :

$$G_{\pm}(u) = \mathcal{P}\{\eta_r^{\pm} \leq u\}, u \geq 0.$$

The SMRW (4.3) can be interpreted as a storage process with arbitrary distribution of time intervals between arrival and departure of goods. The superposition of two renewal processes (4.2) is completely determined **by the Markov renewal process (MRP)** :

$$\mathcal{I}_n, S_n, n \geq 0,$$

which is determined by the following stochastic relations :

$$\mathcal{I}_{n+1} = (\mathcal{I}_n - \alpha_{n+1}^{+})I(\mathcal{I}_n > 0) + (\mathcal{I}_n + \alpha_{n+1}^{-})I(\mathcal{I}_n \leq 0),$$
$$S_{n+1} = S_n + (\mathcal{I}_n \wedge \alpha_{n+1}^{+})I(\mathcal{I}_n > 0) + (-\mathcal{I}_n \wedge \alpha_{n+1}^{-})I(\mathcal{I}_n \leq 0).$$

The imbedded Markov chain (IMC) $\mathcal{I}_n, n \geq 0$, is defined by the transition probabilities :

$$\mathcal{P}\{\mathcal{I}_{n+1} \in dy | \mathcal{I}_n = x > 0\} = P_{+}(x - dy),$$
$$\mathcal{P}\{\mathcal{I}_{n+1} \in dy | \mathcal{I}_n = x \leq 0\} = P_{-}(dy - x).$$

It is easy to verify that there exists a stationary distribution of IMC and their density is defined as follows :

$$\rho_{+}(x) = \rho_{+}p_{+}^{*}(x), x > 0; \qquad \rho_{-}(x) = \rho_{-}p_{-}^{*}(x), x \leq 0,$$

where

$$p_{\pm}^{*}(x) = \overline{P}_{\mp}(\mp x)|p_{\mp}, \qquad \overline{P}_{\pm}(x) := 1 - P_{\pm}(x),$$
$$p_{\pm} := E\alpha_1^{\pm} = \int_0^{\infty} \overline{P}_{\pm}(x)dx, \qquad p := p_{+} + p_{-}, \rho_{\pm} = p_{\mp}/p.$$

The process of superposition (4.2) is a counting process of MRP $\mathcal{I}_n, S_n, n \geq ($

$$\nu(t) = \max\{n : S_n \leq t\}, \ t \geq 0.$$

The imbedded SMRW is defined by :

$$\xi_n = \xi(S_n), n \geq 0.$$

The SMRW (4.3) can be given by :

$$\xi(t) = \xi_{\nu(t)}, t \geq 0.$$

The main fact is that the two component process $\xi_n, \mathcal{I}_n, n \geq 0$, is a Markov chain which can be given by the transition probabilities :

$$\mathcal{P}\{\xi_{n+1} \in u + dv, \mathcal{I}_{n+1} \in dy | \xi_n = u, \mathcal{I}_n = x\} =$$
$$\begin{cases} G_{\pm}(dv)P_{+}(x - dy), x > 0, y \gtrless 0; \\ G_{\pm}(dv)P_{-}(dy - x), x < 0, y \gtrless 0. \end{cases}$$

2. A BOUNDARY PROBLEM FOR SMRW

The SMRW (4.3) is considered with delaying screen on a level $\Gamma > 0$:

$$\xi_{n+1}^{\Gamma} = \min(\xi_n^{\Gamma} + \eta_{n+1}, \Gamma), n \geq 0,$$

$$\xi^{\Gamma}(t) = \xi_{\nu(t)}^{\Gamma}, t \geq 0,$$

where

$$\eta_{n+1} := \eta_{n+1}^{+} I(\mathcal{I}_n > 0) - \eta_{n+1}^{-} I(\mathcal{I}_n \leq 0)$$

and

$$\xi_0^{\Gamma} = \xi_0 \leq \Gamma.$$

The process $\xi^{\Gamma}(t)$ is bounded on top : $\xi^{\Gamma}(t) \leq \Gamma$ for all $t \geq 0$, and $\xi^{\Gamma}(t) = \Gamma$ up to the next negative jump of SMRW (4.3). The crucial fact is that the two component process $\xi_n^{\Gamma}, J_n, n \geq 0$, is a Markov chain also, with some restriction on jumps of the first component. The main boundary characteristic of SMRW with delaying screen is the first passage time :

$$\tau_u^{\Gamma}(x) = \inf\{t : \xi^{\Gamma}(t) \leq 0 | \xi_0^{\Gamma} = u, \mathcal{I}_0 = x\} \tag{4.4}$$

Lemma 1 *The generating function of the first passage time (4.4)*

$$\varphi_{\pm}(u, \pm x) = E[\exp(-\lambda \tau_u^{\Gamma}(x)) | \mathcal{I}_0 \gtrless 0] \tag{4.5}$$

is determined by a solution of the conjugate system of integral equations

$$\varphi_{\pm}(u, x) = \int_0^{\infty} G_{\pm}(dv) \int_0^x P_{\pm}(dt) e^{-\lambda t} \varphi_{\pm}(u \pm v, x - t) +$$

$$+ \int_0^{\infty} G_{\mp}(dv) \int_x^{\infty} P_{\pm}(dt) e^{-\lambda x} \varphi_{\mp}(u \mp v, t - x) \tag{4.6}$$

with additional boundary conditions

$$\varphi_{\pm}(u, x) = 1, u \leq 0,$$

$$\varphi_{\pm}(\Gamma + u, x) = \varphi_{\pm}(\Gamma, x), u \geq 0. \tag{4.7}$$

The proof of Lemma 1 is based on the stochastic relation[1] for the first passage time (4.4) of SMRW (4.3) without delaying screen :

$$\tau_u^{\pm}(x) \doteq \theta_x^{\pm} + \tau_{u \pm \eta_{\pm}}^{\pm}(x - \alpha^{\pm}) I(\alpha^{\pm} < x) + \tau_{u \mp \eta_{\mp}}^{\mp}(\alpha^{\pm} - x) I(\alpha^{\pm} \geq x)$$

with additional boundary condition :

$$\tau_u^{\pm}(x) = 0, u \leq 0.$$

This is, by definition, $\theta_x^{\pm} := \alpha^{\pm} \wedge x$. The existence and uniqueness of a solution of the conjugate system (4.6) can be proved by using the classical methods of upper and lower functions [1,2].

It is hardly to believe that a solution of the system (4.6) can generally be represented in an explicit form, in general. However, there are particular cases when an explicit form of the solution can be obtained [3-13].

3. A SEMI-CONTINUOUS SMRW

A semi-continuous SMRW is defined under a restriction on the distribution function of random variables which are used in the definition of SMRW (4.3).

The semi-continuous lattice SMRW is defined under the following restrictions :

(A) $\mathcal{P}\{\eta_r^+ = 1\} = 1$, and the distribution function $G_-(u)$ is latticed.

(B) $\mathcal{P}\{\alpha_k^- > t\} = e^{-at}, a > 0$, and the distribution function $P_+(t)$ is absolutely continuous.

The explicit form of the generating function (4.5) is rather complicated. Nevertheless, it can be considered limiting behavior of the first passage time (4.4) when the delaying screen is moving off to infinity : $\Gamma \to \infty$.

To formulate limit result we have to introduce some preliminary notations and conditions.The total negative jump between two renewal points of process $\nu_+(t)$ is

$$\gamma = \sum_{r=1}^{\nu_-(\alpha^+)} \eta_r^- .$$

Note that the counting process $\nu_-(t)$ is a Poisson process with intensity a under the restriction (B). Hence by the Kolmogorov-Prokhorov formula,

$$\rho := E\gamma = ap_+g_-,$$

where

$$p_+ := E\alpha^+ = \int_0^\infty \bar{P}_+(t)dt, \qquad g_- := E\eta^- = \int_0^\infty \bar{G}_-(v)dv.$$

It is easy to check that the generating function of γ has the following representation :

$$Ez^\gamma = p_+(a(1 - g_-(z))).$$

In what follows the Cramér condition is supposed to be valid :

$$\sup\{z : |p_+(a(1-g_-(z)))| < \infty\} = z^* > 1,$$

(C)

$$p_+(a(1-g_-(z^*))) > z^*.$$

Under the condition (C) there exists the unique positive solution $z_0 > 1$ of the equation :

$$p_+(a(1-g_-(z_0))) = z_0. \tag{4.8}$$

The imbedded random walk is defined by the sums

$$\chi_n = \sum_{k=1}^{n} \gamma_k - n, n \geq 0, \qquad \chi_0 = 0, \tag{4.9}$$

where $\gamma_k, k \geq 1$, are i.i.d. random variables distributed equally with γ. Under the condition $\rho < 1$ there exists a finite maximum with probability one :

$$\bar{\chi} := \max_{n\geq 0} \chi_n.$$

Theorem 1 ([7,8]) *The limit distribution of the first passage time (4.4) for semi-continuous lattice SMRW (4.3) with delaying screen $\Gamma \to \infty$, in the stationary regime $\rho < 1$, under restrictions (A),(B) and (C) is determined by the following relation*

$$\lim_{\Gamma\to\infty} E\exp[-\lambda z_0^{-\Gamma} \tau_u^\Gamma(x)] = \mathcal{P}\{\bar{\chi} \geq u\} + \mathcal{P}\{\bar{\chi} < u\} \frac{c}{c+\lambda}, \tag{4.10}$$

where the intensity of the limit exponential distribution

$$c = (1-\rho)^2 (z_0 - 1)/z_0 \mu_0 g_+,$$

$$\mu_0 = p'_+(a(1-g_-(z_0))) - 1.$$

A similar result can be obtained under the following restriction (instead of (A))

$$(A') \qquad \mathcal{P}\{\eta_r^+ > u\} = e^{-bu}, \; b > 0,$$

and the distribution function $G_-(u)$ is absolutely continuous, or

$$(A'') \qquad \mathcal{P}\{\eta_r^+ = k\} = (1-p)p^{k-1}, \; k \geq 1,$$

and the distribution function $G_-(u)$ is latticed.

The asymptotical exponentiality of the first passage time (4.4) for SMRW (4.3) in the general case must remain valid. But it is unknown how to calculate the intensity of limit exponential distribution (see, for example [23]).

Remark 1 *A similar limit result was obtained at the first passage time for the compound Poisson process with negative drift in [4].*

Remark 2 *The limit result (4.10) has an heuristic explanation. The atom of the limit distribution* $p = \mathcal{P}\{\bar{\chi} \geq u\}$ *is a measure of trajectories which attain the absorbing barrier without reflecting from the delaying screen.*

Remark 3 *Asymptotical behavior of the first passage time for SMRW with a delaying screen is distinct in three cases:*

$$\rho < 1, \ \rho = 1, \ \rho > 1.$$

For the investigation of SMRW's by another approach see [17,18].

4. QUEUEING SYSTEM $M|G|1|\Gamma$

A uniserial queueing system with a restriction on queue by level $\Gamma > 0$, is set by an exponentially distributed arrival time :

$$\mathcal{P}\{\alpha_k^+ > t\} = e^{-at}, t \leq 0, a > 0,$$

and by an arbitrary distribution function of service time :

$$\mathcal{P}\{\alpha_k^- \leq t\} = P(t), t \geq 0.$$

The queueing process $\xi(t), t \geq 0$, which is the number of customers in the system at time t, can be considered as a semi-continuous lattice SMRW (4.3) with jumps $\eta_r^{\pm} = 1, r \geq 0$, and with the delaying screen on the level $u = 0$. The first time to failure of the customer is :

$$\tau_u^{\Gamma}(x) = \inf\{t : \xi(t) \geq \Gamma | \xi(0) = u, \mathcal{I}_0 = x\}$$

The total positive jump between two renewal points of the renewal process $\nu_-(t)$ is :

$$\gamma = \nu_+(\alpha^-), \rho = E\gamma = ap, p := E\alpha^-.$$

Theorem 1 provides the following :

Corollary 1 *Under the conditions of Theorem 1 in the stationary regime,* $\rho < 1$, *the limit distribution of time to failure in the queueing system* $M|G|1|\Gamma$ *is exponential,*

$$\lim_{\Gamma\to\infty} \mathcal{P}\{z_0^{-\Gamma}\tau_u^{\Gamma}(x) > t\} = e^{-ct}, \quad t \geq 0,$$

with an intensity described by

$$c = (1-\rho)^2(z_0 - 1)/z_0\mu_0 p,$$

$$\mu_0 = p'(a(1-z_0)) - 1.$$

The normalized parameter $z_0 > 1$ is a positive solution of the equation :

$$z_0 = p(a(1-z_0)).$$

The method of potentiality [4] can be applied to investigate in an explicit form the boundary problem with a delaying screen on the level $u = 0$, and with an absorbing screen on the level $\Gamma > 0$, for the imbedded random walk :

$$\chi_n = u + \sum_{k=1}^{n} \xi_k, n \geq 0, \tag{4.11}$$

where $\xi_k, k \geq 1$, are identically distributed with $\xi = \gamma - 1$. It is evidence, that the mentioned above boundary problem for the random walk (4.11) describes the length of queue in the system $M|G|1|\Gamma$ up to the first time to failure of customer,

$$\tau_u^\Gamma = \inf\{n : \chi_n \geq \Gamma | \chi_0 = u\}$$

Introduce the following notions.
The potential of the random walk (4.11) is determined by its generating function :

$$R(z) = \sum_{m=0}^{\infty} z^m R_m = z[p(a(1-z)) - 1]^{-1}.$$

The resolvent on the semi axis of the random walk (4.11) is determined by its generating function :

$$R(z,s) = \sum_{m=0}^{\infty} z^m R_m(s) = [z^{-1}p(a(1-z)) - 1 - s]^{-1}.$$

Theorem 2 ([19]) *The generating function of time to failure in the queueing system $M|G|1|\Gamma$ is represented by the relation :*

$$Ev^{\tau_u^\Gamma} = 1 + s[C_\Gamma(s)R_u(s) + \sum_{m=1}^{u} R_m(s)],$$

where

$$C_\Gamma(s) = R_{\Gamma+1}(s)|[R_\Gamma(s) - R_{\Gamma+1}(s)], s := v^{-1} - 1.$$

The mean value of time to failure is represented by the relation :

$$E\tau^{\Gamma}_{\Gamma-u} = C_{\Gamma}R_u - \sum_{m=1}^{u} R_m, C_{\Gamma} = R_{\Gamma}/[R_{\Gamma+2} - R_{\Gamma+1}].$$

The asymptotic representation for the potential [4] provides an asymptotic analysis as $\Gamma \to \infty$, for the distribution function of time to failure of customer in the queueing system $M|G|1|\Gamma$ for all values of the average drift,

$$\rho = E\alpha^{-}/E\alpha^{+}.$$

5. QUEUEING SYSTEM $G|M|1|\Gamma$

A uniserial queueing system of $G|M|1|\Gamma$ type is described by the distribution function $P(t) = \mathcal{P}\{\alpha^{+} \leq t\}$ of the interval α^{+} between two adjacent arrival moments of customer, and by the exponential distributed service time.

$$\mathcal{P}\{\alpha^{-} > t\} = e^{-at}, t \geq 0, a > 0.$$

The total negative jump between two renewal moments of the counting process $\nu_{+}(t), t \geq 0$, is :

$$\gamma = \nu_{-}(\alpha^{+}), \rho := E\gamma = ap, p := E\alpha^{+}.$$

The imbedded random walk (4.11), with jumps $\xi_k, k \geq 1$, identically distributed with $\xi = \gamma - 1$, is determined by the generating function of jumps :

$$g(z) := Ez = z^{-1}p(a(1-z))$$

Theorem 1 provides the following limit result.

Corollary 2 *Under the conditions of Theorem 1 in the stationary regime, $\rho > 1$, the limit distribution of the normalized time to failure of customer in the system $G|M|1|\Gamma$ is exponential :*

$$\lim_{\Gamma\to\infty} \mathcal{P}\{z_0^{-\Gamma}\tau_u^{\Gamma} > t\} = e^{-ct}, t \geq 0,$$

with an intensity described by ;

$$c = (\rho - 1)^2(1 - z_0^{-1})/\mu_0,$$

$$\mu_0 = p'(a(1 - z_0)) - 1.$$

The normalized parameter $z_0 > 1$, is a positive solution of the equation :

$$z_0 = p(a(1 - z_0)).$$

The boundary problem for the imbedded random walk (4.11) with a delaying screen on the level $u = \Gamma > 0$, and with an absorbing screen on the level $u = 0$, describes the length of queue in the system $G|M|1|\Gamma$ up to the first time to failure of customer. The method of potentiality [4] can be applied to investigate in an explicit form the distribution function of time to failure.

Theorem 3 ([20]) *The generating function of time to failure τ_u^Γ in the system $G|M|1|\Gamma$ is represented by the relation ($z = v^{-1} - 1$):*

$$Ev^{\tau_u^\Gamma} = \left[1 + s\sum_{m=1}^{u} R_m(s)\right]\left[1 + s\sum_{m=1}^{\Gamma} R_m(s)\right]^{-1}.$$

The mean value of time to failure is represented by the relation

$$E\tau_u^\Gamma = a\sum_{m=u+1}^{\Gamma} R_m.$$

The asymptotic representation for the potential $R_m, m \geq 0$, [8], provides the asymptotic analysis of the distribution function of time to failure for all values of the drift parameter $\rho = E\alpha^+/E\alpha^-$ (see [9]). For the investigation of the uniserial $GI|G|1$ queue type system see [11,14].

6. ASYMPTOTIC ANALYSIS OF SMRW IN THE SERIES SCHEME

The asymptotic behavior of the first passage time (4.4) as $\Gamma \to \infty$, in a general scheme is dependent on the average drift at unit time which is defined by the relation :

$$\rho = g_-/p_- - g_+/p_+. \tag{4.12}$$

Note that by the model limit theorem for renewal processes :

$$E\nu_\pm(t)/t \to 1/p_\pm \quad \text{as} \quad t \to \infty.$$

Hence, the mean value of the first and the second sums in (4.3) have the following asymptotic evaluation :

$$E\sum_{r=1}^{\nu_\pm(t)} \eta_r^\pm/t \simeq g_\pm/p_\pm \quad \text{as} \quad t \to \infty.$$

Hence, the average drift at unit time for SMRW (4.3) is evaluated by :

$$E\xi(t) \simeq u - \rho t \quad \text{as} \quad t \to \infty.$$

The average drift at unit time (4.12) determines three variants of behavior of SMRW (4.3) :

(i) positive drift to absorbing screen : $\rho > 0$;

(ii) positive drift to delaying screen : $\rho < 0$;

(iii) zero average drift : $\rho = 0$.

The second variant is the most complicated. The another two variants, $\rho > 0$ and $\rho = 0$ can be investigated by using asymptotic analysis of integral equations in the series scheme with some small series parameter $\varepsilon > 0$, based on a solution of the singular perturbation problems [16].

7. AVERAGE SCHEME. POSITIVE DRIFT, $\rho > 0$

The SMRW (4.3) in the series scheme is considered with the small series parameter $\varepsilon > 0$:

$$\xi_\varepsilon(t) = u + \varepsilon \sum_{r=1}^{\nu_+(t)} \eta_r^+ - \varepsilon \sum_{r=1}^{\nu_-(t)} \eta_r^- . \tag{4.13}$$

The delaying screen is considered on the level $\Gamma > 0$, and the absorbing screen is considered on the level $u = 0$.

Lemma 2 *The generating function of the first passage time*

$$\varphi_\pm^\varepsilon(u, \pm x) = E[\exp(-\varepsilon\lambda\tau_u^\Gamma(x)|x \gtrless 0]$$

is determined by a solution of the conjugate system of integral equations :

$$\varphi_\pm^\varepsilon(u,x) = \int_0^\infty G_\pm(dv) \int_0^x P_\pm(dt) e^{-\varepsilon\lambda t} \varphi_\pm^\varepsilon(u \pm \varepsilon v, t - x) +$$

$$+ \int_0^\infty G_\mp(dv) \int_x^\infty P_\pm(dt) e^{-\varepsilon\lambda x} \varphi_\mp^\varepsilon(u \mp \varepsilon v, t - x) \tag{4.14}$$

with additional boundary conditions

$$\varphi_\pm^\varepsilon(u,x) = 1, u \le 0; \varphi_\pm^\varepsilon(\Gamma + u, x) = \varphi_\pm^\varepsilon(\Gamma, x), u \ge 0. \tag{4.15}$$

The proof of Lemma 2 is similar to the proof of Lemma 1.

Theorem 4 ([21]) *Under the assumption of positive drift to the absorbing screen:*

$$\rho = g_-/p_- - g_+/p_+ > 0,$$

The average scheme is valid :

$$\lim_{\varepsilon\to\infty} E[\exp(-\varepsilon\lambda\tau_u^\Gamma(x))] = \exp(-u/\rho) \tag{4.16}$$

That is

$$\varepsilon\tau_u^\Gamma(x) \Rightarrow u/\rho \quad as \quad \varepsilon \to 0,$$

in distribution and by probability.

8. DIFFUSION APPROXIMATION (ZERO DRIFT, $\rho = 0$)

The SMRW (4.3) is considered in the series scheme with the normalized parameter series $\varepsilon > 0$,

$$\xi_\varepsilon(t) = u/\varepsilon + \varepsilon^2 \sum_{r=1}^{\nu_+(t)} \eta_r^+ - \varepsilon^2 \sum_{r=1}^{\nu_-(t)} \eta_r^- \tag{4.17}$$

The first passage time (4.4) is considered for the SMRW (4.17) with the delaying screen on the level $\Gamma = 1/\varepsilon$, and with the absorbing screen on the level $u = 0$.

Lemma 3 *The generating function of the first passage time :*

$$\varphi_\pm^\varepsilon(u, \pm x) = E[\exp(-\varepsilon^2\lambda\tau_{u/\varepsilon}^{1/\varepsilon}(x))|x \gtrless 0] \tag{4.18}$$

is determined by a solution of the conjugate system of integral equations :

$$\varphi_\pm^\varepsilon(u, x) = \int_0^\infty G_\pm(dv) \int_0^x P_\pm(dt) e^{-\varepsilon^2\lambda t} \varphi_\pm^\varepsilon(u \pm \varepsilon v, x - t) +$$

$$+ \int_0^\infty G_\mp(dv) \int_x^\infty P_\pm(dt) e^{-\varepsilon^2\lambda x} \varphi_\mp(u \mp \varepsilon v, t - x) \tag{4.19}$$

The proof of Lemma 3 is similar to the proof of Lemma 1 taking under consideration of the normalized condition in (4.17) with additional boundary conditions :

$$\varphi_\pm^\varepsilon(u, x) = 1, u \le 0; \varphi_\pm^\varepsilon(1 + u, x) = 0, u \ge 0. \tag{4.20}$$

Theorem 5 *Under the assumption of zero drift :*

$$\rho = g_-/p_- - g_+/p_+ = 0,$$

the diffusion approximation scheme is valid :

$$\lim_{\varepsilon\to 0} E[\exp(-\varepsilon^2\lambda\tau_{u/\varepsilon}^{1/\varepsilon}(x))] = \frac{ch[(1-u)\sqrt{-2\lambda}/\sigma]}{ch[\sqrt{-2\lambda}/\sigma]}. \tag{4.21}$$

The diffusion parameter σ^2 is determined in an explicit form.

9. PROBLEMS OF SINGULAR PERTURBATION IN AVERAGE AND THE DIFFUSION APPROXIMATION SCHEMES

The proofs of Theorem 4 and 5 are based on the asymptotic representations of the operators which define the integral equations (4.14) and (4.19).

Lemma 4 *The conjugate system of integral equations (4.14) for the generating function of the first passage time has the following asymptotic representation :*

$$[I - P + \varepsilon Q_1(x) + \varepsilon\theta(\varepsilon)]\varphi^\varepsilon(u, x) = 0. \tag{4.22}$$

The operator

$$P\varphi_\pm(x) := \int_0^x P_\pm(dt)\varphi_\pm(x-t) + \int_x^\infty P_\pm(dt)\varphi_\mp(t-x) \tag{4.23}$$

defines the transition probability of the IMC $\mathcal{I}_n$, $n \geq 0$.

The perturbing operator :

$$Q_1(x)\varphi_\pm(u) := a_\pm(x)\varphi'(u) + \lambda b_\pm(x)\varphi(u). \tag{4.24}$$

The operator $\theta(\varepsilon)$ satisfies the negligible condition

$$\|\theta(\varepsilon)\varphi\| \to 0 \quad \text{as} \quad \varepsilon \to 0 \tag{4.25}$$

on a smooth function φ.

The stationary distribution of the IMC with the transition probabilities (4.23) provides the normal solvability [16] of the operator $Q_0 := I - P$. A solution of the singular perturbation problem in the average scheme can be applied to the equation (4.22). The average equation for the limit generating function in Theorem 4 is the following :

$$\rho\varphi'(u) + \lambda\varphi(u) = 0.$$

Taking into account the boundary condition (4.15),the limit generating function is represented as follows :

$$\varphi(u) = \exp(-\lambda u/\rho)$$

Lemma 5 *The conjugate system of integral equations (4.19) for the generating function of the first passage time has the following asymptotic representation:*

$$[I - P + \varepsilon Q_1(x) + \varepsilon^2 Q_2(x) + \varepsilon^2\theta(\varepsilon)]\varphi^\varepsilon = 0. \tag{4.26}$$

The perturbing operators $Q_k(x), k = 1, 2$, are defined by the relations :

$$Q_1(x)\varphi_{\pm}(u) := a_{\pm}(x)\varphi'_{\pm}(u),$$

$$Q_2(x)\varphi_{\pm}(u) := c_{\pm}(x)\varphi''_{\pm}(u) + \lambda b_{\pm}(x)\varphi_{\pm}(u),$$

with the given functions $a(x), b(x)$ and $c(x)$, and also the function $c(x)$ satisfies the balance condition :

$$\int_0^{\infty} \rho_+(x)a_+(x)dx + \int_0^{\infty} \rho_-(x)a_-(x)dx = 0. \tag{4.27}$$

The operator $\theta(\varepsilon)$ satisfies the negligible condition (4.25).

The solution of the boundary problem (4.20) for the equation (4.26) is realized by a solution of singular perturbation problem in diffusion approximation scheme [16]. The averaged equation for the limit generating function in Theorem 5 has the following form :

$$\frac{1}{2}\sigma^2\varphi''(u) + \lambda\varphi(u) = 0. \tag{4.28}$$

Taking into account the boundary conditions :

$$\varphi(0) = 1, \quad \varphi'(0) = 0,$$

the solution of (4.28) is represented by right-hand side of (4.21).

The verification of the limit results in Theorems 4 and 5 is realized by the method of upper and lower functions [1,2] modified to the boundary conditions (4.15) and (4.20).

Notes

1. The equality $\doteq$ means the equal distribution of left and right sides.

References

[1] Kolmogorov A.N. (1933) *Ueber die Grenzwertsatze der Wahrscheinlichkeitsrechnung.* Izv. Acad. of Sc. 366-372.

[2] Khinchin, A.Ya. (1936) *Asymptotic Laws of Theory of Probability.* Moskow, ONTI, 96p. (in Russian).

[3] Borovkov, A.A., Rogozin, B.A. (1964) *Boundary problems for some two dimensional random walks.* Theor. Prob. and Appl. v.9, N3, 401-430 (in Russian).

[4] Korolyuk, V.S. (1975) *Boundary Problems for Compound Poison Processes.* Kiev, Naukova Dumka, - 139 p. (in Russian).

[5] Korolyuk, V.S., Suprun, V.N., Shurenkov, V.M. (1976) *Method of potential in boundary problems for processes with independent increments and with jumps of one sign.* Theor. Prob. and Appl. v.21, N2, 253-259.

[6] Pirdzanov, B., Pirliev, B. (1984) *Random walk with Poisson drift.* Ukr. Math. J. v.36, N2, 265-269 (in Russian).

[7] Korolyuk, V.S., Pirliev, B. (1984) *Random walk on semiaxis on superposition of two renewal processes.* Ukr. Math. J. v.36, N4, 483-486 (in Russian).

[8] Nepesova, J., Pirdzanov, B. (1986) *Limit distribution of the first passage time of random walk with reflection on superposition two renewal processes.* Ukr. Math. J., v.38, N2, 195-199 (in Russian).

[9] Korolyuk, V.S., Bratiichuk, N.S., Pirdzanov, B. (1987) *Boundary Problems for Random Walks.* Ashkhabad, Ylym, - 258 p. (in Russian).

[10] Bratiichuk, N.S., Pirdzanov, B. (1990) *Limit theorem for the first passage time and the first overjump by the regenerative semi-Markov walk.* Proceeding of Inst. Appl. Math. and Mech., 41-48 (in Russian).

[11] Bratiichuk, N.S., Pirdzanov, B. (1991) *Busy period of $GI|G|1$ queueing system.* DAN of Ukraine, N9, 50-53 (in Russian).

[12] Pirdzanov, B. (1990) *Semi-Markov walks on superposition of two renewal processes.* Ukr. Math. J. v.42, N11, 1550-1508. (in Russian).

[13] Korolyuk, V.S. (1996) *Ruin problems. Explicit and asymptotic approaches.* Theory of Stoch. Processes, 2 (18), N1-2, 4-14.

[14] Bratiichuk, N.S. (1995) *Limit theorems for some characteristics of system $GI|G|1$.* Exploring Stoch. Laws, Proceeding, VSP, 77-90.

[15] Bratiichuk, N.S., Korolyuk, V.S. (1998) *Ruin problems for semi-Markov walks.* Abstract of VII Vilnus Conference.

[16] Korolyuk, V.S., Turbin, A.F. (1982) *Markov Renewal Processes in Reliability Analysis of Systems.* Kiev,Naukova Dumka.-236 p. (in Russian).

[17] Gusak, D.V. (1995) *About intersection of a level by processes which are given by a sums of random value of terms.* Ukr. Math. J. v.47, N7, 897-914. (in Ukraine).

[18] Nasyrova T. (1984) *Processes of Semi-Markov Walks.* Baky, Elym.-170 p. (in Russian).

[19] Khusanov, M. (1977) *Analysis of distribution of the length queue maximum in queueing system $M|G|1|\infty$.* News of Acad. S. of Uzbekistan, N4, 29-33. (in Russian).

[20] Butko,T., Korolyuk, V.S. (1984) *Method of potential in the investigation of a queueing system* $GI|M|1|\infty$. Kiev, Inst. Math., 16-27. (in Russian).

[21] Korolyuk, V.S., Muzrapov, S., Drapkin, L. (1995) *Sojourn time of the semi- Markov random walk with positive drift on a semiaxis.* DAN of Ukraine, N1, 27-29. (in Russian).

[22] Korolyuk, V.S., Korolyuk, V.V. (1998) *Stochastic Models of Systems*, Kluwer Academic Puplishers. – 196p.

[23] Iscoe, I. and McDonald, D. *Asymptotics of exit times for Markov jump processes I*, (1994) Ann. of Prob., v.22, N1, 372-397.

Chapter 5

DIFFUSION APPROXIMATION FOR PROCESSES WITH SEMI-MARKOV SWITCHES AND APPLICATIONS IN QUEUEING MODELS

Vladimir V. Anisimov
Bilkent University, Turkey and
Kiev University, Ukraine

Abstract Stochastic processes with semi-Markov switches (or in semi-Markov environment) and general Switching processes are considered. In case of asymptotically ergodic environment functional Averaging Principle and Diffusion Approximation types theorems for trajectory of the process are proved. In case of asymptotically consolidated environment a convergence to a solution of a differential or stochastic differential equation with Markov switches is studied. Applications to the analysis of random movements with fast semi-Markov switches and semi-Markov queueing systems in case of heavy traffic conditions are considered.

Keywords: Semi-Markov process, switching process, averaging principle, diffusion approximation, consolidation of states, queueing models, random walks.

1. INTRODUCTION

In various models appearing at study of complex stochastic systems such as state-dependent queueing systems and networks, information and computer systems, production and manufacturing systems, etc., we come to a necessity to consider systems working in different scales of time (slow and fast) and such that their local transition characteristics can be dependent on a current value of some another stochastic process (external random environment, discrete interference of chance, stochastic failures and switches or in general some functional on a trajectory of a system).

An operation of a wide range of these systems can be described in terms of so called Switching Stochastic Processes and in particular in terms of processes with semi-Markov switches.

The main property of a Switching Process (*SP*) is that the character of its operation varies spontaneously (switches) at certain epochs of time which can be random functionals of a previous trajectory.

SP's appear at study of queueing systems and networks, branching and migration processes in a random environment, at the analysis of stochastic dynamical systems with random perturbations, random movements and other various applications.

Taking into account a high dimension and a complex structure, exact analytic solutions for these processes can be obtained only for special rare cases, and methods of a direct stochastic simulation work usually slow and do not give a possibility of parametric investigation of a system. Therefore asymptotic methods play the basic role at the investigation and approximate analytic modelling.

Different asymptotic approaches for various classes of complex stochastic systems are considered in books of Buslenko et al. (1973), Kovalenko (1980), Anisimov et al. (1987), Basharin et al. (1989), and papers of Harrison (1995), Harrison and Williams (1996), Mandelbaum and Pats (1998).

In the paper we give a general description of *SP*'s, consider some important subclasses of *SP*'s paying the main attention to processes with semi-Markov switches (*PSMS*), investigate results of Averaging Principle (*AP*) and Diffusion approximation (*DA*) types and consider models of asymptotic decreasing dimension and consolidation of the state space for *PSMS*.

Applications to the analysis of random movement and state-dependent queueing models in semi-Markov environment in cases when the states of the environment can be asymptotically averaged or the environment allows an asymptotic consolidation of its state space are considered.

2. SWITCHING STOCHASTIC PROCESSES

2.1 PRELIMINARY REMARKS

SP's are described as two-component processes $(x(t), \zeta(t))$, $t \geq 0$, with the property existing a sequence of epochs $t_1 < t_2 < \ldots$ such that on each interval $[t_k, t_{k+1})$, $x(t) = x(t_k)$ and the behavior of the process $\zeta(t)$ depends on the value $(x(t_k), \zeta(t_k))$ only. The epochs t_k are switching times and $x(t)$ is the discrete switching component (see Anisimov, 1977, 1978, 1988a).

SP's can be described in terms of constructive characteristics and they are very suitable in analyzing and asymptotic investigation of complex stochastic systems with "rare" and "fast" switches (Anisimov, 1978, 1988a, 1994-1996).

We mention that switching times may be determined by external factors (for instance, in the case when a system is operating in some random environment) and also by inner and interconnected factors. In general switching times may be some random functionals of the previous trajectory of the system.

According to A.N. Kolmogorov, *SP*'s are the special class of random processes with discrete interference of chance or processes with discrete component. A wide range of processes with discrete component have been studied by different authors: Markov processes homogeneous on the 2nd component (Ežov and Skorokhod, 1969), processes with independent increments and semi-Markov switches (Anisimov, 1973, 1978), piecewise Markov aggregates (Buslenko et. al., 1973), Markov processes with semi-Markov interference of chance (Gikhman and Skorokhod, 1973), and Markov and semi-Markov evolutions (Griego and Hersh, 1969; Hersh, 1974; Kertz, 1978ab; Kurtz, 1972, 1973; Papanicolaou and Hersh, 1972; Pinsky, 1975; Korolyuk and Swishchuk, 1986, 1994).

Law of Large Numbers and CLT for special classes of random evolutions were proved by many authors (Griego and Hersh, 1969; Hersh and Papanicolaou, 1972; Kurtz, 1973, Kertz, 1978ab; Pinsky, 1975; Anisimov, 1973; Korolyuk and Turbin, 1978; Watkins, 1984; Korolyuk and Swishchuk, 1986, 1994). These results are mostly devoted to the analysis of processes with independent increments in a Markov or semi-Markov environment.

Limit theorems for general scheme of *SP*'s were studied in the author's papers in the following directions.

Theorems about convergence of a one *SP* to another are proved in the class of *SP*'s when the number of switches does not tend to infinity ("rare" switches) (see Anisimov, 1978, 1988ab). As in usual a limiting *SP* has a more simple structure and depends on less number of parameters, these results give us the possibility to decrease dimension asymptotically and consolidate in some sense the state space of *SP*.

On the base of these results the theory of asymptotic consolidation (merging) of state space and decreasing dimension for Markov and semi-Markov processes (homogeneous as well as non-homogeneous in time) was constructed (Anisimov, 1973, 1988a).

Asymptotic consolidation of states in particular means the following. Suppose that some Markov process (*MP*) or semi-Markov process (*SMP*) has transition probabilities of different orders and its state space can be divided to regions such that transition probabilities between them are small in some sense and the states in each region asymptotically communicate. Then under rather general conditions additive functionals on the process can be weakly approximated by processes with independent increments in Markov or semi-Markov environment with number of states equals to the number of regions.

Several results devoted to the asymptotic analysis of integral functionals and flows of rare events on trajectories of *SMP*'s operating in different scales of time are obtained in (Anisimov, 1973, 1977, 1978a). Applications to the asymptotic analysis of queueing models and multiprocessor computer systems in conditions of fast service can be found in Anisimov (1988a, 1996a), Anisimov et al. (1987), Anisimov and Sztrik (1989), Sztrik and Kouvatsos (1991).

The next direction of the investigations devoted to the case of fast switches (number of switches tends to infinity). In that case, if the increments on each switching interval are small, it is reasonable to expect taking into account the recurrent character of an operation of the process, that under some general conditions a process trajectory converges to a solution of some ordinary differential equation – *Averaging Principle* (*AP*), and the normed deviation weakly converges to some diffusion process - *Diffusion Approximation* (*DA*). Results of this type for different subclasses of *SP*'s were proved by Anisimov (1992, 1994, 1995), Anisimov and Aliev (1990).

Applications of these results to study an asymptotic behaviour of characteristics for Markov queueing systems and networks under transient conditions and with large number of calls were investigated by Anisimov (1992, 1995, 1996a), Anisimov and Lebedev (1992).

Now we consider a description of some important classes of *SP*'s such as Recurrent Processes of a Semi-Markov type, Processes with Semi-Markov Switches and give a general description of *SP*'s.

2.2 RECURRENT PROCESSES OF A SEMI-MARKOV TYPE

Let $\mathcal{F}_k = \{(\xi_k(\alpha), \tau_k(\alpha)), \alpha \in \mathcal{R}^r\}, k \geq 0$, be jointly independent families of random variables with values in $\mathcal{R}^r \times [0, \infty)$ and S_0 be independent of $\mathcal{F}_k, k \geq 0$ random variable in $\mathcal{R}^r$. We assume the measurability in α of variables introduced concerning σ-algebra $\mathcal{B}_{\mathcal{R}^r}$. Denote

$$t_0 = 0, \quad t_{k+1} = t_k + \tau_k(S_k), \quad S_{k+1} = S_k + \xi_k(S_k), \quad k \geq 0 \qquad (5.1)$$

and put

$$S(t) = S_k \quad \text{as} \quad t_k \leq t < t_{k+1}, \quad t \geq 0. \tag{5.2}$$

Then a process $S(t)$ forms a Recurrent Process of a Semi-Markov type (*RPSM*) (see Anisimov, 1992, Anisimov and Aliev, 1990).
In the homogeneous case (distributions of families $\mathcal{F}_k$ do not depend on the parameter k) the process $S(t)$ is a homogeneous *SMP*. If also distributions of families $\mathcal{F}_k$ do not depend on the parameter α, epochs $t_k, k \geq 0$ form a recurrent flow and $S(t)$ is a generalized renewal process. If variables $\tau_k(\alpha)$ have exponential distributions, the process $S(t)$ is a *MP*.

2.3 RECURRENT PROCESS OF A SEMI-MARKOV TYPE WITH ADDITIONAL MARKOV SWITCHES

Let $\mathcal{F}_k = \{(\xi_k(x, \alpha), \tau_k(x, \alpha)), x \in X, \alpha \in \mathcal{R}^r\}, k \geq 0$ be jointly independent families of random variables with values in $\mathcal{R}^r \times [0, \infty)$, and let $x_l, l \geq 0$ be a *MP*, independent of $\mathcal{F}_k, k \geq 0$, with values in X, (x_0, S_0) be an initial value. We assume here and further the measurability in the pair (x, a) of variables introduced concerning σ-algebra $\mathcal{B}_X \times \mathcal{B}_{R^r}$ and put

$$t_0 = 0, \ t_{k+1} = t_k + \tau_k(x_k, S_k), S_{k+1} = S_k + \xi_k(x_k, S_k), \ k \geq 0, \tag{5.3}$$

$$S(t) = S_k, \quad x(t) = x_k \quad \text{as} \quad t_k \leq t < t_{k+1}, \quad t \geq 0. \tag{5.4}$$

Then the process $(x(t), S(t))$ forms a *RPSM* with additional Markov switches. We assume that *RPSM* is regular, i.e. the component $x(t)$ has a finite number of jumps on each finite interval with probability one. If the distributions of variables $\tau_k(x, \alpha)$ do not depend on parameters (α, k), then the process $x(t)$ is a *SMP*.

2.4 GENERAL CASE OF RPSM

Let $\mathcal{F}_k = \{(\xi_k(x, \alpha), \tau_k(x, \alpha), \beta_k(x, \alpha)), x \in X, \alpha \in \mathcal{R}^r\}, k \geq 0$ be jointly independent families of random variables with values in $\mathcal{R}^r \times [0, \infty) \times X$, X be some measurable space, (x_0, S_0) be an initial value. We put

$$t_0 = 0, \quad t_{k+1} = t_k + \tau_k(x_k, S_k),$$

$$S_{k+1} = S_k + \xi_k(x_k, S_k), \quad x_{k+1} = \beta_k(x_k, S_k), \quad k \geq 0, \tag{5.5}$$

$$S(t) = S_k,\ x(t) = x_k \quad \text{as} \quad t_k \le t < t_{k+1},\quad t \ge 0. \tag{5.6}$$

Then the pair $(x(t), S(t))$, $t \ge 0$ forms a general *RPSM*. We mention that in this case we have feedback between both components $x(t)$ and $S(t)$. In particular, when distributions of the variables $\beta_k(x, \alpha)$ do not depend on the parameter α, the sequence x_k forms a *MP* and we obtain the previous case of *RPSM* with Markov switches.

2.5 PROCESSES WITH SEMI-MARKOV SWITCHES

Now we consider an operation of some random process in a semi-Markov environment. Let $\mathcal{F}_k = \{\zeta_k(t, x, \alpha), t \ge 0, x \in X, \alpha \in \mathcal{R}^r\}$, $k \ge 0$ be jointly independent parametric families of random processes, where $\zeta_k(t, x, \alpha)$ at each fixed k, x, α be a random process with trajectories in Skorokhod space $\mathcal{D}^r_\infty$, and let $x(t)$, $t \ge 0$ be a right-continuous *SMP* in X independent of $\mathcal{F}_k$, $k \ge 0$, S_0 be an initial value. Denote by $0 = t_0 < t_1 < \ldots$ the epochs of sequential jumps for $x(\cdot)$, $x_k = x(t_k)$, $k \ge 0$. We construct a process with semi-Markov switches (or in a semi-Markov environment) as follows: put $S_{k+1} = S_k + \xi_k$, where $\xi_k = \zeta_k(\tau_k, x_k, S_k)$, $\tau_k = t_{k+1} - t_k$, and denote

$$\zeta(t) = S_k + \zeta_k(t - t_k, x_k, S_k) \quad \text{as} \quad t_k \le t < t_{k+1},\quad t \ge 0. \tag{5.7}$$

Then a two-component process $(x(t), \zeta(t))$, $t \ge 0$ is called a Process with Semi-Markov Switches (*PSMS*). Let us introduce also an imbedded process

$$S(t) = S_k \quad \text{as} \quad t_k \le t < t_{k+1}. \tag{5.8}$$

Then two-component process $(x(t), S(t))$ forms a *RPSM* with additional Markov switches.

Suppose that $\{\zeta(t, x),\ t \ge 0\}$ is a family of *MP* and $\zeta(t, x, \alpha)$ denotes the process $\zeta(t, x)$ with initial value α. In that case the process $(x(t), \zeta(t))$ forms a Markov random evolution (when the process $x(t)$ is *MP*) or semi-Markov one (when the process $x(t)$ is *SMP*) (see Hersh, 1974; Kurtz, 1973; Kertz, 1978b; Pinsky, 1975; Korolyuk and Swishchuk, 1994).

2.6 SWITCHING PROCESSES

Now we give a general construction of a Switching Process (*SP*). Let

$$\mathcal{F}_k = \{(\zeta_k(t, x, \alpha), \tau_k(x, \alpha), \beta_k(x, \alpha)), t \ge 0, x \in X, \alpha \in \mathcal{R}^r\},\quad k \ge 0$$

be jointly independent parametric families where $\zeta_k(t, x, \alpha)$ at each fixed k, x, α be a random process in Skorokhod space $\mathcal{D}_\infty{}^r$ and $\tau_k(x, \alpha), \beta_k(x, \alpha)$ be possibly dependent on $\zeta_k(\cdot, x, \alpha)$ random variables, $\tau_k(\cdot) > 0$, $\beta_k(\cdot) \in X$. Let also (x_0, S_0) be independent of $\mathcal{F}_k, k \geq 0$, initial value. We put

$$t_0 = 0, \quad t_{k+1} = t_k + \tau_k(x_k, S_k), \quad S_{k+1} = S_k + \xi_k(x_k, S_k),$$

$$x_{k+1} = \beta_k(x_k, S_k), \quad k \geq 0, \tag{5.9}$$

where $\xi_k(x, \alpha) = \zeta_k(\tau_k(x, \alpha), x, \alpha)$, and set

$$\zeta(t) = S_k + \zeta_k(t - t_k, x_k, S_k), \; x(t) = x_k, \quad \text{as } \; t_k \leq t < t_{k+1}, \; t \geq 0. \tag{5.10}$$

Then a two-component process $(x(t), \zeta(t))$, $t \geq 0$ is called a *SP* (see Anisimov, 1977, 1978). In concrete applications the component $x(\cdot)$ usually means some random environment, and $S(\cdot)$ means the trajectory of the system. We also mention that the general construction of a *SP* allows the dependence (feedback) between both components $x(\cdot)$ and $S(\cdot)$.

2.7 EXAMPLES OF SWITCHING PROCESSES

Now we consider some models of *SP*'s as examples.

PII with semi-Markov switches. Let $x(t), t \geq 0$ be a *SMP* with state space X and let independent of it and jointly independent families of homogeneous processes with independent increments (*PII*) $\{\xi_k(t, x), t \geq 0, x \in X\}, k \geq 0$ and random variables $\{\gamma_k(x), x \in X\}, k \geq 1$ with values in $\mathcal{R}^r$ be given. Suppose for simplicity that distributions of variables introduced do not depend on the index k.

We construct a two-component process $(x(t), \zeta(t)), t \geq 0$ with values in $(X, \mathcal{R}^r)$ as follows. Let (x_0, ζ_0) be some initial value. Denote the epochs of sequential jumps for $x(t)$ by $0 = t_0 < t_1 < t_2$ and let $x_k = x(t_k), k \geq 0$ be the embedded *MP*. Then we put $\zeta(0) = \zeta_0$ and

$$\zeta(t) = \zeta(t_k) + \xi_k(t, x_k) - \xi_k(t_k, x_k), \quad t_k \leq t < t_{k+1},$$

$$\zeta(t_{k+1}) = \zeta(t_{k+1} - 0) + \gamma_{(k+1)}(x_{k+1}), \quad k \geq 0.$$

By the construction on the fixed trajectory of $x(t)$ the process $\zeta(t)$ is operating like a non-homogeneous process with independent increments and with additional jumps in the epochs $t_k, k > 0$ of the sizes $\gamma_k(x_k)$. It is called *PII* with semi-Markov switches (see Anisimov, 1973).

We remark that if the process $x(\cdot)$ is a *MP*, then the pair $(x(t), \zeta(t))$, $t \geq 0$ forms a *PII* with Markov switches (or *MP* homogeneous in the 2nd component (see Ežov and Skorokhod, 1969).

In this way we can construct a Poisson process with semi-Markov switches. Let the family of non-negative functions $\lambda(x)$, $x \in X$ and *SMP* $x(t), t \geq 0$ with values in X be given. Denote by $\Pi_{\lambda(\cdot)}(t)$ a Poisson process with instantaneous value of parameter $\lambda(x(t))$ at time t. Then a two-component process $(x(t), \Pi_{\lambda(\cdot)}(t))$, $t \geq 0$ is a *PSMS*. In particular if $x(t)$ is a *MP* then the process $\Pi_{\lambda(\cdot)}(t)$ is a Markov modulated Poisson process. Processes of this type arise at study of input flows at queueing models in a random environment.

Random movements with SMP switches. Let $\{v(i, \alpha), \alpha \in \mathcal{R}^r\}$, $i = 1, 2, \ldots, m$ be a family of continuous vector-valued functions in $\mathcal{R}^r$, and $x(t), t \geq 0$ be a *SMP* with finite number of states $X = \{1, 2, \ldots, d\}$. We put $\zeta_k(t, i, \alpha) = tv(i, \alpha)$, $t \geq 0$, $i = \overline{1, d}$. Denote by $0 = t_0 < t_1 < \ldots$ times of sequential jumps for $x(t)$, $x_k = x(t_k)$. Then *PSMS* $(x(t), \zeta(t)), t \geq 0$ constructed by the family of processes $\{\zeta_k(t, i, \alpha),\ t \geq 0,\ \overline{1, m}\}$ and switching times $t_k, k \geq 0$ forms a random movement in R^r with semi-Markov switches. If we denote $\nu(t) = \max\{k : k \geq 0, t_k < t\}, \zeta_k = \zeta(t_k)$, then :

$$\zeta(t) = \zeta(0) + \sum_{k=0}^{\nu(t)-1} (t_{k+1} - t_k) v(x_k, \zeta_k) + (t - t_{\nu(t)}) v(x_{\nu(t)}, \zeta_{\nu(t)}).$$

Dynamical systems in semi-Markov environment. Let $\{f(x, \alpha),\ \alpha \in \mathcal{R}^r\}$, $x \in X$ be a family of deterministic functions with values in $\mathcal{R}^r$, $\Gamma_k = \{\gamma_k(x, \alpha), x \in X, \alpha \in \mathcal{R}^r\}$, $k \geq 0$, be jointly independent families of random variables with values in $\mathcal{R}^r$ and $x(t), t \geq 0$ be a *SMP* in X independent of introduced families Γ_k. Put $x_k = x(t_k)$ and denote by $0 = t_0 < t_1 < \ldots$ sequential times of jumps for the process $x(t)$. We introduce the process $\zeta(t)$ as follows : $\zeta(0) = \zeta_0$ and

$$d\zeta(t) = f(x_k, \zeta(t))dt, \quad t_k \leq t < t_{k+1},$$

$$\zeta(t_{k+1} + 0) = \zeta(t_{k+1} - 0) + \gamma_k(x_k, \zeta(t_{k+1} - 0)), \quad k \geq 0.$$

Then the process $\zeta(t)$ forms a dynamical system with semi-Markov switches.

Stochastic differential equations with semi-Markov switches. Let $\{c(x, a), b(x, a), x \in X, \alpha \in R\}$ be deterministic vector and matrix-valued functions of dimensions r and $r \times r$ respectively, $w(t), t \geq 0$ be a standard Wiener process in $\mathcal{R}^r$ and let $x(t), t \geq 0$ be a *SMP* independent of $w(\cdot)$. We introduce the process $\zeta(t)$ as a solution of the following stochastic differential equation :

$$\zeta(0) = \zeta_0, \quad d\zeta(t) = c(x(t), \zeta(t))dt + b(x(t), \zeta(t))dw(t),$$

where at each $x \in X$ the coefficients $c(x, a)$ and $b(x, a)$ satisfy the conditions of the existence and uniqueness theorem. Then the pair $(x(t), \zeta(t))$, $t \geq 0$ forms a *PSMS.* It is also possible to describe a feedback between components. Another example can be Markov continuous time space-dependent branching processes (see Anisimov, 1996b).

Switching state-dependent queueing models. A class of *SP*'s gives a possibility to describe various classes of stochastic queueing models such as state-dependent queueing systems and networks $SM_Q/M_Q/m/\infty$, $M_{SM,Q}/M_{SM,Q}/l/k$, $(M_{SM,Q}/M_{SM,Q}/m_i/k_i)^r$, with batch Markov or semi-Markov input, finite number of nodes, different types of calls, impatient calls and possibly of a random size (volume of information or necessary job), batch state-dependent service, which are switched by some external semi-Markov environment and current values of queues, also retrial queueing models, etc.
For these models switching times are usually times of any changes in the system (Markov models), times of jumps of the environment (in case of external semi-Markov environment), times of exit from some regions for the process generated by queue, waiting times, etc.

3. AVERAGING PRINCIPLE AND DIFFUSION APPROXIMATION FOR RPSM

We study limit theorems for *RPSM* in the triangular scheme for the case of fast switches. This means that we consider the process on the interval $[0, nT]$, $n \to \infty$ and characteristics of the process depend on the parameter n in such a way that the number of switches on each interval $[na, nb]$, $0 < a < b < T$ tends, by probability, to infinity. Then, under natural assumptions, the normed trajectory of $S_n(nt)$ uniformly converges by probability to some function which is the solution of an ordinary differential equation, and normed difference between trajectory and this solution weakly converges in Skorokhod space $\mathcal{D}_T$ to some diffusion process.

Let us consider *AP* and *DA* type theorems for simple *RPSM*, because these results have various applications in queueing models (see Anisimov, 1995, 1996a; Anisimov and Lebedev, 1992).

Let for each $n = 1, 2, \ldots$ $\mathcal{F}_{nk} = \left\{ ((\xi_{nk}(\alpha)), \tau_{nk}(\alpha)), \alpha \in \mathcal{R}^r \right\}$, $k \geq 0$ be jointly independent families of random variables taking values in $\mathcal{R}^r \times [0, \infty)$, with distributions do not depend on index k, and let S_{n0} be independent of $\mathcal{F}_{nk}, k \geq 0$ initial value in $\mathcal{R}^r$. Put

$$t_{n0} = 0, \quad t_{nk+1} = t_{nk} + \tau_{nk}(S_{nk}), \quad S_{nk+1} = S_{nk} + \xi_{nk}(S_{nk}), \ k \geq 0, \tag{5.11}$$

$$S_n(t) = S_{nk} \ \text{ as } \ t_{nk} \leq t < t_{nk+1}, \ t \geq 0.$$

Assume that there exist functions $m_n(\alpha) = E\tau_{n1}(n\alpha)$, $b_n(\alpha) = E\xi_{n1}(n\alpha)$.

Theorem 1 *(Averaging principle)*
Suppose that for any $N > 0$

$$\lim_{L \to \infty} \limsup_{n \to \infty} \sup_{|\alpha| < N} \left\{ E\tau_{n1}(n\alpha)\chi(\tau_{n1}(n\alpha) > L) + E|\xi_{n1}(n\alpha)|\chi(|\xi_{n1}(n\alpha)| > L) \right\} = 0, \tag{5.12}$$

as $\max(|\alpha_1|, |\alpha_2|) < N$,

$$|m_n(\alpha_1) - m_n(\alpha_2)| + |b_n(\alpha_1) - b_n(\alpha_2)| < C_N|\alpha_1 - \alpha_2| + \alpha_n(N), \tag{5.13}$$

where C_N are some bounded constants, $\alpha_n(N) \to 0$ uniformly in $|\alpha_1| < N, |\alpha_2| < N$, and there exist functions $m(a) > 0$, $b(a)$ and a proper random variable s_0 such that as $n \to \infty$, $n^{-1}S_{n0} \xrightarrow{P} s_0$, and for any $\alpha \in \mathcal{R}^r$

$$m_n(\alpha) \to m(\alpha) > 0, \ b_n(\alpha) \to b(\alpha). \tag{5.14}$$

Then

$$\sup_{0 \leq t \leq T} |n^{-1}S_n(nt) - s(t)| \xrightarrow{P} 0, \tag{5.15}$$

where

$$s(0) = s_0, \ \ \mathrm{d}s(t) = m(s(t))^{-1}b(s(t))\mathrm{d}t, \tag{5.16}$$

and T is any positive number such that $y(+\infty) > T$ with probability one, where

$$y(t) = \int_0^t m(\eta(u))\mathrm{d}u, \tag{5.17}$$

$$\eta(0) = s_0, \ \ \mathrm{d}\eta(u) = b(\eta(u))\mathrm{d}u \tag{5.18}$$

(it is supposed that a solution of equation (5.18) exists on each interval and is unique).

Now we consider a convergence of the process $\gamma_n(t) = n^{-1/2}(S_n(nt) - ns(t))$, $t \in [0,T]$ to some diffusion process. Denote

$$\tilde{b}_n(\alpha) = m_n(\alpha)^{-1} b_n(\alpha), \quad \tilde{b}(\alpha) = m(\alpha)^{-1} b(\alpha),$$

$$\rho_n(\alpha) = \xi_{n1}(n\alpha) - b_n(\alpha) - \tilde{b}(\alpha)(\tau_{n1}(n\alpha) - m_n(\alpha)),$$

$$q_n(\alpha, z) = \sqrt{n}\Big(\tilde{b}_n(\alpha + \frac{1}{\sqrt{n}} z) - \tilde{b}(\alpha)\Big), \quad D_n^2(\alpha) = E\rho_n(\alpha)\rho_n(\alpha)^*$$

(we denote the conjugate vector by the symbol *).

Theorem 2 *(Diffusion approximation)*
Let conditions (5.13)-(5.14) be satisfied where in (5.13) $\sqrt{n}\alpha_n(N) \to 0$, *there exist continuous vector-valued function* $q(\alpha, z)$ *and matrix-valued function* $D^2(\alpha)$ *such that in any domain* $|\alpha| < N$ $|q(\alpha, z)| < C_N(1+|z|)$, *and uniformly in* $|\alpha| < N$ *at each fixed* z

$$\sqrt{n}\Big(\tilde{b}_n(\alpha + n^{-1/2} z) - \tilde{b}(\alpha)\Big) \to q(\alpha, z), \tag{5.19}$$

$$D_n^2(\alpha) \to D^2(\alpha), \tag{5.20}$$

$\gamma_n(0) \stackrel{W}{\Rightarrow} \gamma_0$, *and for any* $N > 0$

$$\lim_{L\to\infty} \limsup_{n\to\infty} \sup_{|\alpha|<Nn} \Big\{E\tau_{n1}^2(\alpha)\chi(\tau_{n1}(\alpha) > L) + E|\xi_{n1}(\alpha)|^2 \chi(|\xi_{n1}(\alpha)| > L)\Big\} = 0. \tag{5.21}$$

Then the sequence of the processes $\gamma_n(t)$ *J-converges on any interval [0,T] such that* $y(+\infty) > T$ *to the diffusion process* $\gamma(t)$ *which satisfies the following stochastic differential equation solution of which exists and is unique:* $\gamma(0) = \gamma_0$,

$$d\gamma(t) = q(s(t), \gamma(t))dt + D(s(t))m(s(t))^{-1/2} dw(t), \tag{5.22}$$

where $s(\cdot)$ *satisfies equation (5.16) (J-convergence denotes a weak convergence of measures in Skorokhod space* D_T*.)*

Proof of Theorems 1, 2. Let us introduce sequences $\eta_{nk} = n^{-1}S_{nk}$, $y_{nk} = n^{-1}t_{nk}, k \geq 0$ and processes $\eta_n(u) = \eta_{nk}$, $y(u) = y_{nk}$ as $n^{-1}k \leq u < n^{-1}(k+1), u \geq 0$. Put $\nu_n(t) = \min\{k : k > 0,\ t_{nk+1} > nt\}$, $\mu_n(t) = \inf\{u : u > 0,\ y_n(u) > t\}$. By definition, $y_n(n^{-1}\nu_n(t)) \leq t <$

$y_n(n^{-1}\nu_n(t)+1))$ and $\mu_n(t) = n^{-1}(\nu_n(t)+1)$. As far as $S_n(nt) = S_{n\nu_n(t)}$, we have a representation

$$n^{-1}S_n(nt) = \eta_n(n^{-1}\nu_n(t)) = \eta_n(\mu_n(t) - 1/n).$$

Thus, *RPSM* $n^{-1}S_n(nt)$ is constructed as a superposition of two processes: $\eta_n(t)$ and $\mu_n(t)$. First we'll study the behaviour of the processes $\eta_n(t)$ and $y_n(t)$, then $\mu_n(t)$ and their superposition. According to (5.11), we can write the relations $\eta_{nk+1} = \eta_{nk} + n^{-1}b_n(\eta_{nk}) + \varphi_{nk}$, $y_{nk+1} = y_{nk} + n^{-1}m_n(\eta_{nk}) + \psi_{nk}$, $k \geq 0$, where $\varphi_{nk} = n^{-1}(\xi_{nk}(n\eta_{nk}) - b_n(\eta_{nk}))$, $\psi_{nk} = n^{-1}(\tau_{nk}(n\eta_{nk}) - m_n(\eta_{nk}))$.

Sequences φ_{nk} and $\psi_{nk}, k \geq 0$ are martingale differences with respect to the sequence of σ-algebras σ_{nk} generated by variables $\{\eta_{ni}, i \leq k\}$. Assume that condition (5.12) holds uniformly in $\alpha \in \mathcal{R}^r$. Then, using the result of Grigelionis (1973), it's not difficult to prove that for any $t > 0$, $\max_{m \leq nt} |\sum_{k=0}^{m} \varphi_{nk}| \xrightarrow{P} 0$. Further applying results of Gikhman and Skorokhod (1978) and using relation (5.14) we obtain

$$\sup_{u \leq t} |\eta_n(u) - \eta(u)| \xrightarrow{P} 0, \quad \sup_{u \leq t} |y_n(u) - y(u)| \xrightarrow{P} 0 \tag{5.23}$$

(see (5.17)(5.18)). As far as $m(a) > 0$, the process $y(t)$ increases strictly monotonically. Thus, the process $y^{-1}(t) = \mu(t)$ exists for such t that $y(+\infty) > t$ with probability one, is continuous and

$$\sup_{u \leq t} |\mu_n(u) - \mu(u)| \xrightarrow{P} 0. \tag{5.24}$$

Using the result of Billingsley (1977) about U-convergence of a superposition of random functions and relation (5.23), we obtain (5.15). Finally, we remark that $\Pr\Big\{\sup_{u \leq t} |s(t)| > N\Big\} \xrightarrow{P} 0$ as $N \to \infty$. Thus it is sufficient to check all conditions in each bounded region $|\alpha| \leq N$. Theorem 1 is proved.

Further denote $v_{nk} = \gamma_n(y_{nk})$, $\tilde{s}_{nk} = s(y_{nk}), k \geq 0$, and suppose for simplicity that s_0 is a nonrandom variable. As far as relation (5.15) holds, the trajectory η_{nk} , $k = 0, 1, \ldots, nT$ belongs to some bounded region with probability close to one. Thus, it is enough to check all conditions only in each bounded region. We have by the construction

$$v_{nk+1} = v_{nk} + n^{-1/2}\Big(\xi_{nk}(n\eta_{nk}) - n(\tilde{s}_{nk+1} - \tilde{s}_{nk})\Big).$$

Using Lagrange formula and relation (5.11), we obtain that

$$\tilde{s}_{nk+1} - \tilde{s}_{nk} = n^{-1}\tilde{b}(\tilde{s}_{nk})\tau_{nk} + \delta_{nk}^{(1)} = n^{-1}\tilde{b}(\tilde{s}_{nk})m_n(\eta_{nk}) + \delta_{nk}^{(1)} + \delta_{nk}^{(2)},$$

where $\tau_{nk} = \tau_{nk}(n\eta_{nk})$, $|\delta_{nk}^{(1)}| \le Cn^{-2}\tau_{nk}^2$ and $E|\delta_{nk}^{(2)}|^2 \le Cn^{-2}$.
After transformation we obtain that

$$v_{nk+1} = v_{nk} + n^{-1}m_n(\eta_{nk})q_n(\tilde{s}_{nk}, v_{nk}) + n^{-1/2}\alpha_{nk} + \delta_{nk}^{(3)}, \qquad (5.25)$$

where $\alpha_{nk} = \xi_{nk}(n\eta_{nk}) - b_n(\eta_{nk}) - \tilde{b}(\tilde{s}_{nk})(\tau_{nk} - m_n(\eta_{nk}))$, $\mathbf{E}|\delta_{nk}^{(3)}|^2 \le n^{-3/2}C$. It is not difficult to prove that $\max_{k\le nT} |\sum_{i=o}^{k} \delta_{ni}^{(3)}| \xrightarrow{\mathrm{P}} 0$. If $n \to \infty$, $k/n \to t$ and $v_{nk} = z$, then according to Theorem 1 $\eta_{n,[nt]} \xrightarrow{\mathrm{P}} \eta(t)$, and $\tilde{s}_{n,[nt]} \to s(y(t)) = \eta(\mu(y(t))) = \eta(t)$. It means that a coefficient at $1/n$ in the right-hand side of (5.25) tends in probability to the value $m(\eta(t))q(\eta(t), z)$. Further, $\mathbf{E}[\alpha_{nk}/\eta_{nk}] = 0$, $\mathbf{E}[\alpha_{nk}\alpha_{nk}^*/\eta_{nk} = \alpha] \to D(\alpha)^2$ and, according to (5.21) variables $|\alpha_{nk}|^2$ are uniformly integrable in each bounded region. Let us introduce a random process $v_n(t) = v_{nk}$ as $k/n \le u < (k+1)/n$, $u \ge 0$. Then from representation (5.25) and results of Gikhman and Skorokhod (1975), it follows that the sequence of processes $v_n(u)$ J-converges on the interval $[0, T]$ to a diffusion process $v(u)$ satisfying the following stochastic differential equation: $v(0) = \gamma_0$,

$$\mathrm{d}v(u) = m(\eta(u))q(\eta(u), v(u))\mathrm{d}u + D(\eta(u))\mathrm{d}w(u). \qquad (5.26)$$

We remark that at $\frac{1}{n}t_{nk} \le t < \frac{1}{n}t_{nk+1}$

$$|s(t) - s(\frac{1}{n}t_{nk})| \le \frac{1}{n}\tau_{nk} \sup_{\frac{1}{n}t_{nk}\le u\le \frac{1}{n}t_{nk+1}} |\tilde{b}(s(u))|.$$

Thus as $\mu_n(T) < \mu(T) + \varepsilon$,

$$\sup_{0\le t\le T} |\gamma_n(t) - v_n(\mu_n(t) - \frac{1}{n})| \le \frac{1}{\sqrt{n}} C_T \max_{k\le n(\mu(T)+\varepsilon)} \tau_{nk}, \qquad (5.27)$$

where $C_T = \sup_{u\le\mu(T)+\varepsilon} |\tilde{b}(s(u))|$. It is not so hard to prove (see Anisimov, 1995) that for any $C > 0$ $\max_{k\le nC} n^{-1/2}\tau_{nk} \xrightarrow{\mathrm{P}} 0$. Finally, we obtain that

$$\sup_{0\le t\le T} |\gamma_n(t) - v_n(\mu_n(t) - 1/n)| \xrightarrow{\mathrm{P}} 0.$$

But the sequence of processes $v_n(\mu_n(t) - 1/n)$ J-converges to the process $v(\mu(t)) = \gamma(t)$. As far as $\mu'(t) = m(s(t))^{-1}$, we calculate the stochastic differential for process $\gamma(t)$ using the formula $\mathrm{d}w(\mu(t)) \sim \sqrt{\mu'(t)}\mathrm{d}w(t)$ and obtain equation (5.22). Theorem 2 is proved.

In conclusion of this section let us consider an important case when process $S_n(t)$ is a homogeneous *MP*. Suppose that $S_n(t)$ is a regular step-wise process and there exist intensities of transition probabilities

$q_n(\alpha, A), \alpha \in \mathcal{R}^r, A \in B^r_{\mathcal{R}}, \alpha \neq A$ such that $q_n(\alpha) = q_n(\alpha, \mathcal{R}^r \setminus \{\alpha\}) < \infty$ for any $\alpha \in \mathcal{R}^r$. We introduce independent families of random variables $\{\xi_{nk}(\alpha), \alpha \in \mathcal{R}^r\}, k \geq 0$ and $\{\tau_{nk}(\alpha), \alpha \in R\}, k \geq 0$ with values in $\mathcal{R}^r$ and $[0, \infty)$ respectively and such that $\tau_{nk}(n\alpha)$ has exponential distribution with parameter $q_n(\alpha)$ and $\Pr\Big\{\xi_{nk}(n\alpha) \in A\Big\} = q_n(\alpha)^{-1} q_n(\alpha, A + \alpha), \alpha \neq A$, where $A + \alpha = \{z : z - \alpha \in A\}$. It is clear that $RPSM$ which is defined by families $(\zeta_{nk}(\alpha), \tau_{nk}(\alpha))$ is equivalent to our MP $S_n(t)$. Denote $m_n(\alpha) = q_n(\alpha)^{-1}, D^2_n(\alpha) = \mathbf{E}\xi_{n1}(n\alpha)\xi_{n1}(n\alpha)^*$ and keep other notations.

Corollary 1 *If conditions of Theorems 1, 2 hold, then the relation (5.15) takes place and the sequence of processes $\gamma_n(t)$ weakly converges to the diffusion process $\gamma(t)$ satisfying equation (5.22).*

We remark that in this case conditions (5.12) and (5.21) for variables $\tau_{n1}(\alpha)$ are automatically satisfied.

4. PROCESSES WITH SEMI-MARKOV SWITCHES

Consider now AP and DA type theorems for $PSMS$. Let for each $n > 0$, $\mathcal{F}_{nk} = \{\zeta_{nk}(t, x, \alpha),\ t \geq 0,\ x \in X,\ \alpha \in \mathcal{R}^r\},\ \ k \geq 0$ be jointly independent families of random processes in D^r_∞, $x_n(t)$, $t \geq 0$ be a SMP in X independent of $\mathcal{F}_{nk}$, S_{n0} be an initial value. Let also $0 = t_{n0} < t_{n1} < \cdots$ be the epochs of sequential jumps of $x_n(\cdot)$, $x_{nk} = x_n(t_{nk})$, $k \geq 0$. We construct a $PSMS$ according to formula (5.7): put $S_{nk+1} = S_{nk} + \xi_{nk}$, where $\xi_{nk} = \zeta_{nk}(\tau_{nk},\ x_{nk},\ S_{nk})$, $\tau_{nk} = t_{nk+1} - t_{nk}$, and denote

$$\zeta_n(t) = S_{nk} + \zeta_{nk}(t - t_{nk}, x_{nk}, S_{nk}) \quad \text{as} \quad t_{nk} \leq t < t_{nk+1},\ t \geq 0. \quad (5.28)$$

Then the process $(x_n(t),\ \zeta_n(t))$, $t \geq 0$ is a $PSMS$.

At first we study an AP for the switched component $\zeta_n(\cdot)$. Consider for simplicity a homogeneous case (distributions of processes $\zeta_{nk}(\cdot)$ do not depend on the index $k \geq 0$). Let $\tau_n(x)$ be a sojourn time in the state x for SMP $x_n(\cdot)$. Denote for each $x \in X$, $\alpha \in \mathcal{R}^r$

$$\xi_n(x, \alpha) = \zeta_{n1}(\tau_n(x), x, \alpha), \quad g_n(x, \alpha) = \sup_{t < \tau_n(x)} |\zeta_{n1}(t, x, \alpha)|.$$

4.1 ASYMPTOTICALLY MIXING ENVIRONMENT

Suppose that MP x_{nk}, $k \geq 0$ has at each $n \geq 0$ a stationary measure $\pi_n(A)$, $A \in \mathcal{B}_X$ and denote $m_n(x) = \mathbf{E}\,\tau_n(x)$, $b_n(x,\alpha) = \mathbf{E}\,\xi_n(x, n\alpha)$,

$$m_n = \int_X m_n(x)\pi_n(\mathrm{d}x), \quad b_n(\alpha) = \int_X b_n(x,\alpha)\pi_n(\mathrm{d}x),$$

$$\alpha_n(k) = \sup_{A,B\in\mathcal{B}_X, i\geq 0} |\mathbf{P}\{x_{ni} \in A, x_{ni+k} \in B\} - \mathbf{P}\{x_{ni} \in A\}\,\mathbf{P}\{x_{ni+k} \in B\}|.$$

Theorem 3 *Suppose that* $n^{-1}S_{n0} \xrightarrow{\mathrm{P}} s_0$, *there exists a sequence of integers* r_n *such that*

$$n^{-1}r_n \to O, \ \sup_{k\geq r_n} \alpha_n(k) \to O, \tag{5.29}$$

for any $N > 0, \varepsilon > 0$

$$\lim_{n\to\infty} \sup_{|\alpha|<N} \sup_x n\mathbf{P}\{n^{-1}g_n(x,\alpha) > \epsilon\} = 0, \tag{5.30}$$

$$\lim_{L\to\infty} \limsup_{n\to\infty} \sup_{|\alpha|<N} \sup_x \{\mathbf{E}\tau_{n1}(x)\chi(\tau_{n1}(x) > L) + \mathbf{E}\,|\xi_{n1}(x,n\alpha)|\chi(|\xi(x,n\alpha)| > L)\} = 0,$$

for any x *as* $\max(|\alpha_1|, |\alpha_2|) < N$ $|b_n(x,\alpha_1) - b_n(x,\alpha_2)| < C_N|\alpha_1 - \alpha_2| + \alpha_n(N)$, *where* C_N *are some constants,* $\alpha_n(N) \to 0$ *uniformly on* $|\alpha_1| < N$, $|\alpha_2| < N$, *also there exists a function* $b(\alpha)$ *and a constant* m *such that for any* $\alpha \in \mathcal{R}^r$ $b_n(\alpha) \to b(\alpha)$, $m_n \to m > 0$. *Then for any* $T > 0$

$$\sup_{0\leq t\leq T} |n^{-1}\zeta_n(nt) - s(t)| \xrightarrow{\mathrm{P}} 0, \tag{5.31}$$

where

$$s(0) = s_0, \quad \mathrm{d}s(t) = m^{-1}b(s(t))\,\mathrm{d}t \tag{5.32}$$

(it is supposed that a solution of the equation (5.32) exists on each interval and is unique).

Remark 1 *Condition (5.29) covers also more general situations than only the case when the process* x_{nk} *is ergodic in the limit. For instance a state space can form n-S-set (see Anisimov, 1973, 1996a).*

Consider a *DA* for the sequence of processes $\gamma_n(t) = n^{-1/2}(\zeta_n(nt) - ns(t))$. Introduce a uniformly strong mixing coefficient for the process x_{nk}: $\varphi_n(r) = \sup_{x,y,A} |\mathbf{P}\{x_{nr} \in A/x_{no} = x\} - P\{x_{nr} \in A/x_{no} = y\}|, r >

0. Put $\tilde{b}_n(\alpha) = b_n(\alpha)m_n^{-1}$, $\tilde{b}(\alpha) = b(\alpha)m^{-1}$, $\rho_{nk}(x,\alpha) = \xi_{nk}(x, n\alpha) - b_n(x,\alpha) - \tilde{b}(\alpha)(\tau_{nk}(x) - m_n(x))$, $D_n(x,\alpha)^2 = \mathbf{E}\rho_{n1}(x,\alpha)\rho_{n1}(x,\alpha)^*$, and $\gamma_n(x,\alpha) = b_n(x,\alpha) - b_n(\alpha) - \tilde{b}(\alpha)(m_n(x) - m_n)$.

Theorem 4 *Suppose that* $\gamma_n(0) \overset{\mathrm{w}}{\Rightarrow} \gamma_0$, *there exist fixed* $r > 0$ *and* $q \in [0,1)$ *such that* $\varphi_n(r) \le q$, $n > 0$, *conditions of Theorem 3 hold where* $\sqrt{n}\alpha_n(N) \to 0$, *and for any* $N > 0$ *the following conditions are satisfied:*

$$\lim_{n\to\infty} \sup_{|\alpha|<N} \sup_x n\mathbf{P}\left\{n^{-1/2} g_n(x,\alpha) > \varepsilon\right\} = 0, \ \forall \varepsilon > 0; \tag{5.33}$$

$$\lim_{L\to\infty} \lim_{n\to\infty} \sup_{|\alpha|<N} \sup_x \left\{\mathbf{E}\tau_{n1}(x)^2 \chi(\tau_{n1}(x) > L) + \mathbf{E}|\xi_{n1}(x,n\alpha)|^2 \chi(|\xi_{n1}(x,n\alpha)| > L)\right\} = 0;$$

$|D_n(x,\alpha_1)^2 - D_n(x,\alpha_2)^2| \le C_N|\alpha_1 - \alpha_2| + \alpha_n(N)$, as $\max(|\alpha_1|, |\alpha_2|) < N$, *where* $\alpha_n(N) \to 0$ *uniformly in* $|\alpha_1| < N$, $|\alpha_2| < N$;
there exist continuous vector-valued function $q(\alpha, z)$ *and matrix-valued functions* $D(\alpha)$ *and* $B(\alpha)$ *such that in any domain* $|\alpha| < N$ $|q(\alpha,z)| < C_N(1 + |z|)$, *uniformly in* $|\alpha| < N$ *at each fixed* z

$$\sqrt{n}\left(\tilde{b}_n(\alpha + n^{-1/2}z) - \tilde{b}(\alpha)\right) \to q(\alpha, z);$$

at any $\alpha \in \mathcal{R}^m$

$$D_n(\alpha)^2 = \int_X D_n(x,\alpha)^2 \pi_n(\mathrm{d}x) \to D(\alpha)^2,$$

$$B_n^{(1)}(\alpha)^2 + B_n^{(2)}(\alpha)^2 + (B_n^{(2)}(\alpha)^*)^2 \to B(\alpha)^2,$$

where $B_n^{(1)}(\alpha)^2 = \int_X \gamma_n(x,\alpha)\gamma_n(x,\alpha)^* \pi_n(\mathrm{d}x)$, *and*

$$B_n^{(2)}(\alpha)^2 = \sum_{k\ge 1} \mathbf{E}\,\gamma_n(x_{n0},\alpha)\gamma_n(x_{nk},\alpha)^*, \tag{5.34}$$

where $\mathbf{P}\{x_{n0} \in A\} = \pi_n(A)$, $A \in \mathcal{B}_X$. *Then the sequence of processes* $\gamma_n(t)$ *J-converges to the diffusion process* $\gamma(t) : \gamma(0) = \gamma_0$,

$$\mathrm{d}\gamma(t) = q(s(t), \gamma(t))\,\mathrm{d}t + m^{-\frac{1}{2}}(D(s(t))^2 + B(s(t))^2)^{\frac{1}{2}}\mathrm{d}w(t), \tag{5.35}$$

where $w(t)$ *is a standard Wiener process in* $\mathcal{R}^r$ *and the solution of (5.35) exists and is unique.*

The proof of Theorems 3, 4 follows the same scheme as the proof of Theorems 1, 2 and uses the results about the convergence of stochastic recurrent sequences in Markov environment to solutions of stochastic differential equations (see Anisimov and Yarachkovskiy, 1986). More details can be found in Anisimov (1994).
These results also can be extended on non-homogeneous in time models (see Anisimov, 1995).

4.2 ASYMPTOTICALLY CONSOLIDATED ENVIRONMENT

Now we consider the case when condition (5.29) is not true. It means that states of the environment do not asymptotically communicate. Suppose for simplicity that *MP* x_{nk} has a finite state space $X = \{1, 2, \ldots, d\}$. We keep the previous notations. Let the following representation holds:

$$X = \bigcup j \in Y X_j, \quad \text{where } X_{j_1} \bigcap X_{j_2} = \emptyset \text{ at } j_1 \neq j_2, \tag{5.36}$$

and also one-step transition probabilities $p_n(i, l) = \Pr\{x_{n1} = l/x_{n0} = i\}$ are represented in the form

$$p_n(i, l) = p_n^{(0)}(i, l) + n^{-1} h_n(i, l), \quad i, l = \overline{1, d}, \tag{5.37}$$

where $\limsup_{n\to\infty} \max_{i,l} |h_n(i, l)| < C$, and for any $j \in Y$ $p_n^{(0)}(i, l) \equiv 0$ at $i \in X_j$, $l \notin X_j$.

For each $j \in Y$ denote by $x_{nk}^{(j)}, k \geq 0$ an auxiliary *MP* with state space X_j and transition probabilities $p_n^{(0)}(i, l)$, $i, l \in X_j$. Suppose that at each j the process $x_{nk}^{(j)}$ satisfies condition (5.29) and denote by $\pi_n^{(j)}(i), i \in X_j$ its stationary distribution. Further for any $j \in Y$, $m \in Y$, $j \neq m$ we put

$$\lambda_n(j, m) = \sum_{i \in X_j} \pi_n^{(j)}(i) \sum_{l \in X_m} h_n(i, l),$$

$$\widehat{m}_n(j) = \sum_{i \in X_j} \pi_n^{(j)}(i) m_n(i), \quad \widehat{b}_n(j, \alpha) = \sum_{i \in X_j} \pi_n^{(j)}(i) b_n(i, \alpha).$$

Suppose that there exist values $\lambda(j, m), \widehat{m}(j)$ and continuous functions $\widehat{b}(j, \alpha)$ such that for any $\alpha \in \mathcal{R}^r, j, m \in Y$, $j \neq m$

$$\lambda_n(j, m) \to \lambda(j, m), \; \widehat{m}_n(j) \to \widehat{m}(j) > 0, \; \widehat{b}_n(j, \alpha) \to \widehat{b}(j, \alpha).$$

Denote by $y(t, j_0)$ a *MP* with values in Y, intensities of transition probabilities $\lambda(j, m)/\widehat{m}(j)$, $j, m \in Y, j \neq m$ and the initial value j_0. Denote also by $z(t, j_0, s_0)$ a solution of differential equation: $z(0, j_0, s_0) = s_0$,

$$dz(t, j_0, s_0) = \widehat{m}(y(t, j_0))^{-1} \, \widehat{b}(y(t, j_0), z(t, j_0, s_0)) dt. \tag{5.38}$$

Let us introduce a consolidated process $\widehat{x}_n(t) = j$ as $x_n(t) \in X_j$, $t \geq 0$.

Theorem 5 *Suppose that at our assumptions* $\Pr(x_n(0) \in X_{j_0}) \to 1$ *as* $n \to \infty$*, relations (5.36),(5.37) are true and corresponding conditions of* *regularity for variables* $\tau_n(\cdot), \xi_n(\cdot), g_n(\cdot)$ *given in Theorem 3 hold. Then the sequence of processes* $(\widehat{x}_n(t), n^{-1}\zeta_n(t))$ *J-converges on each interval* $[0, T]$ *to the process* $(y(t, j_0), z(t, j_0, s_0))$.

The proof is based on limit theorems for *SP*'s in the case of rare switches (see Anisimov, 1978, 1988ab). The main steps are as follows. The process $(\widehat{x}_n(t), n^{-1}\zeta_n(t)$ is represented as a *SP* for which switching times are the times of sequential jumps between regions X_j. Then on the interval between two jumps the process $n^{-1}\zeta_n(t)$ behaves as a process in asymptotically quasi-ergodic Markov environment and on the base of results of Theorem 3 it converges to a solution of differential equation with coefficients averaged by stationary measure in corresponding region. Further an interval of time between two sequential switches asymptotically has an exponential distribution with parameter which is obtained by averaging in stationary measure of normed transition probabilities from a region (see Anisimov, 1973, 1988a). Thus the limiting process can be described as a solution of a differential equation with Markov switches. In the case that $\widehat{b}_0(j, \alpha) \equiv 0$, it is also possible to prove a *DA* for $\zeta_n(t)$. We mention that in this case a class of limiting processes belongs to the class of dynamical systems or diffusion processes with Markov switches (see section 2.7).

5. APPLICATIONS

5.1 RANDOM MOVEMENTS

Consider *AP* and *DA* for a random movement with semi-Markov switches described in the section 2.7. Suppose that sojourn times of *SMP* $x(t)$ depend on parameter n in such a way that $\tau_n(i) = n^{-1}\tau(i)$. Assume that 2nd moments exist and denote $\mathbf{E}\tau(i) = m(i), \mathbf{Var}\,\tau(i) = \sigma^2(i), i = \overline{1, d}$.

1) At first consider an ergodic case. Suppose that the embedded *MP* x_k doesn't depend on parameter n and is irreducible. Denote by $\pi_i, i = \overline{1, d}$ its stationary distribution. Let $m = \sum_{i=1}^{d} m(i)\pi_i > 0,\ b(\alpha) = \sum_{i=1}^{d} v(i, \alpha)m(i)\pi_i$. At stationary conditions $(\mathbf{P}\{x_0 = i\} = \pi_i, i = \overline{1, d})$ we denote $B^{(2)}(\alpha)^2 = \sum_{k\geq 1} \mathbf{E}\, m(x_0)m(x_k)(v(x_0, \alpha) - m^{-1}b(\alpha))(v(x_k, \alpha) - m^{-1}b(\alpha))^*$, $B(\alpha)^2 = \sum_{i=i}^{m} \pi_i m(i)^2(v(i, \alpha) - m^{-1}b(\alpha))(v(i, \alpha) - m^{-1}b(\alpha))^* + B^{(2)}(\alpha)^2 + (B^{(2)}(\alpha)^*)^2$, $D(\alpha)^2 = \sum_{i=i}^{d} \pi_i(v(i, \alpha) - m^{-1}b(\alpha))(v(i, \alpha) - m^{-1}b(\alpha))^*\sigma(i)^2$.

Statement 1 *Let functions $v(i, \alpha)$ be locally Lipschitz and have no more than linear growth. Then for any $T > 0$,*

$$\sup_{0 \le t \le T} |\zeta_n(t) - s(t)| \xrightarrow{P} 0,$$

where $s(\cdot)$ satisfies equation (5.32), and the sequence $\sqrt{n}(\zeta_n(t) - s(t))$ J-converges to the diffusion process satisfying equation (5.35) with $q(\alpha, z) = m^{-1} b'(\alpha) z$.

The proof directly follows from the results of Theorems 3, 4.

2) Further suppose that the embedded *MP* also depends on the parameter n in such a way that conditions (5.36),(5.37) hold. For simplicity suppose that each region X_j forms in a limit one essential class. Let $x_k^{(j)}$ be an auxiliary *MP* in X_j with limiting transition probabilities and stationary distribution $\pi^{(j)}(i), i \in X_j$. At any $j \in Y$ denote

$$\widehat{m}(j) = \sum_{i \in X_j} m(i) \pi^{(j)}(i), \quad \widehat{b}(j, \alpha) = \sum_{i \in X_j} v(i, \alpha) m(i) \pi^{(j)}(i). \qquad (5.39)$$

Let $y(t, j_0)$ be the *MP* introduced in Theorem 5.

Statement 2 *Suppose that at our assumptions* $\Pr(x_n(0) \in X_{j_0}) \to 1$ *as $n \to \infty$, at any $j \in Y$, $\widehat{m}(j) > 0$ and functions $\widehat{b}(j, \alpha)$ are locally Lipschitz and have no more than linear growth. Then the sequence $\zeta_n(t)$ J-converges on each interval* $[0, T]$ *to the process $z(t, j_0, s_0)$ (see (5.38)).*

3) Consider now the case when in (5.39) $\widehat{b}(j, \alpha) \equiv 0$. For each region X_j put $\widehat{D}(j)^2 = \sum_{i \in X_j} v(i, 0) v(i, 0)^* \sigma(i)^2 \pi^{(j)}(i)$, $\widehat{B}^{(1)}(j)^2 = \sum_{i \in X_j} m(i)^2 \times v(i, 0) v(i, 0)^* \pi^{(j)}(i)$, in stationary conditions ($\Pr(x_0^{(j)} = i) = \pi^{(j)}(i), i \in X_j$) define

$$\widehat{B}^{(2)}(j)^2 = \sum_{k \ge 1} \mathbf{E}\, m(x_0^{(j)}) m(x_k^{(j)}) v(x_0^{(j)}, 0) v(x_k^{(j)}, 0)^*,$$

and denote $\widehat{C}(j)^2 = \widehat{D}(j)^2 + \widehat{B}^{(1)}(j)^2 + \widehat{B}^{(2)}(j)^2 + (\widehat{B}^{(2)}(j)^*)^2$.

Statement 3 *At conditions of Statement 2 the sequence $\sqrt{n}\zeta_n(t)$ J-converges to the process $\gamma(t, j_0, s_0)$ which can be represented as follows :*

$$\gamma(t, j_0, s_0) = \int_0^t \widehat{m}(y(t, j_0))^{-1/2} \widehat{C}(y(t, j_0))\, dw(t).$$

This is the Wiener process with Markov switches.

5.2 SEMI-MARKOV STATE-DEPENDENT QUEUEING MODELS

The results obtained can be effectively applied to the analysis of overloading state-dependent semi-Markov queueing models. Consider as an example a queueing system $SM/M_{SM,Q}/1/\infty$. Let $x(t),\ t \geq 0$ be a *SMP* with values in X. Denote by $\tau(x)$ a sojourn time in the state x. Let non-negative function $\mu(x,\alpha),\ x \in X,\ \alpha \geq 0$, be given. There is one server and infinitely many places for waiting. At first consider the model when calls enter the system one at a time at the epochs of jumps $t_1 < t_2 < \ldots$ of the process $x(t)$. Put $x_k = x(t_k + 0)$. If a call enters the system at time t_k and the number of calls in the system becomes equal to Q, then the intensity of service on the interval $[t_k, t_{k+1})$ is $\mu(x_k, n^{-1}Q)$. After service the call leaves the system. Let Q_{n0} be an initial number of calls, and $Q_n(t)$ be a number of calls in the system at time t.

1) At first consider the case when the embedded *MP* $x_k,\ k \geq 0$ doesn't depend on parameter n and is uniformly ergodic with stationary measure $\pi(A),\ A \in \mathcal{B}_X$. We put $m(x) = \mathbf{E}\tau(x)$, $m = \int_X m(x)\pi(dx)$, $c(\alpha) = \int_X \mu(x,\alpha)m(x)\pi(dx)$, $b(\alpha) = (1-c(\alpha))m^{-1}$, $g(x,\alpha) = 1 - m(x)(1-c(\alpha)+\mu(x,\alpha)m)m^{-1}$, $G(\alpha) = c'(\alpha)$, $d^2(x) = \mathbf{Var}\,\tau(x)$, $d^2 = \int_X d^2(x)\pi(dx)$, $e_1(\alpha) = \int_X \mu^2(x,\alpha)d^2(x)\pi(dx)$, $e_2(\alpha) = \int_X \mu(x,\alpha)d^2(x)$ $\pi(dx)$ and $D^2(\alpha) = c(\alpha)+e_1(\alpha)+2(1-c(\alpha))m^{-1}e_2(\alpha)+(1-c(\alpha))^2m^{-2}d^2$.

Statement 4 *Suppose that $m > 0$, the function $\mu(x,\alpha)$ is locally Lipschitz with respect to α uniformly in $x \in X$, the function $c(\alpha)$ has no more then linear growth and $n^{-1}Q_n(0)\xrightarrow{P}s_0 > 0$. Then the relation (5.31) holds with $\zeta_n(nt) = Q_n(nt)$, where $ds(t) = m^{-1}(1-c(s(t))dt$, $s(0) = s_0$, and T is any positive value such that $s(t) > 0,\ t \in [0,T]$. Suppose in addition that variables $\tau(x)^2$ are uniformly integrable, the function $c(\alpha)$ is continuously differentiable, $n^{-1/2}(Q_n(0) - s_0)\overset{W}{\Rightarrow}\gamma_0$, and*

$$B^2(a) = \mathbf{E}\Big(g(x_0,\alpha)^2 + 2\sum_{k=1}^{\infty} g(x_0,\alpha)g(x_k,\alpha)\Big),$$

where $P\{x_0 \in A\} = \pi(A), A \in \mathcal{B}_X$. Then the sequence of processes $\gamma_n(t) = n^{-1/2}(Q_n(nt) - ns(t))$ J-converges on the interval $[0,T]$ to the diffusion process $\gamma(t) : \gamma(0) = \gamma_0$,

$$d\gamma(t) = -m^{-1}G(s(t))\gamma(t)dt + m^{-1/2}\left(D^2(s(t)) + B^2(s(t))\right)^{1/2}dw(t).$$

Proof. At first we represent a queue in the system as a *PSMS*. In our case epochs t_k are switching times and variable $\xi_{nk}(x, n\alpha)$ can be represented in a form : $\xi_{n1}(x, n\alpha) = 1 - \Pi_{\mu(x,\alpha)}(\tau(x))$, where $\Pi_\lambda(t)$ is a Poisson process with parameter λ. It is easy to see that $\mathbf{E}\xi_1(x, n\alpha) = 1 - \mu(x,\alpha)m(x)$ and using result of Theorem 3 it is not difficult to obtain *AP*. Further we can simply calculate another characteristics and obtain *DA* using result of Theorem 4. We mention also that the process of changing queue is monotone on each interval $[t_k, t_{k+1})$. Thus U-convergence of embedded *RPSM* to a limit process automatically implies U-convergence of *PSMS* that is conditions (5.30), (5.33) are automatically satisfied. This finally proves Statement 4.

We remark that condition $s(t) > 0,\ t \in [0,T]$ is in fact a heavy traffic condition. For instance it is always true if $c(\alpha) < 1,\ \alpha > 0$.

2) Now suppose that the embedded *MP* $x_k,\ k \geq 0$ also depends on parameter n in such a way that conditions (5.36), (5.37) hold. For simplicity we consider the case of a finite state space X. Suppose that each region X_j forms in a limit one essential class and denote by $\pi^{(j)}(i), i \in X_j$ its stationary distribution. At any $j \in Y$ denote

$$\widehat{m}(j) = \sum_{i \in X_j} m(i)\pi^{(j)}(i), \quad \widehat{c}(j,\alpha) = \sum_{i \in X_j} \mu(i,\alpha)m(i)\pi^{(j)}(i). \tag{5.40}$$

Let $y(t, j_0)$ be a *MP* introduced in Theorem 5.

Statement 5 *If at our assumptions* $\Pr(x_n(0) \in X_{j_0}) \to 1$ *as* $n \to \infty$, *at any* $j \in Y$ $\widehat{m}(j) > 0$ *and functions* $\widehat{c}(j,\alpha)$ *are locally Lipschitz and have no more than linear growth, then the sequence* $n^{-1}Q_n(nt)$ *J-converges on the interval* $[0,T]$ *to the process* $q(t, j_0, s_0)$ *such that* $q(0, j_0, s_0) = s_0$ *and*

$$dq(t, j_0, s_0) = \widehat{m}(y(t, j_0))^{-1}\Big(1 - \widehat{c}(y(t, j_0), q(t, j_0, s_0))\Big)dt,$$

and T *is any positive value such that* $q(t, j_0, s_0) > 0$ *for all* $t \in [0,T]$ *with probability one.*

5.3 MARKOV MODELS WITH SEMI-MARKOV SWITCHES

Consider now a queueing system $M_{SM,Q}/M_{SM,Q}/1/\infty$. Let $x(t),\ t \geq 0$ be a *SMP* with values in $X = \{1, 2, \ldots, d\}$ and sojourn times $\tau(i)$. Let the family of non-negative functions $\{\lambda(i,\alpha),\ \mu(i,\alpha),\ \alpha \geq 0\},\ i \in X$ be given. There is one server and infinitely many places for waiting. The instantaneous rates of input flow and service depend on the state of $x(\cdot)$, value of the queue and parameter n in the following way: if at time t, $x(t) = i$ and $Q_n(t) = Q$, then an input rate is $\lambda(i, n^{-1}Q)$ and service rate is $\mu(i, n^{-1}Q)$. Calls enter the system one at a time. We mention that here times t_k are also switching times but at these times we have no additional jumps of input flow and finishing service.

1) At first consider the case when the embedded *MP* $x_k,\ k \geq 0$ doesn't depend on parameter n and is irreducible with stationary distribution $\pi_i,\ i \in X$. We keep the previous notations for values $m(i)$ and m and put $b(\alpha) = \sum_i(\lambda(i,\alpha) - \mu(i,\alpha))m(i)\pi_i$.

Statement 6 *Suppose that functions* $\lambda(i,\alpha), \mu(i,\alpha)$ *are locally Lipschitz with respect to* $\alpha,\ m > 0,$ *the function* $b(\alpha)$ *has no more then linear growth and* $n^{-1}Q_n(0)\xrightarrow{P}s_0 > 0$. *Then the relation (5.31) holds with* $\zeta_n(nt) = Q_n(nt)$, *where* T *is any positive value such that* $s(t) > 0$ *on the interval* $[0,T]$.

2) Now suppose that the embedded *MP* $x_k,\ k \geq 0$ also depends on parameter n in such a way that conditions (5.36), (5.37) hold. Suppose that each region X_j forms in a limit one essential class and denote by $\pi^{(j)}(i), i \in X_j$ its stationary distribution. At any $j \in Y$ denote $\widehat{m}(j) = \sum_{i\in X_j} m(i)\pi^{(j)}(i),\ \widehat{b}(j,\alpha) = \sum_{i \in X_j}(\lambda(i,\alpha) - \mu(i,\alpha))m(i)\pi^{(j)}(i)$. Let $y(t, j_0)$ be a *MP* introduced in Theorem 5.

Statement 7 *If at our assumptions conditions of Statement 5 are valid (also for functions* $\lambda(i,\alpha)$*), then the sequence* $n^{-1}Q_n(nt)$ *J-converges on the interval* $[0,T]$ *to the process* $q(t, j_0, s_0)$ *such that* $q(0, j_0, s_0) = s_0$ *and*

$$dq(t, j_0, s_0) = \widehat{m}(y(t, j_0))^{-1}\widehat{b}(y(t, j_0), q(t, j_0, s_0))dt.$$

Using the same technique we can apply these results to retrial queues and queueing networks $(SM/M_{SM,Q}/1/\infty)^r$, $(M_{SM,Q}/M_{SM,Q}/1/\infty)^r$ of a semi-Markov type with input and service depending on the state of some *SMP* and current values of queues in the nodes, different types of customers, impatient customers, etc.

Some non-Markov queueing models $G_Q/M_Q/1/\infty$, $SM_Q/M_Q/1/\infty$ and $(G_Q/M_Q/1/\infty)^r$ are considered in (Anisimov, 1992, 1995, 1996a).

Another direction of applications can be branching processes and dynamical systems with stochastic perturbations. For near-critical branching processes with semi-Markov switches and large number of particles an *AP* is proved by Anisimov (1996b), and for dynamical systems with quick semi-Markov perturbations *AP* and *DA* are given in (Anisimov, 1994, 1995).

References

[1] Anisimov, V.V. (1973). Asymptotic consolidation of the states of random processes, *Cybernetics*, **9**, No.3, 494–504.

[2] Anisimov, V.V. (1977). Switching processes, *Cybernetics*, **13**, No. 4, 590–595.

[3] Anisimov, V.V. (1978). Limit theorems for switching processes and their applications, *Cybernetics*, **14**, No. 6, 917–929.

[4] Anisimov, V.V. (1988a). Random Processes with Discrete Component. Limit Theorems, Publ. Kiev Univ., Kiev.

[5] Anisimov, V.V. (1988b). Limit theorems for switching processes, *Theor. Probab. and Math. Statist.*, No. 37, 1–5.

[6] Anisimov, V.V. (1992). Limit theorems for Switching Processes, In *Proceedings of Postgrad. School and Conference Functional Analysis III, 1989, Dubrovnik, Yugoslavia*, No. 40, 1992, 235–262, Var. publ. ser., Aarhus Univ.

[7] Anisimov, V.V. (1994). Limit theorems for processes with semi-Markov switchings and their applications, *Random Oper. & Stoch. Eqv.*, **2**, No. 4, 333–352.

[8] Anisimov, V.V. (1995). Switching processes: averaging principle, diffusion approximation and applications, *Acta Applicandae Mathematicae*, Kluwer, **40**, 95–141.

[9] Anisimov, V.V. (1996a). Asymptotic analysis of switching queueing systems in conditions of low and heavy loading, *Matrix-Analytic Methods in Stochastic Models, S.R. Chakravarthy and A.S. Alfa, eds., Lect. Notes in Pure and Appl. Mathem. Series*, **183**, 241–260, Marcel Dekker, Inc.

[10] Anisimov, V.V. (1996b): Averaging principle for near-critical branching processes with semi-Markov switching, *Theory Probab. and Math. Statist.*, No. 52, 13–26.

[11] Anisimov, V.V., Zakusilo, O.K., Dontchenko, V.S. (1987). The elements of queueing theory and asymptotic analysis of systems. Publ. Visca Scola, Kiev.

[12] Anisimov, V.V. and Aliev, A.O. (1990). Limit theorems for recurrent processes of semi-Markov type, *Theor. Probab. & Math. Statist.*, No. 41, 7–13.

[13] Anisimov, V.V. and Lebedev E.A. (1992). Stochastic Queueing Networks. Markov Models., Publ. Kiev Univ., Kiev.

[14] Anisimov, V.V. and Sztrik, Ja. (1989a). Asymptotic analysis of some complex renewable systems operating in random environments, *European J. of Operational Research*, **41**, 162–168.

[15] Anisimov, V.V. and Jurachkovskiy, A.P. (1986). A limit theorem for stochastic difference schemes with random coefficients, *Theory Probab. and Math. Statist.*, No. 33, 1–9.

[16] Basharin, G.P., Bocharov, P.P. and Kogan, Ja. A. (1989). Analysis of Queues in Computing Networks, Publ. Nauka, Moscow.

[17] Billingsley, P. (1968). *Convergence of Probability Measures.* John Wiley & Sons, New York.

[18] Buslenko, N.P., Kalashnikov, V.V. and Kovalenko, I.N. (1973). *Lectures on the Theory of Complex Systems,* Publ. Sov. Radio, Moscow.

[19] Ezov, I.I. and Skorokhod, A.V. (1969). Markov processes which are homogeneous in the second component, *Theor. Probab. Appl.*, **14**, 679-692.

[20] Gikhman, I.I. and Skorokhod A.V. (1973). *Theory of Random Processes II.*, Publ. Nauka, Moscow.

[21] Gikhman, I.I. and Skorokhod, A.V. (1978). *Theory of Random Processes III.*, Springer-Verlag.

[22] Griego, R. and Hersh, R. (1969). Random evolutions, Markov chains, systems of partial differential equations, In *Proc. Nat. Acad. Sci.*, No 62, 305–308.

[23] Grigelionis, B.I. (1973). The relative compactness of sets of probability measures in $D_{(0,\infty)}(X)$, in *Math. Trans. Acad. Sci. Lithuanian SSR*, **13**.

[24] Harrison, J.M. (1995). Balanced fluid models of multiclass queueing network: a heavy traffic conjecture, *Stochastic Networks, IMA Volumes in Mathematics and its Appl.*, eds. F. Kelly and R. Williams, **71**, 1–20, Springer-Verlag, New York.

[25] Harrison, J.M. and Williams, R.J. (1996). A multiclass closed queueing network with unconventional heavy traffic behavior, *Ann. Appl. Probab.*, **6**, No. 1, 1–47.

[26] Hersh, R. (1974). Random evolutions: survey of results and problems, *Rocky Mount. J. Math.*, 4, No. **3**, 443–475.

[27] Kertz, R. (1978a). Limit theorems for semigroups with perturbed generators with applications to multi-scaled random evolutions, *TAMPS*, No. 27, 215–233.

[28] Kertz R. (1978b). Random evolutions with underlying semi-Markov processes, *Publ. Res. Jnst. Math. Sci.*, **14**, 589–614.

[29] Koroljuk, V.S. and Swishchuk, A.V. (1986). Central limit theorem for semi-Markov random evolutions, *Ukrainian Math. J.*, **38**, 330-335.

[30] Koroljuk, V.S. and Swishchuk, A.V. (1994). *Random evolutions*, Kluwer Acad. Publ.

[31] Koroljuk, V.S. and Turbin, A.F. (1978). *Mathematical foundations of phase consolidations of complex systems*, Publ. Naukova Dumka, Kiev.

[32] Kovalenko, I.N. (1980). *Investigations in the Analysis of Reliability of Complex Systems*, Naukova Dumka, Kiev.

[33] Kurtz, T. (1973). A limit theorem for perturbed operator semigroups with applications to random evolutions, *Journal Funct. Anal.*, No. 12, 55–67.

[34] Mandelbaum, A., Pats, G. (1998). State-dependent stochastic networks. Part I: Approximation and applications, *Annals Appl. Probab.*, **8**, No. 2, 569–647.

[35] Papanicolaou, G. and Hersh, R. (1972). Some limit theorems for stochastic equations and applications, *Indiana Univ. Math. Jour.*, **21**, 815–840.

[36] Pinsky, M. (1975). Random evolutions, *Lecture Notes in Math.*, No. 451, 89–100, Springer, New York.

[37] Sztrik, J. and Kouvatsos, D. (1991). Asymptotic analysis of a heterogeneous multiprocessor system in a randomly changing environment, *IEEE Trans. on Software Engineering*, **17**, No. 10, 1069–1075.

[38] Watkins, J.C. (1984). A central limit problem in random evolutions, *Ann. Prob.*, **12**, No. 2, 480–513.

Chapter 6

APPROXIMATIONS FOR SEMI-MARKOV SINGLE ION CHANNEL MODELS

Susan M. Pitts

University of Cambridge, UK

Abstract We consider a semi-Markov framework for the openings and closures of a single ion channel, incorporating time interval omission, as in [2]. Using a functional approach, we give theoretical results that yield approximation formulae for the relevant semi-Markov kernels, and which may be used to obtain approximation formulae for various other quantities of interest built on these semi-Markov kernels. This work generalises earlier work [8] in an alternating renewal setting, and extends the methodology of the functional approach to multidimensional problems.

Keywords: Time interval omission, Fréchet derivatives.

1. INTRODUCTION

This paper is primarily concerned with obtaining approximation formulae for quantities of interest in semi-Markov models for single ion channel models with time interval omission. It is a generalisation of [8], which deals with a simple alternating renewal model for the open and closed times of single ion channels. We use here the same functional approach to stochastic models, where a quantity of interest is regarded as the output of a functional that acts on some known input quantity. In addition to focusing on approximation formulae for ion channel models, we also open the way for new and wider applications of the functional approach to other stochastic models, in that we develop a technical framework for dealing with models that are multidimensional. In particular, the key step in the functionals under consideration below involves a *multidimensional geometric random sum*. This quantity arises in other contexts (for an application in risk theory see [1]), so that the results indicated here have application to other models built around such random sums. Further, it is hoped that the general framework will provide a setting within which similar approximations can be developed for, e.g. random walks on a Markov chain, Markov modulated queues, etc.

First we give an outline of the semi-Markov ion channel model, as considered in [2] (see also [3]), using their notation. This model describes the state of a single ion channel in a biological membrane. When the channel (molecule) is open, ions may pass through the membrane. Let the state space be $\{1, 2, \ldots, n\}$, and suppose that it splits into two classes, $\mathcal{O} = \{1, 2, \ldots, n_o\}$ of n_O *open* states and $\mathcal{C} = \{n_o+1, \ldots, n\}$ of n_c *closed* states, $n = n_o + n_c$. We model the state of the channel as a semi-Markov process with semi-Markov kernel $\mathbf{G} = (G_{ij})$, $i, j \in \{1, \ldots, n\}$, where for $t \geq 0$, $G_{ij}(t)$ is $\mathrm{P}($next state is j and current sojourn is at most t | current state is $i)$. We assume differentiability, and put $g_{ij} = G'_{ij}$, i.e. $\mathbf{g} = \mathbf{G}'$. Observe that $G_{ij}(\infty)$ is not necessarily equal to one, so that g_{ij} may be a defective density, although we do have $\sum_{j=1}^{n} G_{ij}(\infty) = 1$ for each i. We assume throughout that the embedded discrete-time Markov chain is irreducible.

In practice, the underlying state of the channel is unobserved, and we may only observe whether the channel is open or closed. This observed process is the *aggregated* process. Following [2], we introduce the *gateway* process $\{(J_k, T_k)\}_{k=0}^{\infty}$, where the T_k's are the successive sojourns in the open and closed classes ($T_0 = 0$), and the J_k's record the states by which these classes are entered (J_0 is the state at time zero). This gives another semi-Markov process. Let its semi-Markov kernel be denoted by $\mathbf{F}$ and put $\mathbf{f} = \mathbf{F}'$. This process is itself unobserved, because the J_k's are unobserved, but properties of the aggregated process may be derived from it. The state space for $\mathbf{F}$ is $\{1, \ldots, m_o, n_o + 1, \ldots, n_o + m_c\}$ consisting of $m = m_O + m_c$ elements, where states $1, \ldots, m_o$ are *open entry states* (i.e. the underlying process can enter $\mathcal{O}$ via these states), and states $n_o + 1, \ldots, n_o + m_c$ are *closed entry states*.

A further difficulty arises from practical limitations in obtaining ion channel records, in that sojourns in either class that are smaller that a fixed deadtime $\tau > 0$ are not observed. This is called *time interval omission* or *limited time resolution*. The aggregated process with time interval omission has an associated (unobserved) gateway process $\{(\tilde{J}_k, \tilde{T}_k)\}_{k=0}^{\infty}$, where the $\tilde{T}_k$'s are the lengths of the successive detected sojourns in $\mathcal{O}$ and $\mathcal{C}$ alternately ($\tilde{T}_0 = 0$), and the $\tilde{J}_k$'s are the corresponding entry states ($\tilde{J}_0 = J_0$) (see [2]). This also gives rise to a semi-Markov process, and we denote its semi-Markov kernel by $\tilde{\mathbf{F}}$, with $\tilde{\mathbf{f}} = \tilde{\mathbf{F}}'$. Properties of the aggregated process with time interval omission can be derived from $\tilde{\mathbf{F}}$. As in [2], we make the technical assumption that the number of open gateway states for $\{(\tilde{J}_k, \tilde{T}_k)\}$ is the same as for $\{(J_k, T_k)\}$, and similarly for the number of closed gateway states.

Approximations for quantities of interest derived from $\tilde{\mathbf{f}}$ are needed because, although there exist relationships between Laplace transforms of the appropriate semi-Markov kernels (see [2]), it is not easy to calculate explicit expressions for e.g. $\tilde{\mathbf{f}}$ given a particular $\mathbf{g}$. Hence there has been much interest in approximations in the literature (see for example [6], [7], and the references in these papers). In [8], using a Banach algebra approach in an alternating renewal model of the openings and closures of a single ion channel model, it is shown that various existing approximations can be seen as fitting into the functional framework, and new first order corrections to such existing approximations are obtained. Th semi-Markov framework considered here includes the above alternating renewal model as a special case.

We now summarise briefly the functional approach in a general context. Suppose we have a functional Ξ that maps some input quantity onto an output quantity of interest, output $= \Xi(\text{input})$. A common scenario in applied probability applications is that there are a few special cases for the input for which there are easy explicit expressions for the corresponding output quantities, but that in the general case it is not easy to find such an expression. (Often, as here, we have an expression in terms of transforms.) Given a "hard" input for which we are interested in the resulting output, we find an "easy" input close by (in an appropriate sense that must be made precise), and we obtain the zeroth order approximation

$$\Xi(\text{hard input}) \approx \Xi(\text{easy input}). \tag{6.1}$$

This procedure is justified if the map Ξ is continuous with respect to the relevant topologies. Extending this, provided the map Ξ is differentiable (in an appropriate sense), we can obtain a first order correction to the above approximation,

$$\Xi(\text{hard input}) \approx \Xi(\text{easy input}) + \Xi'_{\text{easy}}(\text{ hard input} - \text{easy input}), \tag{6.2}$$

where we write Ξ'_{easy} for the derivative of Ξ at the easy input. This functional approach has been successfully used in renewal theory [4], [9], and in queueing theory [5]. In all of the applications so far, the quantities forming the input and output of the functionals have been one-dimensional densities, or pairs of densities (or sequences in discrete-time settings). Here we extend the framework to include matrices of functions as input and output quantities.

2. RESULTS

In §2.1 we explain the set-up, and then in §2.2 we show how the functionals are represented. In §2.3 we give our main results.

2.1 THE FRAMEWORK

Let L^1 be the space of complex-valued measurable functions f on $(0,\infty)$ that satisfy $\|f\|_{L^1} = \int_0^\infty |f(t)|dt < \infty$. Let $A = \{(f,\alpha) : f \in L^1, \alpha \in \mathbf{C}\}$. We think of an element of A as corresponding to a complex measure (see Chapter 6 of [11]) on $[0,\infty)$ consisting of an atom of size α at zero and an absolutely continuous part with density f on $(0,\infty)$. To echo this identification we write $f+\alpha\delta_0$ instead of (f,α), where δ_0 is a unit mass at zero. Define a norm on A by $\|f+\alpha\delta_0\| = \|f\|_{L^1} + |\alpha|$, so that $\|f+\alpha\delta_0\|$ gives the total variation of the measure represented by $f+\alpha\delta_0$. Define the convolution $f*g$ of f and g in L^1 by $(f*g)(t) = \int_0^t f(t-x)g(x)\,dx$, and let $(f+\alpha\delta_0)*(g+\beta\delta_0) = f*g+\alpha g+\beta f+\alpha\beta\delta_0$, i.e. $*$ gives the convolution of the two measures. Then $(A, \|\cdot\|, *)$ is a commutative Banach algebra with a unit δ_0. This space is the basic space used in [8]. An element a in A is invertible with inverse $a^{*(-1)}$ in A if $a * a^{*(-1)} = \delta_0$.

Let k be a positive integer, and let $\mathbf{A}_k$ be the set of all $k\times k$ matrices $\mathbf{a} = (a_{ij})$ where a_{ij} is in A, with element-wise definition of addition and scalar multiplication. We give $\mathbf{A}_k$ a norm defined by $\|\mathbf{a}\| = \max_i \sum_j \|a_{ij}\|$, and we define the convolution $\mathbf{a}*\mathbf{b}$ of $\mathbf{a}$ and $\mathbf{b}$ in $\mathbf{A}_k$ by $(\mathbf{a}*\mathbf{b})_{ij} = \sum_l a_{il} * b_{lj}$. Then, using the relevant properties of A, it is easy to show that $(\mathbf{A}_k, \|\cdot\|, *)$ is a (non-commutative) Banach algebra with two-sided identity $\mathbf{e}$ $(= \mathbf{e}_k) = \mathrm{diag}(\delta_0, \ldots, \delta_0)$. In particular, the norm inequality holds, i.e. $\|\mathbf{a}*\mathbf{b}\| \le \|\mathbf{a}\|\,\|\mathbf{b}\|$. An element $\mathbf{a}$ has an inverse $\mathbf{a}^{*(-1)}$ in $\mathbf{A}_k$ if $\mathbf{a}*\mathbf{a}^{*(-1)} = \mathbf{a}^{*(-1)}*\mathbf{a} = \mathbf{e}$.

There is a close connection between Banach algebras and transforms. For f in L^1 let $\hat{f}(\theta) = \int e^{i\theta x} f(x)\,dx$ $(\theta \in \mathbf{R})$ be the Fourier transform of f. For $\theta \in \mathbf{R}$ let $(f+\alpha\delta_0)\hat{}(\theta) = \hat{f}(\theta)+\alpha$, and define $(f+\alpha\delta_0)\hat{}(\infty) = \alpha$. The Gelfand transform of an element $a \in A$ is a map from $\mathbf{R}\cup\{\infty\}$ to $\mathbf{C}$ that takes θ to $\hat{a}(\theta)$. The Banach algebra A is semisimple (see [10] 11.8), which means that an element a of A is identified by $\{\hat{a}(\theta) : \theta \in \mathbf{R}\cup\{\infty\}\}$. Theorem 11.5(d) of [10] implies that $a \in A$ is invertible if and only if $\hat{a}(\theta) \neq 0$ for all θ in $\mathbf{R}\cup\{\infty\}$. For $\mathbf{a} = (a_{ij})$, define $\hat{\mathbf{a}}(\theta)$ to be the matrix $(\hat{a}_{ij}(\theta))$, $\theta \in \mathbf{R}\cup\{\infty\}$. Then $(\mathbf{a}*\mathbf{b})\hat{}(\theta) = \hat{\mathbf{a}}(\theta)\hat{\mathbf{b}}(\theta)$ for all $\theta \in \mathbf{R}\cup\{\infty\}$, where the operation on the right-hand side is matrix multiplication. Further, if $\hat{\mathbf{a}}(\theta) = \hat{\mathbf{b}}(\theta)$ for all $\theta \in \mathbf{R}\cup\{\infty\}$ then the fact that A is semisimple means that we can deduce that $\mathbf{a} = \mathbf{b}$. Transforms provide access to inverses in $\mathbf{A}_k$, in that

$\hat{\mathbf{a}}(\theta)$ is invertible for all $\theta \in \mathbf{R} \cup \{\infty\}$ if and only if $\mathbf{a}$ is invertible in $\mathbf{A}_k$, (6.3)

and in this case $(\mathbf{a}^{*(-1)})\hat{}(\theta) = (\hat{\mathbf{a}}(\theta))^{-1}$.

Where necessary, we extend the definition of $*$ in an obvious way to include convolutions of compatible non-square matrices of elements of A.

2.2 THE FUNCTIONALS

In this subsection we use the arguments underlying the Laplace transform relationships given in [2] to obtain representations, within the framework of the previous subsection, of the functionals dealing with the aggregation step from $\mathbf{g}$ to $\mathbf{f}$ and with the time interval omission step from $\mathbf{f}$ to $\tilde{\mathbf{f}}$.

First we introduce some notation. Partition the kernel $\mathbf{g} \in \mathbf{A}_n$ of the underlying semi-Markov process as

$$\mathbf{g} = \begin{bmatrix} \mathbf{g}_{oo} & \mathbf{g}_{oc} \\ \mathbf{g}_{co} & \mathbf{g}_{cc} \end{bmatrix},$$

and put

$$\mathbf{g}_0 = \begin{bmatrix} \mathbf{g}_{oo} & \mathbf{0} \\ \mathbf{0} & \mathbf{g}_{cc} \end{bmatrix}, \quad \mathbf{g}_1 = \begin{bmatrix} \mathbf{0} & \mathbf{g}_{oc} \\ \mathbf{g}_{co} & \mathbf{0} \end{bmatrix}.$$

Let $\mathbf{l}$ be the $(m_o + m_c) \times (m_o + (n_o - m_o) + m_c + (n_c - m_c))$ matrix given by

$$\mathbf{l} = \begin{bmatrix} \mathbf{e}_{m_o} & \mathbf{0} & \mathbf{0} & \mathbf{0} \\ \mathbf{0} & \mathbf{0} & \mathbf{e}_{m_c} & \mathbf{0} \end{bmatrix},$$

and write $\mathbf{I}$ for the identity matrix, with dimension appropriate to the context. We next obtain an expression for $\mathbf{f}$ in terms of $\mathbf{l}$, $\mathbf{g}_0$ and $\mathbf{g}_1$, derived from Theorem 3.1 in [2]. First note that elementwise

$$0 \le |\hat{\mathbf{g}}_{oo}(\theta)| \le \hat{\mathbf{g}}_{oo}(0).$$

Since $\hat{\mathbf{g}}_{oo}(0)$ is the same as the Laplace transform of $\mathbf{g}_{oo}$ evaluated at zero (they are both the transition matrix for transitions within $\mathcal{O}$), we can directly apply the argument on page 778 in the proof of Theorem 3.1 in [2] to find that $\mathbf{I} - \hat{\mathbf{g}}_{oo}(\theta)$ is invertible , and similarly that $\mathbf{I} - \hat{\mathbf{g}}_{cc}(\theta)$ is invertible, for all θ in $\mathbf{R}$. Hence $\mathbf{I} - \hat{\mathbf{g}}_0(\theta)$ is invertible for all θ in $\mathbf{R}$.

Further, $\mathbf{I} - \hat{\mathbf{g}}_{\mathcal{OO}}(\infty) = \mathbf{I}$ by our density assumptions. Thus $(\mathbf{e} - \mathbf{g}_{\mathrm{O}})\hat{}(\theta)$ is invertible for all θ in $\mathbf{R} \cup \{\infty\}$, and so (6.3) implies that $\mathbf{e} - \mathbf{g}_{\mathrm{O}}$ is invertible in $\mathbf{A}_n$.

Following page 777 of [2], for $i \in \mathcal{O}$ let

$$H_{ij}(t) = \mathrm{P}\big(\text{first open sojourn} \leq t, \text{ next state} = j \mid \text{initial state } i\big),$$

if $j \in \mathcal{C}$, and let $H_{ij}(t) = 0$ for all t if $j \in \mathcal{O}$. Define $H_{ij}(t)$ similarly for $i \in \mathcal{C}$, and put $\mathbf{h} = \mathbf{H}'$. Again we follow [2] and condition on the length of the initial sojourn in $i \in \mathcal{O}$ to obtain

$$\mathbf{h}_{\mathcal{OC}} = \mathbf{g}_{\mathcal{OC}} + \mathbf{g}_{\mathcal{OO}} * \mathbf{h}_{\mathcal{OC}}.$$

Since $\mathbf{e} - \mathbf{g}_{\mathcal{OO}}$ is invertible, this gives

$$\mathbf{h}_{\mathcal{OC}} = \big(\mathbf{e} - \mathbf{g}_{\mathcal{OO}}\big)^{*(-1)} * \mathbf{g}_{\mathcal{OC}}.$$

Similarly we find

$$\mathbf{h}_{\mathcal{CO}} = \big(\mathbf{e} - \mathbf{g}_{\mathcal{CC}}\big)^{*(-1)} * \mathbf{g}_{\mathcal{CO}},$$

and so

$$\mathbf{h} = \big(\mathbf{e} - \mathbf{g}_{\mathrm{O}}\big)^{*(-1)} * \mathbf{g}_{1}.$$

Pre- and post-multiplying by $\mathbf{l}$ and $\mathbf{l}^T$ respectively (^{T} denotes transpose) in order to bring the resulting matrix to $\mathbf{A}_m$, we obtain

$$\mathbf{f} = \mathbf{l} * \big(\mathbf{e} - \mathbf{g}_{\mathrm{O}}\big)^{*(-1)} * \mathbf{g}_{1} * \mathbf{l}^T. \tag{6.4}$$

To help in interpreting (6.4), when $\|\mathbf{g}_{\mathrm{O}}\| < 1$ Theorem 10.7 of [10] implies that we may write $\big(\mathbf{e} - \mathbf{g}_{\mathrm{O}}\big)^{*(-1)} * \mathbf{g}_{1}$ as

$$\sum_{k=0}^{\infty} \mathbf{g}_{\mathrm{O}}^{*k} * \mathbf{g}_{1} = \begin{bmatrix} \mathbf{0} & \sum \mathbf{g}_{\mathcal{OO}}^{*k} * \mathbf{g}_{\mathcal{OC}} \\ \sum \mathbf{g}_{\mathcal{CC}}^{*k} * \mathbf{g}_{\mathcal{CO}} & \mathbf{0} \end{bmatrix},$$

where $\mathbf{g}_{\mathrm{O}}^{*0}$ is defined to be $\mathbf{e}$ and $\mathbf{g}_{\mathrm{O}}^{*k}$ is the (matrix) k-fold convolution of $\mathbf{g}_{\mathrm{O}}$. We interpret $\big(\mathbf{g}_{\mathcal{OO}}^{*k} * \mathbf{g}_{\mathcal{OC}}\big)_{ij}$, $i \in \mathcal{O}$ and $j \in \mathcal{C}$, as the density of a sojourn starting in i and taking k steps within $\mathcal{O}$ and then leaving $\mathcal{O}$ for $j \in \mathcal{C}$.

In (6.4), if $\mathbf{g}_{\mathrm{O}} = \mathbf{0}$ then there is no difference between the underlying and the aggregated processes, and $\mathbf{f} = \mathbf{g}$. Otherwise, we observe that (6.4) can be written in terms of normalised densities as follows. For any state i, let Let $G_{i\mathcal{O}} = \sum_{j\in\mathcal{O}} G_{ij}(\infty)$ and let $G_{i\mathcal{C}} = \sum_{j\in\mathcal{C}} G_{ij}(\infty)$. For

$i \in \mathcal{O}$, if $G_{i\mathcal{O}} > 0$ then there is positive probability that from state i the underlying process jumps to an open state. In this case we define $(\breve{\mathbf{g}}_{oo})_{ij} = (\mathbf{g}_{oo})_{ij}/G_{i\mathcal{O}}$, $j \in \mathcal{O}$; if $G_{i\mathcal{O}} = 0$ then $(\breve{\mathbf{g}}_{oo})_{ij}$ is defined to be zero. Define $\breve{\mathbf{g}}_{cc}$ similarly. Let

$$\breve{\mathbf{g}}_0 = \begin{bmatrix} \breve{\mathbf{g}}_{oo} & \mathbf{0} \\ \mathbf{0} & \breve{\mathbf{g}}_{cc} \end{bmatrix},$$

so that $\|\breve{\mathbf{g}}_0\| = 1$. Put

$$\mathbf{p}_o = \mathrm{diag}(G_{1\mathcal{O}}\delta_0, \ldots, G_{n_o\mathcal{O}}\delta_0),$$

$$\mathbf{p}_c = \mathrm{diag}(G_{n_o+1,\mathcal{C}}\delta_0, \ldots, G_{n\mathcal{C}}\delta_0),$$

and let $\mathbf{p}$ be the diagonal matrix given by

$$\mathbf{p} = \begin{bmatrix} \mathbf{p}_o & \mathbf{0} \\ \mathbf{0} & \mathbf{p}_c \end{bmatrix}.$$

Then $\mathbf{g}_0 = \mathbf{p} * \breve{\mathbf{g}}_0$. Similarly, defining $\breve{\mathbf{g}}_1$ by

$$\breve{\mathbf{g}}_1 = \begin{bmatrix} \mathbf{0} & \breve{\mathbf{g}}_{oc} \\ \breve{\mathbf{g}}_{co} & \mathbf{0} \end{bmatrix},$$

so that $\mathbf{g}_1 = (\mathbf{e} - \mathbf{p}) * \breve{\mathbf{g}}_1$, we find that (6.4) becomes

$$\mathbf{f} = \mathbf{1} * (\mathbf{e} - \mathbf{p} * \breve{\mathbf{g}}_0)^{*(-1)} * (\mathbf{e} - \mathbf{p}) * \breve{\mathbf{g}}_1 * \mathbf{1}^T. \tag{6.5}$$

For the time interval omission semi-Markov kernel, we introduce some more notation, again following [2]. Recall

$$\mathbf{f} = \begin{bmatrix} \mathbf{0} & \mathbf{f}_{oc} \\ \mathbf{f}_{co} & \mathbf{0} \end{bmatrix}.$$

Let $d_i = 1 - \sum_{j=1}^{m} F_{ij}(\tau)$ be the probability that a sojourn starting in state i is greater that the detection limit τ, and let

$$\mathbf{d} = \mathrm{diag}(d_1\delta_0, \ldots, d_m\delta_0) = \begin{bmatrix} \mathbf{d}_o & \mathbf{0} \\ \mathbf{0} & \mathbf{d}_c \end{bmatrix}.$$

Our assumptions imply $d_i > 0$, all i. If $d_i = 1$ for all i then every sojourn is detected, which means that time interval omission is not an issue and $\tilde{\mathbf{f}} = \mathbf{f}$. Assume we are not in this case and so there is at least one state i with $d_i < 1$. If $d_i < 1$, put $f_{ij}^U(t) = f_{ij}(t)/(1 - d_i)$ if $0 < t \le \tau$ (0 otherwise). If $d_i = 1$ then define $f_{ij}^U(t) = 0$ for all t and j. Similarly define $f_{ij}^D(t) = f_{ij}(t)/d_i$ if $\tau < t < \infty$ (0 otherwise). The subscripts U and D refer to "undetected" and "detected" respectively.

Following the probabilistic argument preceding (4.11) in [2] we find

$$\tilde{\mathbf{f}}_{oc} = \mathbf{f}_{oc}^D * \sum_{k=0}^{\infty} \left[(\mathbf{e} - \mathbf{d}_c) * \mathbf{f}_{co}^U * \mathbf{f}_{oc}\right]^{*k} * \mathbf{d}_c,$$

and

$$\tilde{\mathbf{f}}_{co} = \mathbf{f}_{co}^D * \sum_{k=0}^{\infty} \left[(\mathbf{e} - \mathbf{d}_o) * \mathbf{f}_{oc}^U * \mathbf{f}_{co}\right]^{*k} * \mathbf{d}_o.$$

Thus

$$\tilde{\mathbf{f}} = \mathbf{f}^D * \sum_{k=0}^{\infty} [(\mathbf{e} - \mathbf{d}) * \mathbf{f}^U * \mathbf{f}]^{*k} * \mathbf{d}. \tag{6.6}$$

Since $d_i > 0$ for all i, we have $\|\mathbf{e} - \mathbf{d}\| < 1$. It can be checked that $\|\mathbf{f}^U * \mathbf{f}\| = 1$. Hence by Theorem 10.7 in [10], the element $\mathbf{e} - (\mathbf{e} - \mathbf{d}) * \mathbf{f}^U * \mathbf{f}$ is invertible in $\mathbf{A}_m$ with inverse $\sum_{k=0}^{\infty} \left[(\mathbf{e} - \mathbf{d}) * \mathbf{f}^U * \mathbf{f}\right]^{*k}$. Then (6.6) becomes

$$\tilde{\mathbf{f}} = \mathbf{f}^D * \left[\mathbf{e} - (\mathbf{e} - \mathbf{d}) * \mathbf{f}^U * \mathbf{f}\right]^{*(-1)} * \mathbf{d}. \tag{6.7}$$

The core of each of (6.5) and (6.7) is a map $\Psi_{\mathbf{p}}$ taking elements of $\mathbf{A}_k$ to $\mathbf{A}_k$ for appropriate k, of the form

$$\Psi_{\mathbf{p}} : \mathbf{h} \mapsto (\mathbf{e} - \mathbf{p} * \mathbf{h})^{*(-1)} * (\mathbf{e} - \mathbf{p}), \tag{6.8}$$

defined for $\mathbf{h} \in \mathbf{A}_k$ such that $\mathbf{e} - \mathbf{p} * \mathbf{h}$ is invertible in $\mathbf{A}_k$, where $\mathbf{p}$ is a diagonal matrix of elements of the form $p_i\delta_0$ and where $\|\mathbf{h}\| = 1$, with elements of $\mathbf{h}$ (possibly defective) densities. In (6.5) we have $\mathbf{h} = \breve{\mathbf{g}}_o$, and in (6.7) we have $\mathbf{p} = \mathbf{e} - \mathbf{d}$ and $\mathbf{h} = \mathbf{f}^U * \mathbf{f}$. We call the output of this map a *multidimensional geometric random sum*, by analogy with the one-dimensional geometric random sum which is of the form $(\delta_0 - pf)^{*(-1)}(1 - p)$ in A, where f is a proper probability density function and $0 < p < 1$. In both the aggregation functional and the time interval omission functional, the difficult part is the evaluation of this multidimensional geometric random sum.

2.3 DERIVATIVES

Following the explanation of the functional approach in the introduction, we see that the main step in finding approximations is to find an appropriate derivative of the relevant functional. We use the Fréchet derivative, and give here a reminder of its definition.

Definition 1 *Let* $(B_i, \|\cdot\|_i)$, $i = 1, 2$ *be Banach spaces and let* U *be a non-empty open subset of* B_1. *The map* $\Psi : U \to B_2$ *is Fréchet differentiable at* $a \in U$ *if there exists a linear bounded map* $\Psi'_a : B_1 \to B_2$ *such that, given* $\epsilon > 0$, *there exists* $\delta > 0$ *such that* $\|x\|_1 < \delta$ *implies that* $\|\Psi(a+x) - \Psi(a) - \Psi'_a(x)\|_2 < \epsilon\|x\|_1$. *The map* Ψ'_a *is the Fréchet derivative of* Ψ *at* a.

We need two properties of Fréchet differentiation. The first is that if a map is Fréchet differentiable at an element of B_1 then it is continuous there, and the second is that Fréchet differentiation satisfies the chain rule.

The next lemma gives the Fréchet derivative of the multidimensional geometric random sum functional $\Psi_{\mathbf{p}}$.

Lemma 1 *The multidimensional geometric random sum functional* $\Psi_{\mathbf{p}}$ *is Fréchet differentiable at every* $\mathbf{a}$ *in* $\mathbf{A}_k$ *with* $\mathbf{e} - \mathbf{p} * \mathbf{a}$ *invertible, with Fréchet derivative at* $\mathbf{a}$ *given by*

$$(\Psi_{\mathbf{p}})'_{\mathbf{a}}(\mathbf{x}) = \left(\mathbf{e} - \mathbf{p} * \mathbf{a}\right)^{*(-1)} * \mathbf{p} * \mathbf{x} * \Psi_{\mathbf{p}}(\mathbf{a}).$$

Proof. From (6.8), $\Psi_{\mathbf{p}}$ can be decomposed

$$\mathbf{a} \mapsto \mathbf{e} - \mathbf{p} * \mathbf{a} \mapsto \left(\mathbf{e} - \mathbf{p} * \mathbf{a}\right)^{*(-1)} \mapsto \left(\mathbf{e} - \mathbf{p} * \mathbf{a}\right)^{*(-1)} * (\mathbf{e} - \mathbf{p}).$$

The Fréchet derivative at $\mathbf{a}$ of the map taking $\mathbf{a}$ in $\mathbf{A}_k$ to $\mathbf{c} + \mathbf{d} * \mathbf{a}$, where $\mathbf{c}$ and $\mathbf{d}$ are in $\mathbf{A}_k$, is $\mathbf{x} \mapsto \mathbf{d} * \mathbf{x}$. Similarly the derivative at $\mathbf{a}$ of the map taking $\mathbf{a}$ to $\mathbf{a} * \mathbf{d}$ is $\mathbf{x} \mapsto \mathbf{x} * \mathbf{d}$. It follows from Theorem 18.4 in [11] that the map that takes an invertible element $\mathbf{a}$ of $\mathbf{A}_k$ to its inverse $\mathbf{a}^{*(-1)}$ is differentiable at $\mathbf{a}$, with derivative $\mathbf{x} \mapsto -\mathbf{a}^{*(-1)} * \mathbf{x} * \mathbf{a}^{*(-1)}$ there. Applying these and using the chain rule, we find that

$$(\Psi_{\mathbf{p}})'_{\mathbf{a}}(\mathbf{x}) = -\left(\mathbf{e} - \mathbf{p} * \mathbf{a}\right)^{*(-1)} * (-\mathbf{p} * \mathbf{x}) * \left(\mathbf{e} - \mathbf{p} * \mathbf{a}\right)^{*(-1)} * (\mathbf{e} - \mathbf{p}),$$

as required. Content-Length: 10414 X-Lines: 304 Status: RO

Using (6.5), we define the aggregation functional Φ by

$$\Phi(\mathbf{x}) = \mathbf{1} * \left(\mathbf{e} - \mathbf{p} * \mathbf{x}\right)^{*(-1)} * (\mathbf{e} - \mathbf{p}) * \breve{\mathbf{g}}_1 * \mathbf{1}^T,$$

for $\mathbf{x} \in \mathbf{A}_n$ such that $\mathbf{e} - \mathbf{p} * \mathbf{x}$ is invertible.

Theorem 1 *The aggregation functional* Φ *is Fréchet differentiable at* $\mathbf{a}$ *in* $\mathbf{A}_n$ *such that* $\mathbf{e}-\mathbf{p}*\mathbf{a}$ *is invertible, with Fréchet derivative at* $\mathbf{a}$ *given by*

$$\Phi'_{\mathbf{a}}(\mathbf{x}) = \mathbf{1} * (\Psi_{\mathbf{p}})'_{\mathbf{a}}(\mathbf{x}) * \breve{\mathbf{g}}_1 * \mathbf{1}^T.$$

Proof. We decompose Φ as

$$\mathbf{a} \mapsto \Psi_{\mathbf{p}}(\mathbf{a}) \mapsto \Psi_{\mathbf{p}}(\mathbf{a}) * \breve{\mathbf{g}}_1 \mapsto \mathbf{1} * \Psi_{\mathbf{p}}(\mathbf{a}) * \breve{\mathbf{g}}_1 * \mathbf{1}^T.$$

The Fréchet derivative at $\mathbf{a}$ of the map from $\mathbf{A}_n$ to $\mathbf{A}_m$, $\mathbf{a} \mapsto \mathbf{1}*\mathbf{a}*\mathbf{1}^T$ is given by $\mathbf{x} \mapsto \mathbf{1}*\mathbf{x}*\mathbf{1}^T$, which is easily checked to be a linear bounded map. Derivatives of the other maps have already been found in Lemma 1 and its proof. The chain rule then yields the result.

The time interval omission functional is defined by

$$\tilde{\Phi}(\mathbf{x}) = \mathbf{f}^D * \left[\mathbf{e} - (\mathbf{e}-\mathbf{d}) * \mathbf{x} * \mathbf{f}\right]^{*(-1)} * \mathbf{d},$$

for $\mathbf{x} \in \mathbf{A}_m$ with $\mathbf{e} - (\mathbf{e}-\mathbf{d}) * \mathbf{x} * \mathbf{f}$ invertible (see (6.7)).

Theorem 2 *The time interval omission functional* $\tilde{\Phi}$ *is Fréchet differentiable at* $\mathbf{a}$ *in* $\mathbf{A}_m$ *such that* $\mathbf{e}-(\mathbf{e}-\mathbf{d})*\mathbf{a}*\mathbf{f}$ *is invertible, with Fréchet derivative at* $\mathbf{a}$ *given by*

$$\tilde{\Phi}'_{\mathbf{a}}(\mathbf{x}) = \mathbf{f}^D * \left(\Psi_{\mathbf{e}-\mathbf{d}}\right)'_{\mathbf{a}*\mathbf{f}}(\mathbf{x}*\mathbf{f}).$$

Proof. Decompose $\tilde{\Phi}$ as

$$\mathbf{a} \mapsto \mathbf{a}*\mathbf{f} \mapsto \Psi_{\mathbf{e}-\mathbf{d}}(\mathbf{a}*\mathbf{f}) \mapsto \mathbf{f}^D * \Psi_{\mathbf{e}-\mathbf{d}}(\mathbf{a}*\mathbf{f}).$$

The derivatives of all the component maps in this decomposition have already been discussed in earlier proofs. Using the chain rule one last time gives the derivative as stated.

3. APPROXIMATIONS

The functional approach and the differentiability given above lead to zeroth and first order approximation formulae to the output of our functionals using (6.1) and (6.2), the former being justified by the continuity of the functionals implied by their Fréchet differentiability. Although these approximating expressions may seem complicated, we see from Lemma 1 and Theorems 1 and 2 that the difficult steps, i.e. those related to evaluation of the multidimensional geometric random sum, is only carried out at the easy input in all these derivatives. The remaining steps in the evaluation of the approximations, although lengthy, are not intrinsically hard. It is expected that the use of a mathematical package such as Maple or Mathematica, as in [9], would be of use in finding the resulting approximating formulae for the semi-Markov kernels. These would then lead to approximations for other quantities of interest. For example, the density of apparent open times incorporating time interval omission can be obtained from $\tilde{\mathbf{f}}$ using the approach in Section 4.2.2 (page 788) of [2].

Since the alternating renewal model is included within the semi-Markov framework, all the zeroth and first order approximations that appear in [8] would arise as corollaries to our present results. However, some new approximations are also suggested by the semi-Markov setting. We illustrate this briefly by considering one of the models in [8]. This is a three state Markov model with one open state, 0, and two closed states, 1 and 2, taken from [6], page 533. The generator is

$$\begin{pmatrix} -\alpha & s\alpha & (1-s)\alpha \\ \beta_1 & -\beta_1 & 0 \\ \beta_2 & 0 & \beta_2 \end{pmatrix},$$

where $0 < s < 1$. In [8], this first of all had to be approximated by an alternating renewal model with exponentially distributed open sojourns, and with closed sojourns distributed as a mixture of exponential distributions, in order to get the problem into the framework of that paper. Once this was done, we found approximations for the density of apparent open times, by approximating the undetected closures in the alternating renewal model by exponential distributions with the correct mean, and evaluating the required density that would arise for this approximating model. A first order approximation was also obtained. However, with our new framework, we avoid the initial approximation by an alternating renewal model. In this example we have $\mathbf{f} = \mathbf{g}$. A natural easy zeroth order approximation to the time interval omission semi-Markov kernel is given by ignoring the lengths of the undetected closed and open sojourns (see [7]). This would amount to replacing $\mathbf{f}^U$ in (6.7) by the $m \times m$ matrix $\mathbf{a}$, where, for $i \in \mathcal{O}$ and $j \in \mathcal{C}$, or for $i \in \mathcal{C}$ and $j \in \mathcal{O}$, $\mathbf{a}_{ij} = \mathrm{P}\big(\text{next state is } j \mid \text{current state is } i, \text{ current sojourn} \leq \tau\big)\delta_0$, and the entries are otherwise zero. This leads to an easier input for the time interval omission functional, and therefore, using the above results, zeroth and first order approximations to various quantities of interest may be obtained.

Finally, it should be pointed out that the theoretical framework established here also allows for second and higher approximations to be obtained on using a Taylor series expansion involving higher derivatives.

Acknowledgments

The use of a functional approach in a semi-Markov setting for single ion channels was suggested in the helpful remarks of an anonymous referee to [8]. I am grateful for the hospitality and support of the Forshunginstitut für Mathematik at ETH Zürich, where this work was carried out.

References

[1] Asmussen, S. (1989). Risk theory in a Markovian environment, *Scandinavian Actuarial Journal*, 69–100.

[2] Ball, F., Milne, R.K. and Yeo, G.F. (1991). Aggregated semi-Markov processes incorporating time interval omission, *Adv. Appl. Prob.*, **23**, 772–797.

[3] Ball,F., Yeo, G.F., Milne, R.K., Edeson, R.O., Madsen, B.W. and Sansom, M.S.P. (1993). Single ion channel models incorporating aggregation and time interval omission, *Biophys. J.*, **64**, 357–374.

[4] Grübel, R. (1989). Stochastic models as functionals: Some remarks on the renewal case, *J. Appl. Prob.*, **26**, 296–303.

[5] Grübel, R. and Pitts, S.M. (1992). A functional approach to the stationary waiting time and idle period distributions of the $GI/G/1$ queue, *Ann. Probab.*, **20**, 1754–1778.

[6] Hawkes, A.G., Jalali, A. and Colquhoun, D. (1990). The distributions of the apparent open times and shut times in a single channel record when brief events cannot be detected, *Phil. Trans. R. Soc. Lond. A.*, **332**, 511–538.

[7] Milne, R.K., Yeo, G.F., Edeson, R.O. and Madsen, B.W. (1988). Stochastic modelling of a single ion channel: an alternating renewal approach with application to limited time resolution, *Proc. R. Soc. Lond.*, **233**, 247–292.

[8] Pitts, S.M. (1998). A functional approach to approximations for the observed open times in single ion channel models, *Journal of Mathematical Analysis and Applications*, **219**, 260–278.

[9] Politis, K. and Pitts, S.M. (1998). Approximations for solutions of renewal-type equations, *Stoch. Proc. Appl.*, **78**, 195–216.

[10] Rudin, W. (1973). *Functional Analysis*, Tata McGraw-Hill, New Delhi.

[11] Rudin, W. (1986). *Real and Complex Analysis,* 3rd ed. McGraw-Hill, New York.

II
STATISTICAL ESTIMATION

Chapter 7

ASYMPTOTIC BEHAVIOR OF THE LOG-LIKELIHOOD FUNCTION IN STOCHASTIC PROCESSES WHEN BASED ON A RANDOM NUMBER OF RANDOM VARIABLES

George G. Roussas*
University of California, USA

Debasis Bhattacharya
University of California, USA

Abstract A general discrete time stochastic process is considered, and it is assumed that its probability laws are of known functional form, except that they depend on a k-dimensional parameter θ. On the basis of a random number of random variables ν_n, defined on the underlying process, the log-likelihood function Λ_{ν_n} is formed, and its asymptotic behavior, as n tends to ∞, is studied. It is shown that the asymptotic distribution of Λ_{ν_n} is normal, both under a fixed probability measure, as well as under certain contiguous probability measures. The same is shown to be true for a k-dimensional random vector, defined in terms of a certain quadratic mean derivative.

The presence of the random variable ν_n introduces some technical difficulties. These difficulties are resolved by means of certain results discussed in an appendix. The underlying idea involved is that, when in a sequence of random variables the deterministic index is replaced by ν_n, the resulting sequence and the original sequence do not differ substantially in the probability sense.

The paper is concluded with the suggestion of some statistical applications.

*This research was supported in part by a research grant to the first author by the University of California, Davis.

Keywords: Asymptotic expansion, asymptotic normality, contiguity, likelihood function, stochastic processes, stopping times.

1. INTRODUCTION

As is well known, it is of great importance, from a statistical inference point of view, to determine the asymptotic distribution of the log-likelihood function. The same is true for some other random variables associated with the likelihood function (for example, related test statistics proposed from the point of view of power considerations). In both these cases, the determination of the asymptotic distribution is required under a sequence of probability measures, which, in sequential experiments, vary with the random sample size, ν_n. The asymptotic distribution is also useful in determining the asymptotic power of a sequential likelihood ratio test under suitable alternatives. As another application of the asymptotic distribution of the log-likelihood function, we hope that it can be used to prove the efficiency of the sequential maximum likelihood estimator for a general stochastic process in the Wolfowitz sense (in short, W-efficiency). It is interesting to note that W-efficiency is based on coverage probabilities and not on the variance of the asymptotic distributions, and is defined even when the asymptotic distribution of the maximum likelihood estimator does not have finite variance (see Wolfowitz (1965)).

Several studies by various authors, regarding the asymptotic distribution of log-likelihood function, have been made. Asymptotic normality of log-likelihood function for a general stochastic process was obtained in Roussas (1979). Le Cam and Yang (1990) covers the asymptotics related to the log-likelihood functions in considerable detail. In a recent paper, Luschgy (1995) considers the local asymptotic quadraticity of a general stochastic process based on stopping times. The approach is carried out by continuous semi-martingale techniques.

In this paper, we consider the sequence of randomly stopped log-likelihood ratios for a general stochastic process and show that the sequence of randomly stopped log-likelihood ratios converges to a Locally Asymptotically Normal (LAN) law, both under a fixed probability measure, as well as under certain contiguous probability measures. The same is also true for a random vector defined in terms of certain quadratic mean derivatives.

In more technical terms, for $n \geq 0$ integer, let $X_0, X_1, ..., X_n$ be the first $n+1$ random variables (r.v.'s) from a stochastic process, each defined on the probability space $(\Omega, \mathcal{A}, P_\theta)$ and taking values in $(\mathsf{S}, \mathcal{S})$. The parameter θ belongs to an open set Θ in $\Re^k$, S is a Borel subset of a Euclidean space, and $\mathcal{S}$ is the σ-field of Borel subsets of S. It is assumed that the joint probability law of any finite number of such r.v.'s is of known functional form except that it depends on the parameter θ. Let $\mathcal{A}_n$ be the σ-field induced in Ω by the r.v.'s $X_0, X_1, ..., X_n$; that is, $\mathcal{A}_n = \sigma(X_0, X_1, ..., X_n)$, and let $P_{n,\theta}$ be the restriction of P_θ to $\mathcal{A}_n$; that is, $P_{n,\theta} = P_\theta | \mathcal{A}_n$. Let $\{\tau_n\}$, $n \geq 1$, be a sequence of positive real numbers (possibly depending on θ) tending increasingly to ∞, and define θ_{τ_n} by $\theta_{\tau_n} = \theta + h\,\tau_n^{-1/2}$, $h \in \Re^k$. Let $\{\nu_n\}$, $n \geq 1$, be a sequence of stopping times defined on the sequence $\{X_n\}$, $n \geq 0$; that is, ν_n are non-negative integer-valued r.v.'s, tending non-decreasingly to ∞ a.s. $[P_\theta]$, and such that $(\nu_n = m) \in \mathcal{A}_m$, $m \geq 0$, for each n. Let $\mathcal{A}_{\nu_n}$ be the σ-field induced by the r.v.'s $X_0, X_1, ..., X_{\nu_n}$, and denote by $\tilde{P}_{n,\theta}$ the restriction of P_θ to $\mathcal{A}_{\nu_n}$; $\tilde{P}_{n,\theta} = P_\theta \, | \mathcal{A}_{\nu_n}$.

It is assumed that, for each $n \geq 0$, the probability measures $P_{n,\theta}$ and P_{n,θ^*} are mutually absolutely continuous for all θ, θ^* in Θ. Then, let

$$q_n(\theta;\theta^*) = q_n(\mathbf{X}_n;\theta;\theta^*) = \frac{dP_{n,\theta^*}}{dP_{n,\theta}} = L_n(\mathbf{X}_n;\theta;\theta^*) = L_n(\theta;\theta^*) \quad (7.1)$$

be specified versions of the Radon-Nikodym derivatives (or the likelihood functions) involved, where $\mathbf{X}_n = (X_0, X_1, ..., X_n)$. Also, denote the transition probability densities by :

$$\varphi_n^2(\theta;\theta^*) = \varphi_n^2(\mathbf{X}_n;\theta;\theta^*) = \frac{q_n(\theta;\theta^*)}{q_{n-1}(\theta;\theta^*)} = q_n(X_n|\mathbf{X}_{n-1};\theta;\theta^*), \quad n \geq 1. \quad (7.2)$$

Then,

$$\begin{aligned} L_n(\theta;\theta^*) &= q_0(X_0;\theta;\theta^*)\prod_{j=1}^{n} q_j(X_j|\mathbf{X}_{j-1};\theta;\theta^*) \\ &= q_0(\theta;\theta^*)\prod_{j=1}^{n} q_j(\theta;\theta^*) \\ &= q_0(\theta;\theta^*)\prod_{j=1}^{n} \varphi_j^2(\theta;\theta^*)\ (L_0(\theta;\theta^*) = q_0(\theta;\theta^*)), \quad (7.3) \end{aligned}$$

and

$$\Lambda_n(\theta;\theta^*) = \log L_n(\theta;\theta^*) = \log q_0(\theta;\theta^*) + 2\sum_{j=1}^{n} \log \varphi_j(\theta;\theta^*) \quad (7.4)$$

is well-defined with P_θ-probability 1 for all $\theta \in \Theta$. It is shown (see Proposition 2) that $\tilde{P}_{n,\theta}$ and $\tilde{P}_{n,\theta^*}$ are mutually absolutely continuous, and that the likelihood function $d\tilde{P}_{n,\theta^*}/d\tilde{P}_{n,\theta}$, L_{ν_n}, is given by :

$$
\begin{aligned}
L_{\nu_n}(\theta;\theta^*) = \frac{d\tilde{P}_{n,\theta^*}}{d\tilde{P}_{n,\theta}} &= q(X_0;\theta;\theta^*)\prod_{j=1}^{\nu_n} q_j(X_j|\mathbf{X}_{j-1};\theta;\theta^*) \\
&= q_0(\theta;\theta^*)\prod_{j=1}^{\nu_n} q_j(\theta;\theta^*) \\
&= q_0(\theta;\theta^*)\prod_{j=1}^{\nu_n} \varphi_j^2(\theta;\theta^*),
\end{aligned}
\tag{7.5}
$$

so that, with P_θ-probability 1 for all $\theta \in \Theta$,

$$\Lambda_{\nu_n}(\theta;\theta^*) = \log L_{\nu_n}(\theta;\theta^*) = \log q_0(\theta;\theta^*) + 2\sum_{j=1}^{\nu_n} \log \varphi_j(\theta;\theta^*). \tag{7.6}$$

It will be assumed in the following that the random functions $\varphi_j(\theta;\cdot)$, $j \geq 1$, are differentiable in quadratic mean (q.m.), when the probability measure P_θ is used, and let $\dot{\varphi}_j(\theta)$, $j \geq 1$, be the derivatives in q.m. involved. In terms of them, define the k-dimensional random vector $\Delta_{\nu_n}(\theta)$ by :

$$\Delta_{\nu_n}(\theta) = 2\nu_n^{-1/2}\sum_{j=1}^{\nu_n} \dot{\varphi}_j(\theta). \tag{7.7}$$

Recall that

$$\theta_{\tau_n} = \theta + h_n\,\tau_n^{-1/2},\ h_n \to h \in \Re^k, \text{ as } n \to \infty, \tag{7.8}$$

so that, for sufficiently large n, $\theta_{\tau_n} \in \Theta$. Therefore, replacing θ^* by θ_{τ_n} in (7.6), we obtain

$$\Lambda_{\nu_n}(\theta) = \log q_0(\theta;\theta_{\tau_n}) + 2\sum_{j=1}^{\nu_n} \log \varphi_{\tau_n j}(\theta), \tag{7.9}$$

where

$$\Lambda_{\nu_n}(\theta) = \Lambda_{\nu_n}(\theta;\theta_{\tau_n}) \text{ and } \varphi_{\tau_n}(\theta) = \varphi_j(\theta;\theta_{\tau_n}). \tag{7.10}$$

The main part of this paper revolves around the expansion of $\Lambda_{\nu_n}(\theta)$ in $\tilde{P}_{n,\theta}$ and also $\tilde{P}_{n,\theta_{\tau_n}}$-probability, and the establishment of its asymptotic distributions. Also, the proof of the asymptotic normality of the important random vector $\Delta_{\nu_n}(\theta)$. In these derivations, the contiguity of the sequences of the probability measures $\left\{\tilde{P}_{n,\theta}\right\}$ and $\left\{\tilde{P}_{n,\theta_{\tau n}}\right\}$ plays an important role, and is established in Proposition 2. At this point, it should be mentioned in passing that the concept of contiguity itself was introduced and developed by Le Cam (1960). The main results themselves follow after a series of lemmas, Lemmas 1 through 4, have been established. These lemmas are reminiscent of Lemmas 5.1-5.10 in Roussas (1979). However, there is a basic difference between parts of these sets of lemmas; namely, those parts involving random number of r.v.'s. Typically, in this context, one already knows that $\mathcal{L}(W_{\alpha_n}|Q_n) \underset{n\to\infty}{\Rightarrow} Q$, a probability measure, for some sequence of r.v.'s $\{W_{\alpha_n}\}$ and some probability measures Q_n (which in the present case will be either $P_{n,\theta}$ or $P_{n,\theta_{\tau n}}$, and one wishes to show that $\mathcal{L}(W_{\alpha_n}|\tilde{Q}_n) \underset{n\to\infty}{\Rightarrow} Q$ for appropriate probability measures $\tilde{Q}_n$ (which in the present case will be either $\tilde{P}_{n,\theta}$ or $\tilde{P}_{n,\theta_{\tau n}}$). Conditions for this to be, indeed, the case are given in Theorem A1 in the Appendix. The essence of the conditions is that W_{α_n} and W_{ν_n} do not differ wildly in the probability sense. This is expressed by requiring that the difference of W_{α_n} and the maximum of W_m over m in a neighborhood of α_n tends to 0 in probability. Theorem A1 takes on the form of one of Propositions 5-7 for specific forms of W_{α_n}. These theorems and propositions play a decisive role in establishing Lemmas 1-4, and hence the proof of Theorems 1 and 2.

The assumptions under which the results in this paper are obtained are given in Section 2. They are followed by certain comments and examples, where these assumptions have been found to be fulfilled. In the following section, the main results of the paper, Theorems 1 and 2 are stated along with two relevant propositions. Section 4. is devoted to the proof of the main results is taken up in the final section of the paper. The paper is concluded with a brief comment regarding applications to statistical inference about the underlying parameter.

Finally, it should be mentioned once and for all that all limits are taken as $n \to \infty$.

2. ASSUMPTIONS

The main assumptions of the paper are gathered together here for easy reference.

(A1) For each $n \geq 0$ integer, the (finite dimensional) probability measures $\{P_{n,\theta}; \theta \in \Theta\}$ are mutually absolutely continuous.

(A2) (i) For each $\theta \in \Theta$, the random functions $\varphi_j(\theta;\cdot)$, defined in (7.2), are differentiable in *q.m.* $[P_\theta]$ uniformly in $j \geq 1$. That is, there exist k-dimensional random vectors $\dot{\varphi}_j(\theta)$ — the *q.m.* derivatives of $\varphi_j(\theta;\theta^*)$ with respect to θ^* at θ — such that

$$\frac{1}{\lambda}\left|[\varphi_j(\theta+\lambda h)-1]-\lambda h'\dot{\varphi}_j(\theta)\right| \to 0$$

in *q.m.* $[P_\theta]$, as $\lambda \to 0$, uniformly on bounded sets of $h \in \Re^k$ and uniformly in $j \geq 1$; " $'$ "stands for transpose.

(ii) For $j \geq 1$, $\dot{\varphi}_j(\theta)$ is $\mathcal{A}_j \times \mathcal{C}$-measurable, where $\mathcal{A}_j = \sigma(X_0, X_1, \ldots, X_j)$ and $\mathcal{C}$ is the σ-field of Borel subsets of Θ.

(A3) (i) For each $\theta \in \Theta$ and each $t \in \Re^k$, $[t'\dot{\varphi}_j(\theta)]^2$, $j \geq 1$, are uniformly integrable with respect to $[P_\theta]$. That is, uniformly in $j \geq 1$,

$$\int_{\{[t'\dot{\varphi}_j(\theta)]^2>a\}} [t'\dot{\varphi}_j(\theta)]^2 dP_\theta \to 0, \text{ as } a \to \infty.$$

(ii) For each $\theta \in \Theta$ and $n \geq 1$, let the $k \times k$ covariance matrix $\bar{\Gamma}_n(\theta)$ be defined by :

$$\bar{\Gamma}_n(\theta) = \frac{1}{n}\sum_{j=1}^{n}\Gamma_j(\theta), \text{ where } \Gamma_j(\theta) = 4\mathcal{E}_\theta\left[\dot{\varphi}_j(\theta)\dot{\varphi}_j'(\theta)\right]. \qquad (7.11)$$

Then $\bar{\Gamma}_n(\theta) \to \bar{\Gamma}(\theta)$ (in any one of the standard norms) and $\bar{\Gamma}(\theta)$ is positive definite for all $\theta \in \Theta$.

(iii) For each $\theta \in \Theta$ and each $t \in \Re^k$, the WLLN's holds, under P_θ, for the sequence of r.v.'s $\{[t'\dot{\varphi}_j(\theta)]^2\}$, $j \geq 1$.

(iv) For each $\theta \in \Theta$ and each $t \in \Re^k$,

$$\frac{1}{n}\sum_{j=1}^{n}\left[\mathcal{E}_\theta\left\{[t'\dot{\varphi}_j(\theta)]^2|\mathcal{A}_{j-1}\right\} - [t'\dot{\varphi}_j(\theta)]^2\right] \to 0 \text{ in } P_\theta\text{-probability.}$$

(A4) For each $\theta \in \Theta$, let $q_0(\theta;\cdot)$ be defined by (7.1). Then $q_0(\theta;\cdot)$ is $\mathcal{A}_0 \times \mathcal{C}$-measurable and continuous in P_θ-probability. Recall that τ_n are positive real numbers tending increasingly to ∞, and ν_n are stopping times defined on $\{X_n,\ n \geq 0\}$. Then

(A5) For each $\theta \in \Theta$, $\frac{\nu_n}{\tau_n} \to 1$ in P_θ-probability.

In many derivations below, τ_n is to be replaced by its integer part α_n; that is,

$$\alpha_n = [\tau_n]\,, \text{ so that } 0 \leq \alpha_n \uparrow \infty \text{ and } \frac{\alpha_n}{\tau_n} \to 1 \text{ as } n \to \infty. \qquad (7.12)$$

2.1 SOME COMMENTS ON THE ASSUMPTIONS

Assumptions (A1)-(A4) are the same as those used in Roussas (1979), and assumption (A5) is required to deal with the random number of r.v.'s involved; suitable versions of them have been employed in Akritas and Roussas (1979) in a Markovian framework. Assumptions (A1)-(A4) are known to be satisfied in a number of examples (see Examples 3.1-3.4, pages 47-52 in Roussas (1972)). They are also fulfilled in the examples given below. The relevant justification may be found in Stamatelos (1976).

Assumption (A5) is, clearly, required regardless of the framework of dependence considered. The remaining assumptions are modified for specific modes of dependence or lack thereof. Thus, if the underlying process is a strictly stationary and ergodic Markov process, then assumptions (A1) and (A4) remain the same, whereas in (A2)(i), $\varphi_j(\theta;\cdot) = \varphi_1(\theta;\cdot)$ for all j; here $\varphi_1(\theta;\cdot)$ is the square root of the one-step transition probability density function. Also, $\dot{\varphi}_j(\theta) = \dot{\varphi}_1(\theta)$ with probability one for all j, and assumption (A2)(i) is imposed on $\dot{\varphi}_1(\theta)$. Assumptions (A3)(i), (iii) and (iv) are superfluous here, and in (A3)(ii), $\bar{\Gamma}_n(\theta) = \Gamma(\theta) = \Gamma_1(\theta) = \Gamma_j(\theta)$ for all j.

In the independent identically distributed case, the modifications just mentioned apply. In this case, $\varphi_1(\theta;\cdot)$ is, of course, the square root of the common probability density function.

Example 1 *The r.v.'s involved satisfy the relation* $X_n = \alpha X_{n-1} + U_n$, *where* $|\alpha| < 1$ *is known and the r.v.'s* U_n, $n \geq 1$, *are independently distributed as* $N(\theta, 1)$, *and* $X_0 = 0$.

Example 2 *The r.v.'s involved constitute a Gaussian process such that :*

$$\mathcal{E}_\theta X_n = \begin{cases} \theta, & n = 2k+1 \\ 2\theta, & n = 2k \end{cases}, \quad Cov_\theta(X_m, X_n) = c$$

for all m, n $(m \neq n)$, *and* $\sigma^2_\theta(X_n) = 2c$ *for all* n.

Next, assumption (A3)(iii) states that

$$\frac{1}{n}\sum_{j=1}^{n}\{[t'\dot{\varphi}_j(\theta)]^2 - \mathcal{E}_\theta[t'\dot{\varphi}_j(\theta)]^2\} \rightarrow 0 \text{ in } P_\theta\text{-probability.} \qquad (7.13)$$

From (7.11), $\frac{1}{4}t'\bar{\Gamma}_n(\theta)t = \frac{1}{4n}\sum_{j=1}^{n} t'\Gamma_j(\theta)t = \frac{1}{n}\sum_{j=1}^{n} \mathcal{E}_\theta[t'\dot{\varphi}_j(\theta)]^2$, so that, by (A3) (ii),

$$\frac{1}{n}\sum_{j=1}^{n} \mathcal{E}_\theta[t'\dot{\varphi}_j(\theta)]^2 \to \frac{1}{4}t'\bar{\Gamma}(\theta)t. \tag{7.14}$$

On the basis of (7.13) and (7.14), (A3) (iii) may be reformulated as follows :

(A3) (iii$'$) For each $\theta \in \Theta$ and each $t \in \Re^k$,

$$\frac{1}{n}\sum_{j=1}^{n}[t'\dot{\varphi}_j(\theta)]^2 \to \frac{1}{4}t'\bar{\Gamma}(\theta)t \text{ in } P_\theta\text{-probability}. \tag{7.15}$$

Finally, on the basis of (7.15), assumption (A3)(iv) may be reformulated thus :

(A3) (iv$'$) For each $\theta \in \Theta$ and each $t \in \Re^k$,

$$\frac{1}{n}\sum_{j=1}^{n} \mathcal{E}_\theta\left\{[t'\dot{\varphi}_j(\theta)]^2|\mathcal{A}_{j-1}\right\} \to \frac{1}{4}t'\bar{\Gamma}(\theta)t \text{ in } P_\theta\text{-probability.}$$

3. ASYMPTOTIC EXPANSION AND ASYMPTOTIC NORMALITY

The quantities $\Lambda_{\nu_n}(\theta;\theta^*)$ and $\Delta_{\nu_n}(\theta)$ are defined in terms of the r.v.'s $X_0, X_1, ..., X_{\nu_n}$ by (7.6) and (7.7), respectively, and they would be expected to be $\mathcal{A}_{\nu_n}$-measurable. That this is, indeed, the case is part of Proposition 1 below. Also, the probability measures $\tilde{P}_{n,\theta}$ and $\tilde{P}_{n,\theta^*}$, defined on $\mathcal{A}_{\nu_n}$, are mutually absolutely continuous for each θ, θ^* in Θ, as already mentioned, and $d\tilde{P}_{n,\theta^*}/d\tilde{P}_{n,\theta}$ is given by (7.5). These facts constitute the second part of Proposition 1. In the proofs of various statements, it is assumed that assumptions (A1)-(A5) hold without explicitly mentioning it, although not all of them are needed everywhere.

Proposition 1
(i) For each θ, θ^* in Θ and each $n \geq 0$, the quantities $\Lambda_{\nu_n}(\theta;\theta^*)$ and $\Delta_{\nu_n}(\theta)$ defined by (7.6) and (7.7), respectively, are $\mathcal{A}_{\nu_n}$-measurable.
(ii) For each θ, θ^* in Θ and each $n \geq 0$, $\tilde{P}_{n,\theta} \approx \tilde{P}_{n,\theta^*}$ and $d\tilde{P}_{n,\theta^*}/d\tilde{P}_{n,\theta}$ is equal to $L_{\nu_n}(\theta;\theta^*)$ defined in (7.5).

Remark 1 *Before we embark on the proof of the proposition, let us recall a few basic facts. Recall that $\mathcal{A}_n = \sigma(X_0, X_1, ..., X_n)$, and set*

$\mathcal{A}_\infty = \bigvee_{n\geq 0} \mathcal{A}_n$*; that is, the* σ*-field generated by the* $\mathcal{A}_n$*'s. Also, for a stopping time* τ *defined on the sequence* $\{X_n,\ n \geq 0\}$*, define* $\mathcal{A}_\tau$ *by :*

$$\mathcal{A}_\tau = \{A \in \mathcal{A}_\infty;\ A \cap (\tau = m) \in \mathcal{A}_m,\ m \geq 0\}.$$

Then $\mathcal{A}_\tau$ *is a* σ*-field. Furthermore, if for a r.v.* X*,* X_τ *is defined by :* $X_\tau(\omega) = X_{\tau(\omega)}(\omega)$*,* $\omega \in \Omega$*, then* X_τ *is a r.v. A reference for such facts is, for example, Chow and Teicher (1988), page 136.*

Proof of Proposition 1

(i) In the first place, $\Lambda_{\nu_n}(\theta;\theta^*)$ and $\Delta_{\nu_n}(\theta)$ are r.v.'s by the above remark. In arguing the $\mathcal{A}_{\nu_n}$-measurability of $\Lambda_{\nu_n}(\theta;\theta^*)$, write just Λ_{ν_n} for simplicity. For $x \in \Re$,

$$(L_{\nu_n} \leq x) = \sum_{m=0}^{\infty} (L_{\nu_n} \leq x, \nu_n = m) = \sum_{m=0}^{\infty} (L_m \leq x, \nu_n = m).$$

However, $(L_m \leq x) \in \mathcal{A}_m$ and also $(\nu_n = m) \in \mathcal{A}_m$, so that $(L_m \leq x, \nu_n = m) \in \mathcal{A}_m$ and therefore in their sum, or $(L_{\nu_n} \leq x)$, is in $\mathcal{A}_\infty$. Also, for each $m \geq 0$,

$$(L_{\nu_n} \leq x) \cap (\nu_n = m) = (L_m \leq x) \cap (\nu_n = m) = (L_m \leq x, \nu_n = m)$$

is in $\mathcal{A}_m$, so that $(L_{\nu_n} \leq x) \in \mathcal{A}_{\nu_n}$.

For the $\mathcal{A}_{\nu_n}$-measurability of $\Delta_{\nu_n}(\theta)$, it suffices to show that the r.v. $t'\Delta_{\nu_n}$ is $\mathcal{A}_{\nu_n}$-measurable for every $t \in \Re^k$, and this is done as in the case of the r.v. $L_{\nu_n}(\theta;\theta^*)$.

(ii) That $\tilde{P}_{n,\theta} \approx \tilde{P}_{n,\theta^*}$ is an immediate consequence of assumption (A1). Thus, it suffices to show that, for each $A \in \mathcal{A}_{\nu_n}$,

$$\int_A L_{\nu_n}(\theta;\theta^*) d\tilde{P}_{n,\theta} = \tilde{P}_{n,\theta^*}(A).$$

As in part (i), write L_{ν_n} rather than $L_{\nu_n}(\theta;\theta^*)$. Then

$$\int_A L_{\nu_n} d\tilde{P}_{n,\theta} = \int_{\bigcup_{j=0}^{\infty}[A\cap(\nu_n=j)]} L_{\nu_n} d\tilde{P}_{n,\theta}$$

$$= \sum_{j=0}^{\infty} \int_{A\cap(\nu_n=j)} L_{\nu_n} d\tilde{P}_{n,\theta} \text{ (since the integrand is } \geq 0)$$

$$= \sum_{j=0}^{\infty} \int_{A\cap(\nu_n=j)} L_j dP_{j,\theta} \text{ (since } A\cap(\nu_n = j) \in \mathcal{A}_j \text{ and } L_j \text{ is } \mathcal{A}_j\text{-measurable)}$$

$$= \sum_{j=0}^{\infty} \int_{A\cap(\nu_n=j)} dP_{j,\theta^*} \text{ (by assumption (A1) and (7.1))}$$

$$= \sum_{j=0}^{\infty} \int_{A\cap(\nu_n=j)} dP_{\theta^*}$$

$$= \sum_{j=0}^{\infty} P_{\theta^*}\left[A \cap (\nu_n = j)\right]$$

$$= P_{\theta^*}(A) = \tilde{P}_{n,\theta^*}(A) \text{ (since } A \in \mathcal{A}_{\nu_n}).$$

This shows that $\tilde{P}_{n,\theta^*} \ll \tilde{P}_{n,\theta}$. That $\tilde{P}_{n,\theta} \ll \tilde{P}_{n,\theta^*}$ is established similarly. □

Once appropriate measurability of $\Lambda_{\nu_n}(\theta)$ (see (7.9) and (7.10)) and $\Delta_{\nu_n}(\theta)$ (see (7.7)) has been established and it has been shown that $L_{\nu_n}(\theta)$ ($= \exp \Lambda_{\nu_n}(\theta)$) is equal to $d\tilde{P}_{n,\theta_{\tau_n}}/d\tilde{P}_{n,\theta}$, we may turn to the asymptotic behavior of $\Lambda_{\nu_n}(\theta)$ and $\Delta_{\nu_n}(\theta)$. All results stated below will hold under assumptions (A1)-(A5) without mentioning it explicitly. Thus, we have :

Theorem 1 *For every $\theta \in \Theta$ and any $h_n \to h \in \Re^k$,*

(i) $\Lambda_{\nu_n}(\theta) - h_n' \Delta_{\nu_n}(\theta) \to -\frac{1}{2}h'\bar{\Gamma}(\theta)h$ in $\tilde{P}_{n,\theta}$-probability,

(ii) $\mathcal{L}[\Delta_{\nu_n}(\theta)|\tilde{P}_{n,\theta}] \Rightarrow N(0, \bar{\Gamma}(\theta))$,

(iii) $\mathcal{L}[\Lambda_{\nu_n}(\theta)|\tilde{P}_{n,\theta}] \Rightarrow N(-\frac{1}{2}h'\bar{\Gamma}(\theta)h, h'\bar{\Gamma}(\theta)h)$.

Assume for a moment that this theorem is true and proceed to establish the following result.

Proposition 2 *The sequences of probability measures $\left\{\tilde{P}_{n,\theta}\right\}$ and $\left\{\tilde{P}_{n,\theta_{\tau_n}}\right\}$ are contiguous, where θ_{τ_n} is defined in (7.8).*

Proof. Set $\tilde{\mathcal{L}}_n = \mathcal{L}[\Lambda_{\nu_n}(\theta)|\tilde{P}_{n,\theta}]$, and $\sigma^2 = h'\bar{\Gamma}(\theta)h$. Then $\left\{\tilde{\mathcal{L}}_n\right\}$ is relatively compact, since $\tilde{\mathcal{L}}_n \Rightarrow N(-\frac{1}{2}\sigma^2, \sigma^2)$ and

$$\int_{\Re} e^{\lambda} dN(-\frac{1}{2}\sigma^2, \sigma^2) = \frac{1}{\sqrt{2\pi}\sigma} \int_{\Re} e^{\lambda} e^{-\frac{1}{2\sigma^2}(\lambda + \frac{1}{2}\sigma^2)^2} d\lambda = 1,$$

as is seen by calculations (see Corollary 7.1, page 34, in Roussas (1972)). Then the asserted contiguity follows by Theorem 6.1 (iv), page 33, in Roussas (1972). □

Theorem 1 and Proposition 2 will lead to the following results.

Theorem 2 *For every $\theta \in \Theta$ and any $h_n \to h \in \Re^k$,*

(i) $\Lambda_{\nu_n}(\theta) - h_n'\Delta_{\nu_n}(\theta) \to -\frac{1}{2}h'\bar{\Gamma}(\theta)h$ *in $\tilde{P}_{n,\theta_{\tau_n}}$-probability,*

(ii) $\mathcal{L}[\Delta_{\nu_n}(\theta)|\tilde{P}_{n,\theta_{\tau_n}}] \Rightarrow N(\bar{\Gamma}(\theta)h, \bar{\Gamma}(\theta))$,

(iii) $\mathcal{L}[\Lambda_{\nu_n}(\theta)|\tilde{P}_{n,\theta_{\tau_n}}] \Rightarrow N(\frac{1}{2}h'\bar{\Gamma}(\theta)h, h'\bar{\Gamma}(\theta)h)$.

When the log-likelihood function and the random vector defined in (7.7) are indexed by non-random entities, Theorems 1 and 2 are established in Roussas (1979). In the present framework but for stationary Markov processes, Theorem 1 (but not Theorem 2) has been shown in Akritas and Roussas (1979). The results in Theorem 2 were also announced in Akritas et al. (1984) but they were not established. The proof of Theorems 1 and 2 in the present general set-up involves the handling of certain delicate points. Some of the results needed will be presented in the form of an appendix; actually, these results are of interest in their own right. The next section, Section 4., deals with a series of lemmas needed in the proof of the main results. The proof of Theorems 1 and 2 is carried out in Section 5.

4. SOME AUXILIARY RESULTS

In order to give a motivation for the lemmas discussed in this section, let us look, for instance, at Theorem 1, where it is asserted that, for any $h_n \to h \in \Re^k$,

$$\Lambda_{\nu_n}(\theta) - h_n'\Delta_{\nu_n}(\theta) \to -\frac{1}{2}h'\bar{\Gamma}(\theta)h \text{ in } P_\theta\text{-probability.} \tag{7.16}$$

(From this point on, let us agree to refer to the probability measure P_θ rather than $P_{n,\theta}$ and $\tilde{P}_{n,\theta}$. Likewise for other values of θ).

This convergence will be established in two steps. First, by considering, with $\alpha_n = [\tau_n]$ (see (7.12)),

$$\Lambda_{\alpha_n}(\theta) = \log q_0(\theta; \theta_{\tau_n}) + 2\sum_{j=1}^{\alpha_n} \log \varphi_{\tau_n j}(\theta), \tag{7.17}$$

and

$$\Delta_{\alpha_n}(\theta) = 2\alpha_n^{-1/2}\sum_{j=1}^{\alpha_n} \varphi_j(\theta), \tag{7.18}$$

and showing that

Proposition 3 *For every $\theta \in \Theta$ and any $h_n \to h \in \Re^k$,*

$$\Lambda_{\alpha_n}(\theta) - h_n' \Delta_{\alpha_n}(\theta) \to -\frac{1}{2} h' \bar{\Gamma}(\theta) h \text{ in } P_\theta\text{-probability,} \tag{7.19}$$

and second,

Proposition 4 *For every $\theta \in \Theta$ and any $h_n \to h \in \Re^k$,*

$$(\Lambda_{\nu_n}(\theta) - h_n' \Delta_{\nu_n}(\theta)) - (\Lambda_{\alpha_n}(\theta) - h_n' \Delta_{\alpha_n}(\theta)) \to 0 \text{ in } P_\theta\text{-probability.} \tag{7.20}$$

Proof of Proposition 3. Let us make here the following observation, which will be useful in many occasions; namely,

$$\begin{aligned} \theta_{\tau_n} &= \theta + h_n \tau_n^{-1/2} = \theta + h_n \left(\frac{\alpha_n}{\tau_n}\right)^{1/2} \alpha_n^{-1/2} \\ &= \theta + \hat{h}_n \alpha_n^{-1/2}, \quad \hat{h}_n = h_n \left(\frac{\alpha_n}{\tau_n}\right)^{1/2} \to h. \end{aligned} \tag{7.21}$$

Since h_n and $\hat{h}_n$ converge to the same limit (h), relation (7.21) suggests that in expressions such as $q_0(\theta; \theta_{\tau_n})$ (see (7.1)), $\varphi_{\tau_n j}(\theta) = \varphi_j(\theta; \theta_{\tau_n})$ (see (7.10)), and $\Lambda_{\alpha_n}(\theta)$, as defined in (7.16), we may replace θ_{τ_n} by θ_{α_n}, where

$$\theta_{\alpha_n} = \theta + \hat{h}_n \alpha_n^{-1/2}, \ \hat{h}_n \to h. \tag{7.22}$$

Thus, we may work with the quantities

$$\Lambda_{\alpha_n}(\theta) = \log q_0(\theta; \theta_{\alpha_n}) + 2 \sum_{j=1}^{\alpha_n} \log \varphi_{\alpha_n j}(\theta),$$

$$\varphi_{\alpha_n j}(\theta) = \varphi_j(\theta; \theta_{\alpha_n}). \tag{7.23}$$

When α_n is replaced by n in (7.22) and (7.23), convergence (7.19) has been established in Roussas (1979) (see Theorem 1.4.1) under assumptions (A1)-(A4). Since $\{\alpha_n\} \subseteq \{n\}$, it follows that (7.19) also holds. □

Remark 2 *Actually, in Theorem 4.1 in Roussas (1979), it is shown that $\Lambda_n(\theta) - h' \Delta_n(\theta) \to -\frac{1}{2} h' \bar{\Gamma}(\theta) h$ in P_θ-probability rather than $\Lambda_n(\theta) - h_n' \Delta_n(\theta) \to -\frac{1}{2} h' \bar{\Gamma}(\theta) h$ in P_θ-probability. However, this is inconsequential, since $\mathcal{L}[\Delta_n(\theta) | P_\theta] \Rightarrow N(0, \bar{\Gamma}(\theta))$ (see Theorem 4.2 in the reference just cited) and $h_n \to h$.*

It is the proof of Proposition 4 that the auxiliary results below are needed. All results not involving stopping times hold under assumptions (A1)-(A4); those involving stopping times also require assumption (A5). Since θ, although arbitrarily chosen in Θ, remains fixed throughout, we may omit it for notational convenience. Thus, we shall write $\varphi_{\alpha_n j}$, $\dot{\varphi}_j$, $\bar{\Gamma}$, Λ_{α_n}, Δ_{α_n}, Λ_{ν_n} and Δ_{ν_n} rather than $\varphi_{\alpha_n j}(\theta)$, $\dot{\varphi}_j(\theta)$, $\bar{\Gamma}(\theta)$, $\Lambda_{\alpha_n}(\theta)$, $\Delta_{\alpha_n}(\theta)$, $\Lambda_{\nu_n}(\theta)$ and $\Delta_{\nu_n}(\theta)$, respectively. Also, in expressions of the form $\Lambda_{\alpha_n} - h_n'\Delta_{\alpha_n} \to -\frac{1}{2}h'\bar{\Gamma}h$ in P_θ-probability, we shall replace h_n by h, as explained in Remark 4.1. Thus, instead of (7.20), it suffices to show that

$$(\Lambda_{\nu_n} - h'\Delta_{\nu_n}) - (\Lambda_{\alpha_n} - h'\Delta_{\alpha_n}) \to 0 \text{ in } P_\theta\text{-probability.} \tag{7.24}$$

However, $\log q_0(\theta;\theta_{\alpha_n}) \to 0$ in P_θ-probability, by assumption (A4). Therefore, by way of (7.23), convergence (7.24) is reduced to

$$\left(2\sum_{j=1}^{\nu_n}\log\varphi_{\alpha_n j} - h'\Delta_{\nu_n}\right) - \left(2\sum_{j=1}^{\alpha_n}\log\varphi_{\alpha_n j} - h'\Delta_{\alpha_n}\right) \to 0 \text{ in } P_\theta\text{-probability.} \tag{7.25}$$

Lemma 1 *Let the r.v.'s W_{α_n} and W_{ν_n} be defined by :*

$$W_{\alpha_n} = \max\left(|\varphi_{\alpha_n j} - 1|;\, 1 \le j \le \alpha_n\right), W_{\nu_n} = \max\left(|\varphi_{\alpha_n j} - 1|;\, 1 \le j \le \nu_n\right).$$

Then, in P_θ-probability :

(i) $W_{\alpha_n} \to 0$, *(ii)* $W_{\nu_n} \to 0$.

Proof. (i) It follows from Lemma 5.2 in Roussas (1979). (ii) For m satisfying the inequalities $(1-\delta)\alpha_n < m < (1-\delta)\alpha_n$, some $0 < \delta < 1$, define W_m by : $W_m = \max\left(|\varphi_{\alpha_n j} - 1|;\, 1 \le j \le m\right)$. On account of part (i), the proof will follow, if we would establish the convergence

$$P_\theta\left\{\max|W_m - W_{\alpha_n}|;\, (1-\delta)\alpha_n < m < (1+\delta)\alpha_n\right\} \to 0.$$

This is so by Theorem A1 applied with β_n and r_n being replaced by α_n and ν_n, respectively. Clearly,

$$\begin{aligned}
&\max\left\{|W_m - W_{\alpha_n}|;\, (1-\delta)\alpha_n < m < (1+\delta)\alpha_n\right\} \\
&\le \max\left\{|W_m - W_{\alpha_n}|;\, (1-\delta)\alpha_n < m \le \alpha_n\right\} \\
&+ \max\left\{|W_m - W_{\alpha_n}|;\, \alpha_n < m < (1+\delta)\alpha_n\right\},
\end{aligned}$$

and, for $(1-\delta)\alpha_n < m \le \alpha_n$, $|W_m - W_{\alpha_n}| = W_{\alpha_n} - W_m \le W_{\alpha_n}$, so that

$$\max\{|W_m - W_{\alpha_n}|;\ (1-\delta)\alpha_n < m \le \alpha_n\} \le W_{\alpha_n} \xrightarrow{P_\theta} 0. \qquad (7.26)$$

For $\alpha_n < m < (1+\delta)\alpha_n$, set $1+\delta = \gamma$, $\gamma\alpha_n = c_n$, and $[c_n] = \gamma_n$. Also, it would be convenient to write $\theta_{\alpha_{n,h}}$ for any parameter point $\theta + h_n\alpha_n^{-1/2}$ with $h_n \to h$. Then

$$\begin{aligned}
\theta_{\alpha_{n,h}} &= \theta + h_n\left(\gamma\cdot\frac{\gamma_n}{c_n}\right)^{1/2}\cdot\gamma_n^{-1/2} = \theta + t_n\gamma_n^{-1/2}\\
t_n &= h_n\left(\gamma\cdot\frac{\gamma_n}{c_n}\right)^{1/2} \to h\gamma^{1/2} = t.
\end{aligned}$$

Thus, $\theta_{\alpha_{n,h}} = \theta_{\gamma_{n,t}}$ with γ_n and t as above. For the purposes of this proof, also set

$$\varphi_{\alpha_{n,h,j}} = \varphi_j(\theta;\theta_{\alpha_{n,h}}),\ \varphi_{\gamma_{n,t,j}} = \varphi_j(\theta;\theta_{\gamma_{n,t}}),$$

and

$$\begin{aligned}
\hat{W}_m &= \max\left(\left|\varphi_{\gamma_{n,t,j}} - 1\right|;\ 1 \le j \le m\right)\\
&= \max\left(\left|\varphi_{\alpha_{n,h,j}} - 1\right|;\ 1 \le j \le m\right)\\
&= W_m,\ 1 \le m \le \gamma_n.
\end{aligned}$$

Then, clearly,

$$\max\left(\left|\varphi_{\alpha_{n,j}} - 1\right|;\ 1 \le j \le (1+\delta)\alpha_n\right) \le \max\left(\left|\varphi_{\alpha_{n,h,j}} - 1\right|;\ 1 \le j \le \gamma_n\right),$$

$$\max\left(\left|\varphi_{\gamma_{n,t,j}} - 1\right|;\ 1 \le j \le \gamma_n\right) = \hat{W}_{\gamma_n} \to 0 \text{ in } P_\theta\text{-probability}$$

by part (i). Also, for $\alpha_n < m < \gamma_n$,

$$|W_m - W_{\alpha_n}| = \left|\hat{W}_m - W_{\alpha_n}\right| \le \hat{W}_m + W_{\alpha_n} \le \hat{W}_{\gamma_n} + W_{\alpha_n}.$$

Hence

$$\begin{aligned}
&\max\{|W_m - W_{\alpha_n}|;\ \alpha_n < m < (1+\delta)\,\alpha_n\}\\
&\qquad\le \max\left(|W_m - W_{\alpha_n}|;\ \alpha_n < m \le \gamma_n\right)\\
&\qquad\qquad\le \hat{W}_{\gamma_n} + W_{\alpha_n} \to 0 \text{ in } P_\theta\text{-probability.} \qquad (7.27)
\end{aligned}$$

Convergences (7.26) and (7.27) complete the proof of part (ii). □

The lemma just established allows the expansion of $\log \varphi_{\alpha_n j}$ and $\log \varphi_{\nu_n j}$, uniformly in j, over an event of P_θ-probability $< \varepsilon, \varepsilon > 0$. That is, uniformly in j,

$$\log \varphi_{\alpha_n j} = (\varphi_{\alpha_n j} - 1) - \frac{1}{2}(\varphi_{\alpha_n j} - 1)^2 + c_{nj}(\varphi_{\alpha_n j} - 1)^3, \quad |c_{nj}| \leq 3,$$

so that

$$2\sum_{j=1}^{\alpha_n} \log \varphi_{\alpha_n j} = 2\sum_{j=1}^{\alpha_n}(\varphi_{\alpha_n j} - 1) - \sum_{j=1}^{\alpha_n}(\varphi_{\alpha_n j} - 1)^2 + \sum_{j=1}^{\alpha_n} c_{nj}(\varphi_{\alpha_n j} - 1)^3, \tag{7.28}$$

and

$$2\sum_{j=1}^{\nu_n} \log \varphi_{\alpha_n j} = 2\sum_{j=1}^{\nu_n}(\varphi_{\alpha_n j} - 1) - \sum_{j=1}^{\nu_n}(\varphi_{\alpha_n j} - 1)^2 + \sum_{j=1}^{\nu_n} c_{nj}(\varphi_{\alpha_n j} - 1)^3. \tag{7.29}$$

We may proceed with the following result.

Lemma 2 *Set* $V_{\alpha_n} = \sum_{j=1}^{\alpha_n} |\varphi_{\alpha_n j} - 1|^3$ *and* $V_{\nu_n} = \sum_{j=1}^{\nu_n} |\varphi_{\alpha_n j} - 1|^3$.
Then, in P_θ-probability :

(i) $V_{\alpha_n} \to 0$, *(ii)* $V_{\nu_n} \to 0$.

Proof. (i) It is contained in the proof of Lemma 5.3 in Roussas (1979) (see relation (5.12) applied with α_n rather than n).
(ii) In Theorem A1 in the Appendix, replace β_n, r_n and W_{β_n} by α_n, ν_n and V_{α_n}, respectively. Then V_{α_n} satisfies condition (C1) on account of part (i) (with Q being the probability measure concentrated at 0). Then the theorem will hold for V_{ν_n} (that is, (ii) will hold), provided condition (C2) is verified. To this end, in Proposition 5 in the Appendix, set :

$$X_{\alpha_n j} = |\varphi_{\alpha_n j} - 1|^3,\ j \geq 1, \text{ and } V_{\alpha_n m} = \sum_{j=1}^{m} X_{\alpha_n j}.$$

Then (7.A.2) becomes

$$P_\theta\left(|V_{\alpha_n m} - V_{\alpha_n}| < \varepsilon \text{ for all } m \text{ such that } |m - \alpha_n| < \delta\alpha_n\right)$$

$$\geq 1 - P_\theta\left\{\sum_{j=[(1-\delta)\alpha_n]+2}^{\alpha_n} X_{\alpha_n j} \geq \varepsilon\right\} - P_\theta\left\{\sum_{j=\alpha_n+1}^{[(1-\delta)\alpha_n]} X_{\alpha_n j} \geq \varepsilon\right\}.$$

However,

$$\sum_{j=[(1-\delta)\alpha_n]+2}^{\alpha_n} X_{\alpha_n j} = \sum_{j=1}^{\alpha_n} X_{\alpha_n j} - \sum_{j=1}^{[(1-\delta)\alpha_n]+1} X_{\alpha_n j}, \tag{7.30}$$

and the first term on the right-hand side in (7.30) tends to 0 in P_θ-probability by part (i). Next, set $\gamma_n = [(1-\delta)\,\alpha_n] + 1$, so that γ_n/α_n tends to $(1-\delta)$, and recall that

$$X_{\alpha_n j} = |\varphi_{\alpha_n j} - 1|^3 = |\varphi_j(\theta; \theta_{\alpha_n}) - 1|^3, \quad \theta_{\alpha_n} = \theta + \hat{h}_n \alpha_n^{-1/2}, \hat{h}_n \to h.$$

But $\theta_{\alpha_n} = \theta + \bar{h}_n \gamma_n^{-1/2}$, $\bar{h}_n = \hat{h}_n \left(\gamma_{n/\alpha_n}\right)^{1/2} \to h(1-\delta)^{1/2} = \bar{h}$. Thus, in the derivations, θ_{α_n} may be replaced by $\theta_{\gamma_n} = \theta + \bar{h}_n \gamma_n^{-1/2}$, $\bar{h}_n \to \bar{h}$, and $X_{\alpha_n j} = X_{\gamma_n j}$. Therefore, by part (i) again,

$$\sum_{j=1}^{[(1-\delta)\alpha_n]+1} X_{\alpha_n j} = \sum_{j=1}^{\gamma_n} X_{\gamma_n j} \to 0 \text{ in } P_\theta\text{-probability.}$$

That $\sum_{j=\alpha_n+1}^{[(1+\delta)\alpha_n]} X_{\alpha_n j} \to 0$ in P_θ-probability follows in the same way. The proof is completed. □

In (7.25), replace the sums over j of $\log \varphi_{\alpha_n j}$, given in (7.28) and (7.29), and use the lemma just established in order to conclude that (7.25) may be replaced by

$$\left[2\sum_{j=1}^{\nu_n}(\varphi_{\alpha_n j} - 1) - \sum_{j=1}^{\nu_n}(\varphi_{\alpha_n j} - 1)^2 - h'\Delta_{\nu_n}\right]$$

$$- \left[2\sum_{j=1}^{\alpha_n}(\varphi_{\alpha_n j} - 1) - \sum_{j=1}^{\alpha_n}(\varphi_{\alpha_n j} - 1)^2 - h'\Delta_{\nu_n}\right] \to 0 \text{ in } P_\theta\text{-probability.} \tag{7.31}$$

It will be shown below that

$$\sum_{j=1}^{\alpha_n}(\varphi_{\alpha_n j} - 1)^2 - \sum_{j=1}^{\nu_n}(\varphi_{\alpha_n j} - 1)^2 \to 0 \text{ in } P_\theta\text{-probability,} \tag{7.32}$$

in which case, (7.31) is reduced to :

$$\left[2\sum_{j=1}^{\nu_n}(\varphi_{\alpha_n j}-1)-h'\Delta_{\nu_n}\right]$$
$$-\left[2\sum_{j=1}^{\alpha_n}(\varphi_{\alpha_n j}-1)-h'\Delta_{\alpha_n}\right]\to 0 \text{ in } P_\theta\text{-probability.} \quad (7.33)$$

It will also be proved below that

$$\left[2\sum_{j=1}^{\nu_n}(\varphi_{\alpha_n j}-1)-\left(\frac{\nu_n}{\alpha_n}\right)^{1/2}h'\Delta_{\nu_n}\right]$$
$$-\left[2\sum_{j=1}^{\alpha_n}(\varphi_{\alpha_n j}-1)-h'\Delta_{\alpha_n}\right]\to 0 \text{ in } P_\theta\text{-probability.} \quad (7.34)$$

Then (7.33) may be further reduced to

$$\left(\frac{\nu_n}{\alpha_n}\right)^{1/2}h'\Delta_{\nu_n}-h'\Delta_{\nu_n}\to 0 \text{ in } P_\theta\text{-probability.} \quad (7.35)$$

Relations (7.32), (7.34), and (7.35), which need to be justified, are now formulated as lemmas.

Lemma 3 *In P_θ-probability,*

(i) $\sum_{j=1}^{\alpha_n}(\varphi_{\alpha_n j}-1)^2-\sum_{j=1}^{\nu_n}(\varphi_{\alpha_n j}-1)^2\to 0,$

(ii) $\sum_{j=1}^{\nu_n}[(\varphi_{\alpha_n j}-1)-\alpha_n^{-1/2}h'\dot{\varphi}_j]-\sum_{j=1}^{\alpha_n}[(\varphi_{\alpha_n j}-1)-\alpha_n^{-1/2}h'\dot{\varphi}_j]\to 0.$

Lemma 4 *In P_θ-probability,*

$$2\alpha_n^{-1/2}\sum_{j=1}^{\nu_n}h'\dot{\varphi}_j-2\nu_n^{-1/2}\sum_{j=1}^{\nu_n}h'\dot{\varphi}_j\to 0.$$

Proof of Lemma 3. (i) First, observe that the uniform integrability assumption (A3)(i) implies uniform boundedness of $\mathcal{E}_\theta\,(h'\dot{\varphi}_j)^2$; that is, $\mathcal{E}_\theta\,(h'\dot{\varphi}_j)^2\le M$ for all j (see, for example (e), page 162, in Loève (1963)). Next, the differentiability in q.m. assumption (A2)(i) implies that

$$\mathcal{E}_\theta\left[\alpha_n^{1/2}(\varphi_{\alpha_n j}-1)\right]^2=\alpha_n\mathcal{E}_\theta(\varphi_{\alpha_n j}-1)^2\to\mathcal{E}_\theta\left(h'\dot{\varphi}_j\right)^2 \text{ uniformly in } j.$$

Since $\mathcal{E}_\theta\,(h'\dot{\varphi}_j)^2 \le M$ for all j, we also have $\alpha_n\mathcal{E}_\theta(\varphi_{\alpha_n j}-1)^2 \le M$ for all j, where M is a generic positive finite constant.

At this point, go to Proposition 6 and set $X_{\alpha_n j} = (\varphi_{\alpha_n j}-1)^2$, $j \ge 1$. Also, replace β_n and r_n by α_n and ν_n, respectively. By the fact that $\mathcal{E}_\theta X_{\alpha_n j} \le M/\alpha_n$ for all j, inequality (7.A.9) becomes

$$P_\theta\left(\left|\sum_{j=1}^{\alpha_n} X_{\alpha_n j} - \sum_{j=1}^{\nu_n} X_{\alpha_n j}\right| \ge \varepsilon\right) \le \varepsilon(1+2M),$$

so that

$$\sum_{j=1}^{\alpha_n} X_{\alpha_n j} - \sum_{j=1}^{\nu_n} X_{\alpha_n j} = \sum_{j=1}^{\alpha_n}(\varphi_{\alpha_n j}-1)^2 - \sum_{j=1}^{\nu_n}(\varphi_{\alpha_n j}-1)^2 \to 0 \text{ in } P_\theta\text{-probability.}$$

(ii) Again, in Proposition 6 in the Appendix, set

$$X_{\alpha_n j} = (\varphi_{\alpha_n j}-1) - \alpha_n^{-1/2}h'\dot{\varphi}_j,\ j \ge 1.$$

Then

$$\mathcal{E}_\theta\left|X_{\alpha_n j}\right| = \mathcal{E}\left|(\varphi_{\alpha_n j}-1) - \alpha_n^{-1/2}h'\dot{\varphi}_j\right| \le 2\mathcal{E}_\theta(\varphi_{\alpha_n j}-1)^2 + \alpha_n^{-1}\mathcal{E}_\theta\left|h'\dot{\varphi}_j\right|^2$$

by the c_r-inequality with $r = 2$ (see, for example, Loève (1963), page 155). As already pointed out in part (i),

$$\mathcal{E}_\theta(\varphi_{\alpha_n j}-1)^2 \le \frac{M}{\alpha_n} \text{ and } \mathcal{E}_\theta\left(h'\dot{\varphi}_j\right)^2 \le M, \text{ for all } j,$$

so that $\mathcal{E}_\theta\left|X_{\alpha_n j}\right| \le M\alpha_n^{-1}$ for all j. Therefore inequality (7.A.9) becomes

$$P_\theta\left(\left|\sum_{j=1}^{\alpha_n} X_{\alpha_n j} - \sum_{j=1}^{\nu_n} X_{\alpha_n j}\right| \ge \varepsilon\right) < \varepsilon(1+M),$$

where throughout M is a generic constant. This completes the proof of part (ii) and hence of the lemma. □

Proof of Lemma 4. In Proposition 7 in the Appendix, replace β_n and r_n by α_n and ν_n, respectively, and set

$$X_{\alpha_n j} = 2\alpha_n^{-1/2}h'\dot{\varphi}_j,\ j \ge 1.$$

By Lemma 5.6 in Roussas (1979), $\mathcal{E}_\theta(h'\dot{\varphi}_j|\mathcal{A}_{j-1}) = 0$ a.s. $[P_\theta]$, and also $\mathcal{E}_\theta\,(h'\dot{\varphi}_j) = 0$. In addition, $\mathcal{E}_\theta\,(h'\dot{\varphi}_j)^2 \le M$ for all j, as was pointed out in the proof of Lemma 3(i), and so that $\sigma_\theta^2(h'\dot{\varphi}_j) = \mathcal{E}_\theta\,(h'\dot{\varphi}_j)^2 \le M$ for all

j, and hence $\sigma^2_\theta(X_{\alpha_n j}) = 4\alpha_n^{-1}\mathcal{E}_\theta(h'\dot{\varphi}_j)^2 \le M\alpha_n^{-1}$ for all j. Therefore, inequality (7.A.16) becomes

$$P_\theta\left(\left|\sum_{j=1}^{\alpha_n} X_{\alpha_n j} - \sum_{j=1}^{\nu_n} X_{\alpha_n j}\right| \ge \varepsilon\right) < \varepsilon(1+M).$$

Thus,

$$2\alpha_n^{-1/2}\sum_{j=1}^{\alpha_n} h'\dot{\varphi}_j - 2\alpha_n^{-1/2}\sum_{j=1}^{\nu_n} h'\dot{\varphi}_j \to 0 \text{ in } P_\theta\text{-probability.} \tag{7.36}$$

However, by Theorem 4.2 in Roussas (1979), applied with n replaced by α_n,

$$\mathcal{L}\left(2\alpha_n^{-1/2}\sum_{j=1}^{\alpha_n} h'\dot{\varphi}_j | P_\theta\right) \Rightarrow N\left(0, h'\bar{\Gamma}h\right),$$

and hence, by (7.36),

$$\mathcal{L}\left(2\alpha_n^{-1/2}\sum_{j=1}^{\nu_n} h'\dot{\varphi}_j | P_\theta\right) \Rightarrow N\left(0, h'\bar{\Gamma}h\right); \tag{7.37}$$

thus $2\alpha_n^{-1/2}\sum_{j=1}^{\nu_n} h'\dot{\varphi}_j$ is bounded in P_θ-probability. Furthermore,

$$2\alpha_n^{-1/2}\sum_{j=1}^{\nu_n} h'\dot{\varphi}_j - 2\nu_n^{-1/2}\sum_{j=1}^{\nu_n} h'\dot{\varphi}_j = [1 - \left(\frac{\alpha_n}{\nu_n}\right)^{1/2}](2\alpha_n^{-1/2}\sum_{j=1}^{\nu_n} h'\dot{\varphi}_j),$$

which tends to 0 in P_θ-probability. This completes the proof of the lemma. □

5. PROOF OF THEOREMS

Proof of Theorem 1. (i) Omitting θ in most expressions below, and using P_θ rather than $\tilde{P}_{n,\theta}$, this part amounts to proving that

$$\Lambda_{\nu_n} - h_n'\Delta_{\nu_n} \to -\frac{1}{2}h'\bar{\Gamma}h \text{ in } P_\theta\text{-probability.} \tag{7.38}$$

According to the successive reductions obtained in the previous section, instead of (7.38), it suffices to show (7.26) or (7.27) or (7.31) or (7.33) or (7.35). However, (7.35) is the same as the expression in Lemma 4.

This completes the justification of this part. (ii) With the simplifications mentioned above, we wish to show that

$$\mathcal{L}\left(\Delta_{\nu_n}|P_\theta\right) \Rightarrow N(0,\bar{\Gamma}),$$

or

$$\mathcal{L}\left(t'\Delta_{\nu_n}|P_\theta\right) \Rightarrow N(0,t'\bar{\Gamma}t) \text{ for every } t \in \Re^k, \tag{7.39}$$

or

$$\mathcal{L}\left(2\nu_n^{-1/2}\sum_{j=1}^{\nu_n} t'\dot{\varphi}_j|P_\theta\right) \Rightarrow N\left(0,t'\bar{\Gamma}t\right), t \in \Re^k.$$

This follows from Lemma 4 and relation (7.37).
(iii) Here, we wish to show that

$$\mathcal{L}\left(\Lambda_{\nu_n}|P_\theta\right) \Rightarrow N\left(-\frac{1}{2}h'\bar{\Gamma}h, h'\bar{\Gamma}h\right).$$

This is an immediate consequence of parts (i) and (ii), expressed in the forms (7.38) and (7.39). □

Proof of Theorem 2. Throughout the proof of the theorem, write P_θ and $P_{\theta_{\tau n}}$ instead of $\tilde{P}_{n,\theta}$ and $\tilde{P}_{n,\theta_{\tau n}}$, respectively. (i) Here, we wish to show that

$$\Lambda_{\nu_n} - h_n'\Delta_{\nu_n} \to -\frac{1}{2}h'\bar{\Gamma}h \text{ in } P_{\theta_{\tau n}}\text{-probability}.$$

However, this is an immediate consequence of Theorem 1 (i) and the contiguity established in Proposition 2. (ii) We wish to show that $\mathcal{L}\left(\Delta_{\nu_n}|P_{\theta_{\tau n}}\right) \Rightarrow N(\bar{\Gamma}h,\bar{\Gamma})$, or, equivalently,

$$\mathcal{L}\left(t'\Delta_{\nu_n}|P_{\theta_{\tau n}}\right) \Rightarrow N(t'\bar{\Gamma}h, t'\bar{\Gamma}t) \text{ for every } t \in \Re^k. \tag{7.40}$$

By Theorem 4.6 in Roussas (1979), $\mathcal{L}\left(\Delta_{\alpha_n}|P_{\theta_{\tau n}}\right) \Rightarrow N(\bar{\Gamma}h,\bar{\Gamma})$, or

$$\mathcal{L}\left(t'\Delta_{\alpha_n}|P_{\theta_{\tau n}}\right) \Rightarrow N(t'\bar{\Gamma}h, t'\bar{\Gamma}t),\ t \in \Re^k. \tag{7.41}$$

Next, the convergence in (7.36) also holds under $P_{\theta_{\tau n}}$ because of the underlying contiguity. In a slightly different form then, one has, for every $t \in \Re^k$,

$$t'\Delta_{\alpha_n} - \left(\frac{\nu_n}{\alpha_n}\right)^{1/2} t'\Delta_{\nu_n} \to 0 \text{ in } P_{\theta_{\tau n}}\text{-probability}. \tag{7.42}$$

Thus, from (7.41) and (7.42),

$$\mathcal{L}\left[\left(\frac{\nu_n}{\alpha_n}\right)^{1/2} t'\Delta_{\nu_n}|P_{\theta_{\tau_n}}\right] \Rightarrow N(t'\bar{\Gamma}h, t'\bar{\Gamma}t),\ t \in \Re^k, \tag{7.43}$$

which implies, in particular, that

$$\left\{\left(\frac{\nu_n}{\alpha_n}\right)^{1/2} t'\Delta_{\nu_n}\right\} \text{ is bounded in } P_{\theta_{\tau_n}}\text{-probability.} \tag{7.44}$$

Therefore

$$\left(\frac{\nu_n}{\alpha_n}\right)^{1/2} t'\Delta_{\nu_n} - t'\Delta_{\nu_n}$$
$$= \left(\frac{\nu_n}{\alpha_n}\right)^{1/2} t'\Delta_{\nu_n}\left[1 - \left(\frac{\alpha_n}{\nu_n}\right)^{1/2}\right] \to 0 \text{ in } P_{\theta_{\tau_n}}\text{-probability} \tag{7.45}$$

This is so because of (7.44) and the fact that $\frac{\alpha_n}{\nu_n}$ converges to 1 in $P_{\theta_{\tau_n}}$-probability. This convergence is a consequence of assumption (A5), relation (7.12), and the contiguity established in Proposition 2. The proof of (7.40) is concluded by means of (7.43) and (7.45). (ii) In (7.40), t can, clearly, be replaced by $h_n \to h$ to obtain

$$\mathcal{L}\left(h_n'\Delta_{\nu_n}|P_{\theta_{\tau_n}}\right) \Rightarrow N(h'\bar{\Gamma}h, h'\bar{\Gamma}h). \tag{7.46}$$

Then relation (7.46) and part (i) of the theorem provide the desired result. □

On the basis of Theorems 1 and 2 obtained here, certain testing hypotheses problems and asymptotic efficiency of estimates may be addressed. Such problems were discussed for fixed sample size in Roussas (1972), pages 99-105 and pages 147-166. Details may be presented elsewhere.

References

[1] Akritas, M.G. and Roussas, G.G. (1979). Asymptotic expansion of the log-likelihood function based on stopping times defined on a Markov process, *Ann. Inst. Statist. Math.*, **31**, 21-38.

[2] Akritas, M.G., Roussas, G.G. and Stamatelos, G.D. (1984). Asymptotic expansion of the log-likelihood function based on stopping times defined on stochastic processes, *Mathematical Structures – Computational Mathematics – Mathematical Modelling*, **2**, 90-96, Sofia.

[3] Anscombe, F. J. (1952). Large sample theory of sequential estimation, *Proc. Camb. Phil. Soc.*, **48**, 600-607.

[4] Chow, Y.S. and Teicher, H (19..). *Probability Theory*, 3rd Ed., Springer-Verlag.

[5] Le Cam, L. (1960). Locally asymptotically normal families of distributions, *Univ. California Publ. Statist.*, **3**, 37-98.

[6] Le Cam, L. and Yang, G.L. (1990). *Asymptotics in Statistics, Some Basic Concepts*, Springer Verlag.

[7] Luschgy, H. (1995). Local asymptotic quadraticity of stochastic process models based on stopping times, *Stochastic Process. Appl.*, **57**, 305-317.

[8] Loève, M. (1963). *Probability Theory*, 3rd Ed., VanNostrand, Princeton.

[9] Roussas, G.G. (1972). *Contiguity of Probability Measures : Some applications in statistics*, Cambridge University Press, Cambridge.

[10] Roussas, G.G. (1979). Asymptotic distribution of the log-likelihood function for stochastic processes, *Z. Wahrscheinlichkeitstheorie verw. Gebiete,* **47**, 31-46.

[11] Stamatelos, G.D. (1976). Asymptotic distribution of the log-likelihood function for stochastic processes : Some examples, *Bull. Soc. Math. Grèce,* **17**, 92-116.

[12] Wolfowitz, J. (1965). Asymptotic efficiency of the maximum likelihood estimator. *Theor. Probability Appl.*,**10**, 247-260.

Appendix

The results stated and/or proved in this appendix are indirectly instrumental in establishing Theorems 1 and 2, in addition to having independent interest. Theorem A1 is a slightly modified version of Theorem 1 of Anscombe (1952).

For its formulation, two conditions to be stated below are required. To this end, let W_n, $n = 1, 2, \ldots$, be real-valued r.v.'s defined on the probability space $(\Omega, \mathcal{A}, P)$, and let $\{\beta_n\}$, $n \geq 1$, be a nondecreasing sequence of positive integers tending to ∞. Then suppose that the following conditions are satisfied; namely,

(C1) $\mathcal{L}(W_{\beta_n}|P) \Rightarrow Q$ a probability measure in $\Re$.

(C2) For every $\varepsilon > 0$ and some $\delta > 0$,

$$P(|W_m - W_{\beta_n}| < \varepsilon; \text{ for all } m \text{ such that } |m - \beta_n| < \delta\beta_n)) \to 1;$$

equivalently,

$$P[\max(|W_m - W_{\beta_n}|;\, m \text{ such that } |m - \beta_n| < \delta\beta_n) \geq \varepsilon] \to 0.$$

We may now formulate the result referred to above.

Theorem A1 *Consider the sequence of r.v.'s $\{W_n\}$, $n \geq 1$, defined on the probability space $(\Omega, \mathcal{A}, P)$ and satisfying conditions (C1) and (C2) above. Let also $\{r_n\}$, $n \geq 1$, be a sequence of positive integer-valued r.v.'s such that $(r_n/\beta_n) \to 1$ in probability. Then*

$$\mathcal{L}(W_{r_n}|P) \Rightarrow Q.$$

Proof. See Theorem 4.1 in Akritas and Roussas (1979). □

Proposition 5 *For each $n = 1, 2, ...$, let X_{nj}, $j \geq 1$, be nonnegative r.v.'s defined on the probability space $(\Omega, \mathcal{A}, P)$. Let $\{\beta_n\}$, $n \geq 1$, be a non-decreasing sequence of positive integers tending to ∞, and for any positive integer m, define W_{nm} by*

$$W_{nm} = \sum_{j=1}^{m} X_{nj}. \tag{7.A.1}$$

Then, for any $\varepsilon > 0$ and any $0 < \delta < 1$,

$$P(|W_{nm} - W_{n\beta_n}| < \varepsilon; \textit{ for all } m \textit{ such that } |m - \beta_n| < \delta\beta_n)$$

$$\geq 1 - P\left\{\sum_{j=[(1-\delta)\beta_n]+2}^{\beta_n} X_{nj} \geq \varepsilon\right\} - P\left\{\sum_{j=\beta_n+1}^{[(1+\delta)\beta_n]} X_{nj} \geq \varepsilon\right\}. \tag{7.A.2}$$

In particular, if, for each n, the r.v.'s X_{nj}, $j \geq 1$, are stationary, then (7.A.2) becomes

$$\begin{aligned} &P(|W_{nm} - W_{n\beta_n}| < \varepsilon; \textit{ for all } m \textit{ such that } |m - \beta_n| < \delta\beta_n) \\ &\qquad \geq 1 - 2P\left(\sum_{j=1}^{[\delta\beta_n]} X_{nj} \geq \varepsilon\right). \end{aligned} \tag{7.A.3}$$

Proof. The inequality $|m - \beta_n| < \delta\beta_n$ implies the inequalities

$$[(1-\delta)\,\beta_n] < m \leq [(1+\delta)\,\beta_n]. \tag{7.A.4}$$

Then

$$\begin{aligned} &P(|W_{nm} - W_{n\beta_n}| < \varepsilon; \text{ for all } m \text{ such that } |m - \beta_n| < \delta\beta_n) \\ = \;& 1 - P(|W_{nm} - W_{n\beta_n}| \geq \varepsilon \\ &\qquad\qquad \text{for at least one } m \text{ such that } |m - \beta_n| < \delta\beta_n) \\ = \;& 1 - P[\max(|W_{nm} - W_{n\beta_n}|) \geq \varepsilon; \\ &\qquad\qquad m \text{ with } |m - \beta_n| < \delta\beta_n]. \end{aligned} \tag{7.A.5}$$

But

$$P[\max(|W_{nm} - W_{n\beta_n}|) \geq \varepsilon;\ m \text{ with } |m - \beta_n| < \delta\beta_n]$$

$$\leq P\{\max(|W_{nm} - W_{n\beta_n}|) \geq \varepsilon;\ m \text{ with } [(1-\delta)\beta_n] < m \leq [1+\delta)\beta_n]\}$$
(by (7.A.4))

$$\leq P\{\max(|W_{nm} - W_{n\beta_n}|) \geq \varepsilon;\ m \text{ with } [(1-\delta)\beta_n] < m < \beta_n\}$$

$$+P\{\max(|W_{nm} - W_{n\beta_n}|) \geq \varepsilon;\ m \text{ with } \beta_n \leq m \leq\ [(1+\delta)\beta_n]\}. \tag{7.A.6}$$

The first maximum on the right-hand side of (7.A.6) is taken over the terms :

$$X_{n,[(1-\delta)\beta_n]+2} + \ldots + X_{n,\beta_n}, X_{n,[(1-\delta)\beta_n]+3} + \ldots + X_{n,\beta_n}, \ldots, X_{n,\beta_n},$$

and therefore is equal to the first sum. Likewise, the second maximum on the same side of (7.A.6) is taken over the terms :

$$0, X_{n,\beta_n+1}, \ldots, X_{n,\beta_n+1} + \ldots + X_{n,[(1+\delta)\beta_n]},$$

and therefore is equal to the last sum above. Therefore inequality (7.A.6) becomes

$$P[\max(|W_{nm} - W_{n\beta_n}|) \geq \varepsilon;\ m \text{ with } |m - \beta_n| < \delta\beta_n]$$

$$\leq P\left\{\sum_{j=[(1-\delta)\beta_n]+2}^{\beta_n} X_{nj} \geq \varepsilon\right\} + P\left\{\sum_{j=\beta_n+1}^{[(1+\delta)\beta_n]} X_{nj} \geq \varepsilon\right\}. \tag{7.A.7}$$

Relations (7.A.5) and (7.A.7) establish inequality (7.A.2).

The first sum on the right-hand side of (7.A.7) consists of $\beta_n - [(1-\delta)\beta_n] - 1$ terms, and the second sum consists of $[(1+\delta)\beta_n] - \beta_n$ terms. Since $x \geq 0$ implies $x - 1 < [x] \leq x$, the numbers of these terms satisfy the inequalities :

$$\delta\beta_n - 1 \leq \beta_n - [(1-\delta)\beta_n] - 1 < \delta\beta_n \text{ and}$$

$$\delta\beta_n - 1 < [(1+\delta)\beta_n] - \beta_n < \delta\beta_n.$$

The first pair of inequalities imply $\beta_n - [(1-\delta)\beta_n] - 1 \leq [\delta\beta_n]$, and the second pair of inequalities imply $[(1+\delta)]\beta_n - \beta_n = [\delta\beta_n]$. Thus, in either case there are $\leq [\delta\beta_n]$ terms in each sum. Then, under stationarity the right-hand side of (7.A.7) is less than or equal to $2P(\sum_{j=1}^{[\delta\beta_n]} X_{nj} \geq \varepsilon)$, and thus inequality (7.A.2) yields (7.A.3). □

The following two properties, stated as Propositions 6 and 7, will prove useful in establishing certain results in Section 4. In addition, they are of considerable independent interest. Although some discussion of them may be found in Akritas and Roussas (1979), it was deemed appropriate to have them stated and proved here in a slightly modified form, and with some more details.

Proposition 6 *For each $n = 1, 2, ...$, let X_{nj}, $j \geq 1$, be r.v.'s defined on the probability space $(\Omega, \mathcal{A}, P)$ and let r_n, $n \geq 1$, be positive integer-valued r.v.'s such that $\frac{r_n}{\beta_n} \to c$, a finite positive constant, where $\{\beta_n\}$, $n \geq 1$, is a nondecreasing sequence of positive integers tending to ∞. Then, for every $\varepsilon > 0$, there exists a positive integer $N = N(\varepsilon)$ such that $n > N$ implies that*

$$P\left(\left|\sum_{j=1}^{\beta_n} X_{nj} - \sum_{j=1}^{r_n} X_{nj}\right| \geq \varepsilon\right) = P(A_n) < \varepsilon + \varepsilon^{-1} \sum_{j=[(c-\varepsilon^2)\beta_n]+2}^{[(c+\varepsilon^2)\beta_n]} \mathcal{E}|X_{nj}|. \tag{7.A.8}$$

In particular, if for each n, $\mathcal{E}|X_{nj}| \leq M_n$, $j \geq 1$, then the above inequality becomes

$$P\left(\left|\sum_{j=1}^{\beta_n} X_{nj} - \sum_{j=1}^{r_n} X_{nj}\right| \geq \varepsilon\right) = P(A_n) < \varepsilon(1 + 2\beta_n M_n). \tag{7.A.9}$$

Proof. In the first place,

$$P\left(\left|\frac{r_n}{\beta_n} - c\right| \geq \varepsilon^2\right) = P(|r_n - c\beta_n| \geq \varepsilon^2\beta_n) < \varepsilon, \; n > N = N(\varepsilon). \tag{7.A.10}$$

Next,

$$\begin{aligned}
\left(\left|\tfrac{r_n}{\beta_n} - c\right| < \varepsilon^2\right) &= (|r_n - c\beta_n| < \varepsilon^2\beta_n) \\
&= \{(c-\varepsilon^2)\beta_n < r_n < (c+\varepsilon^2)\beta_n\} \\
&\subseteq \{[(c-\varepsilon^2)\beta_n] < r_n \leq [(c+\varepsilon^2)\beta_n]\} \\
&= \{[(c-\varepsilon^2)\beta_n] < r_n < \beta_n\} + \{\beta_n \leq r_n \leq [(c+\varepsilon^2)\beta_n]\}.
\end{aligned}$$

Therefore, for $n > N$,

$$\begin{aligned}
P(A_n) &= P[A_n \cap (|r_n - c\beta_n| \geq \varepsilon^2\beta_n)] + P[A_n \cap (|r_n - c\beta_n| < \varepsilon^2\beta_n)] \\
&\leq P(|r_n - c\beta_n| \geq \varepsilon^2\beta_n) + P(A_n \cap \{[(c-\varepsilon^2)\beta_n] < r_n < \beta_n\}) \\
&+ P(A_n \cap \{\beta_n \leq r_n \leq [(c+\varepsilon^2)\beta_n]\}).
\end{aligned} \tag{7.A.11}$$

Then, on account of (7.A.10), relation (7.A.11) becomes

$$P(A_n) < \varepsilon + p_{n1} + p_{n2}, \tag{7.A.12}$$

where

$$\begin{aligned} p_{n1} &= P\left(A_n \cap \{[(c-\varepsilon^2)\beta_n] < r_n < \beta_n\}\right) \\ p_{n2} &= P\left(A_n \cap \{\beta_n \le r_n \le [(c+\varepsilon^2)\beta_n]\}\right). \end{aligned}$$

However,

$$\begin{aligned} p_{n1} &= P\left\{A_n \cap \bigcup_{i=[(c-\varepsilon^2)\beta_n]+1}^{\beta_n-1} (r_n = i)\right\} \\ &= P\left\{\bigcup_{i=[(c-\varepsilon^2)\beta_n]+1}^{\beta_n-1} A_n \cap (r_n = i)\right\} \\ &= P\left\{\bigcup_{m=[(c-\varepsilon^2)\beta_n]+2}^{\beta_n} \left(\left|\sum_{j=m}^{\beta_n} X_{nj}\right| \ge \varepsilon\right)\right\} \\ &= P(\max\{\left|\sum_{j=m}^{\beta_n} X_{nj}\right| \ge \varepsilon; [(c-\varepsilon^2)\beta_n] + 2 \le m \le \beta_n\}) \\ &\le P(\max\{\sum_{j=m}^{\beta_n} |X_{nj}| \ge \varepsilon; [(c-\varepsilon^2)\beta_n] + 2 \le m \le \beta_n\}) \\ &= P\left\{\sum_{j=[(c-\varepsilon^2)\beta_n]+2}^{\beta_n} |X_{nj}| \ge \varepsilon\right\} \le \varepsilon^{-1} \sum_{j=[(c-\varepsilon^2)\beta_n]+2}^{\beta_n} \mathcal{E}\,|X_{nj}|; \end{aligned}$$

that is,

$$p_{n1} \le \varepsilon^{-1} \sum_{j=[(c-\varepsilon^2)\beta_n]+2}^{\beta_n} \mathcal{E}\,|X_{nj}|. \tag{7.A.13}$$

Next,

$$\begin{aligned} p_{n2} &= P\left\{A_n \cap \bigcup_{i=\beta_n}^{[(c+\varepsilon^2)\beta_n]} (r_n = i)\right\} \\ &= P\left\{\bigcup_{i=\beta_n}^{[(c+\varepsilon^2)\beta_n]} A_n \cap (r_n = i)\right\} \\ &= P\left\{\bigcup_{m=\beta_{n+1}}^{[(c+\varepsilon^2)\beta_n]} \left(\left|\sum_{j=\beta_n+1}^{m} X_{nj}\right| \ge \varepsilon\right)\right\} \\ &= P\left(\max\left\{\sum_{j=\beta_n+1}^{m} |X_{nj}| \ge \varepsilon; \beta_n + 1 \le m \le [(c+\varepsilon^2)\beta_n]\right\}\right) \\ &\le P\left(\max\left\{\sum_{j=\beta_n+1}^{m} |X_{nj}| \ge \varepsilon; \beta_n + 1 \le m \le [(c+\varepsilon^2)\beta_n]\right\}\right) \end{aligned}$$

$$= P\left\{\textstyle\sum_{j=\beta_n+1}^{[(c+\varepsilon^2)\beta_n]} |X_{nj}| \geq \varepsilon\right\} \leq \varepsilon^{-1} \textstyle\sum_{j=\beta_n+1}^{[(c+\varepsilon^2)\beta_n]} \mathcal{E}\,|X_{nj}|\,;$$

that is,

$$p_{n2} \leq \varepsilon^{-1} \sum_{j=\beta_n+1}^{[(c+\varepsilon^2)\beta_n]} \mathcal{E}\,|X_{nj}|\,. \qquad (7.A.14)$$

Then, by means of (7.A.13) and (7.A.14), inequality (7.A.12) becomes

$$P(A_n) < \varepsilon + \varepsilon^{-1} \sum_{j=[(c-\varepsilon^2)\beta_n]+2}^{[(c+\varepsilon^2)\beta_n]} \mathcal{E}\,|X_{nj}|\,, \quad n > N,$$

which is what relation (7.A.8) asserts. Relation (7.A.9) follows from (7.A.8) because

$$\left[\left(c+\varepsilon^2\right)\beta_n\right] - \left\{\left[\left(c-\varepsilon^2\right)\beta_n\right]+2\right\}+1 = \left[\left(c+\varepsilon^2\right)\beta_n\right] - \left[\left(c-\varepsilon^2\right)\beta_n\right] - 1$$

$$\leq \left(c+\varepsilon^2\right)\beta_n - \left[\left(c-\varepsilon^2\right)\beta_n\right] - 1 \leq 2\varepsilon^2\beta_n.$$

□

Proposition 7 *For each $n = 1, 2, \ldots$, let X_{nj}, $j \geq 1$, be r.v.'s defined on the probability space $(\Omega, \mathcal{A}, P)$ and such that*

$$\mathcal{E}(X_{n,j+1}|X_{n1}, \ldots, X_{nj}) = 0 \ a.s. \ [P]\,, \ \mathcal{E}X_{n1} = 0.$$

Also, let $\{\beta_n\}$, $n \geq 1$, be a nondecreasing sequence of positive integers tending to ∞. Furthermore, let $\frac{r_n}{\beta_n} \to 1$ in probability, where r_n, $n \geq 1$, are positive integer-valued r.v.'s. Then, for every $\varepsilon > 0$, there exists a positive integer $N = N(\varepsilon)$ such that $n > N$ implies

$$P\left(\left|\sum_{j=1}^{\beta_n} X_{nj} - \sum_{j=1}^{r_n} X_{nj}\right| \geq \varepsilon\right) = P(A_n) < \varepsilon + \varepsilon^{-2} \sum_{j=[(c+\varepsilon^3)\beta_n]+2}^{[(c+\varepsilon^3)\beta_n]} \sigma^2\left(X_{nj}\right). \qquad (7.A.15)$$

In particular, if for each n, $\sigma^2(X_{nj}) \leq M_n$, $j \geq 1$, then the above inequality becomes

$$P\left(\left|\sum_{j=1}^{\beta_n} X_{nj} - \sum_{j=1}^{r_n} X_{nj}\right| \geq \varepsilon\right) = P(A_n) < \varepsilon\left(1 + 2\beta_n M_n\right). \qquad (7.A.16)$$

Proof. For $\varepsilon > 0$, there exists $N(\varepsilon) = N$ such that $n > N$ implies

$$P\left(\left|\frac{r_n}{\beta_n} - c\right| \geq \varepsilon^3\right) = P\left(|r_n - c\beta_n| \geq \varepsilon^3 \beta_n\right) < \varepsilon, \qquad (7.A.17)$$

whereas

$$\begin{aligned}
\left(\left|\tfrac{r_n}{\beta_n} - c\right| \geq \varepsilon^3\right) &= \left(|r_n - c\beta_n| < \varepsilon^3 \beta_n\right) \\
&\subseteq \left\{\left[\left(c - \varepsilon^3\right)\beta_n\right] < r_n \leq \left[\left(c + \varepsilon^3\right)\beta_n\right]\right\} \\
&= \left\{\left[\left(c - \varepsilon^3\right)\beta_n\right] < r_n < \beta_n\right\} \\
&+ \left\{\beta_n \leq r_n \leq \left[\left(c + \varepsilon^3\right)\beta_n\right]\right\},
\end{aligned}$$

so that, for $n > N$,

$$\begin{aligned}
P(A_n) &= P[A_n \cap \left(|r_n - c\beta_n| \geq \varepsilon^3\beta_n\right)] + P\left[A_n \cap \left(|r_n - c\beta_n| < \varepsilon^3\beta_n\right)\right] \\
&\leq P\left(|r_n - c\beta_n| \geq \varepsilon^3\beta_n\right) + P\left(A_n \cap \left\{\left[\left(c - \varepsilon^3\right)\beta_n\right] < r_n < \beta_n\right\}\right) \\
&+ P\left(A_n \cap \left\{\beta_n \leq r_n \leq \left[\left(c + \varepsilon^3\right)\beta_n\right]\right\}\right) \\
&< \varepsilon + p_{n1} + p_{n2} \text{ (by (7.A.17))},
\end{aligned} \qquad (7.A.18)$$

where

$$\begin{aligned}
p_{n1} &= P\left(A_n \cap \left\{\left[\left(c - \varepsilon^3\right)\beta_n\right] < r_n < \beta_n\right\}\right), \\
p_{n2} &= P\left(A_n \cap \left\{\beta_n \leq r_n \leq \left[\left(c + \varepsilon^3\right)\beta_n\right]\right\}\right).
\end{aligned}$$

However,

$$\begin{aligned}
p_{n1} &= P\left\{A_n \cap \bigcup_{i=[(c-\varepsilon^3)\beta_n]+1}^{\beta_{n-1}} (r_n = i)\right\} \\
&= P\left\{\bigcup_{i=[(c-\varepsilon^3)\beta_n]+1}^{\beta_{n-1}} A_n \cap (r_n = i)\right\} \\
&= P\left\{\bigcup_{m=[(c-\varepsilon^3)\beta_n]+2}^{\beta_n} \left(\left|\sum_{j=m}^{\beta_n} X_{nj}\right| \leq \varepsilon\right)\right\} \\
&= P\left(\max\left\{\left|\sum_{j=m}^{\beta_n} X_{nj}\right| \geq \varepsilon; \left[\left(c - \varepsilon^3\right)\beta_n\right] + 2 \leq m \leq \beta_n\right\}\right).
\end{aligned} \qquad (7.A.19)$$

The assumptions made allow the application of the extended Kolmogorov inequality (see, page 386, in Loève (1963)) to the last expressions on the right-hand side of (7.A.19) to produce the bound $\varepsilon^{-2} \bigcup_{j=[(c-\varepsilon^3)\beta_n]+2}^{\beta_n} \sigma^2(X_{nj})$. That is,

$$p_{n1} \leq \varepsilon^{-2} \bigcup_{j=[(c-\varepsilon^3)\beta_n]+2}^{\beta_n} \sigma^2(X_{nj}). \tag{7.A.20}$$

Similarly,

$$\begin{aligned}
p_{n2} &= P\left\{A_n \cap \bigcup_{i=\beta_n}^{[(c+\varepsilon^3)\beta_n]} (r_n = i)\right\} \\
&= P\left\{\bigcup_{i=\beta_n}^{[(c+\varepsilon^3)\beta_n]} A_n \cap (r_n = i)\right\} \\
&= P\left\{\bigcup_{m=\beta_n+1}^{[(c+\varepsilon^3)\beta_n]} \left(\left|\sum_{j=\beta_n+1}^{m} X_{nj}\right| \geq \varepsilon\right)\right\} \\
&= P\left(\max\left\{\left|\sum_{j=\beta_n+1}^{m} X_{nj}\right| \geq \varepsilon; \beta_n + 1 \leq m \leq \left[\left(c+\varepsilon^3\right)\beta_n\right]\right\}\right).
\end{aligned} \tag{7.A.21}$$

Applying the above cited inequality to the last expression on the right-hand side of (7.A.21), we obtain

$$p_{n2} \leq \varepsilon^{-2} \bigcup_{j=\beta_n+1}^{[(c+\varepsilon^3)\beta_n]} \sigma^2(X_{nj}). \tag{7.A.22}$$

Relations (7.A.20) and (7.A.22) yield

$$p_{n1} + p_{n2} \leq \varepsilon^{-2} \bigcup_{j=[(c-\varepsilon^3)\beta_n]+2}^{[(c+\varepsilon^3)\beta_n]} \sigma^2(X_{nj}). \tag{7.A.23}$$

Then (7.A.15) follows by way of (7.A.18) and (7.A.23). Inequality (7.A.16) follows from (7.A.15) and the fact that the number of variances involved are

$$\left[\left(c+\varepsilon^3\right)\beta_n\right] - \left[\left(c-\varepsilon^3\right)\beta_n\right] - 1 \leq \left(c+\varepsilon^3\right)\beta_n - \left(c-\varepsilon^3\right)\beta_n = 2\varepsilon^3\beta_n. \square$$

Chapter 8

SOME ASYMPTOTIC RESULTS AND EXPONENTIAL APPROXIMATION IN SEMI-MARKOV MODELS

George G. Roussas*
University of California, USA

Debasis Bhattacharya
University of California, USA

Abstract Consider a continuous-time parameter semi-Markov process $\{X_t, t \geq 0\}$ with the random variables involved defined on the probability space $(\Omega, \mathcal{A}, P_\theta)$ and taking values in the finite set $\{1, 2, \ldots, m\}$; θ is a k-dimensional parameter belonging in an open set Θ the parameter space. For $T > 0$ let $P_{T,\theta}$ be the restriction of P_θ to the σ-field generated by the random variables $\{X_t, 0 \leq t \leq T\}$. Under suitable regularity conditions, the log-likelihood ratio, $\log(dP_{T,\theta^*}/dP_{T,\theta})$, exists for all θ and θ^*, and for parameter points θ^* tending to θ at a specified rate, as $T \to \infty$, the corresponding log-likelihood ratio assumes a certain expansion in the probability sense. The asymptotic distribution of the log-likelihood ratio is also determined, as is the asymptotic distribution of a k-dimensional random vector closely related to the log-likelihood ratio in the sense of probability. These results hold both under a sequence of probability measures depending on a fixed parameter point, as well as under a sequence of probability measures contiguous to the former sequence. Also, it is shown that, in the neighborhood of a parameter point θ, the underlying family of probability measures may be approximated, in the sup-norm sense, by an exponential family of probability measures. These results may be used for carrying out statistical inference with asymptotically optimal properties.

*This research was supported in part by a research grant to the first author by the University of California, Davis.

Keywords: Asymptotic expansion, asymptotic normality, contiguity, semi-Markov process, stopping time.

1. INTRODUCTION, NOTATION, AND SOME AUXILIARY RESULTS

The starting point in many attempts to model a discrete or a continuous time process has been a suitable Markovian process. Markov modeling has the implicit assumption that the duration of stay in a state is geometric in the discrete case, and exponential in the continuous case. That is, the rate of leaving a state is constant, irrespective of how long the process has been there. It has been observed, however, that in many practical situations the probability of transition from one state to another does depend strongly on the length of stay in the former state. It has also been observed that in some cases there is a cumulative inertia effect, whereby the probability of leaving a state decreases as the length of stay increases.

In all such cases, Markovian modeling would not be adequate. On the other hand, a semi-Markov process would adequately describe transitions in situations described above. Consequently, semi-Markovian modeling has become quite popular in a host of diverse areas, ranging from social sciences, especially in manpower planning or mobility studies; to analysis of survival data; to health care planning; to earthquake occurrences; to reliability theory etc. A wealth of relevant information may be found in Janssen (1986). Here, we consider a continuous-time parameter semi-Markov process (S.M.P.) $\{X(t), t \in [0, \infty)\}$, where the random variables (r.v.'s) $X(t), t \geq 0$, are defined on the probability space $(\Omega, \mathcal{A}, P_\theta)$ and take values in $\{1, 2, \ldots, m\}$ $(m < \infty)$, and θ belongs in an open subset Θ of $\Re^k (k \geq 1)$. The process moves from one state to another state, according to a countable state Markov chain, and stays in a given state a random length of time, whose distribution may depend on the present state as well as the next, and need not be exponential. For $T > 0$, let $\mathcal{A}_T$ be the σ-field induced by the r.v.'s $X(t), t \in [0, T]$; that is, $\mathcal{A}_T = \sigma\{X(t), t \in [0, T]\}$, and let $P_{T,\theta}$ be the restriction of P_θ to $\mathcal{A}_T$; $P_{T,\theta} = P_\theta | \mathcal{A}_T$. In the sequel, it will be assumed that $P_{T,\theta} \approx P_{T,\theta^*}$ for any pair θ, θ^* in Θ and all $T \geq 0$, where "$\approx$" denotes mutual absolute continuity of the measures involved. Set $\theta_T = \theta + h_T[c(\theta)T]^{-1/2}$, where $h_T \to h \in \Re^k$, as $T \to \infty$, and $c(\theta)$ is a certain positive quantity defined below (see relation (8.13), and let $\Lambda_T(\theta)$ be the log-likelihood function

$\log(dP_{T,\theta_T}/dP_{T,\theta})$; that is,

$$\Lambda_T(\theta) = \log(dP_{T,\theta_T}/dP_{T,\theta}), \tag{8.1}$$

which is well-defined with $P_{T,\theta}$-probability 1 for all $\theta \in \Theta$.

Our aim is to study the asymptotic behavior of $\Lambda_T(\theta)$, as $T \to \infty$, as well as to obtain an approximation (in the sup-norm sense) of the family of families of probability measures $\{P_{T,\theta+h_T[c(\theta)T]^{-1/2}}; h_T \in \Re^k\}, T \geq 0$, in the neighborhood of θ, by an exponential family of families of probability measures. The results obtained could then be used for making statistical inference about the parameter θ with asymptotically optimal properties. In a recent paper, Greenwood and Wefelmeyer (1996) considered S.M.P. with general state space, and studied empirical estimates of linear functions of certain distributions, as well as other estimates such as the jump frequency, the proportion of time spent in a set etc. All estimation takes place in a non-parametric framework. Results obtained here are also discussed in Akritas and Roussas (1980) except for the exponential approximation stated above. Conditions in the reference just cited are somewhat weakened (for example, assumption (A5) in page 78 is dropped), and also some derivations, such as that of Theorem 5.4 in page 77, are significantly elaborated on. In addition to the basic reference Akritas and Roussas (1980), another two references, Roussas and Bhattacharya (1998a,b), will provide all background results needed presently.

The r.v.'s $X(t), t \geq 0$, may be represented in terms of a 2-dimensional Markov process and a stopping time, as indicated below (see also Pyke (1961)). Then the problem of studying issues related to the initial process becomes to a large extent a problem of studying a randomly stopped Markov process. The notation and assumptions to be used here are drawn from Akritas and Roussas (1980). In this respect, the borrowing is quite extensive in order to make the present paper as self-sufficient as possible. To this end, for each $\theta \in \Theta$ and $i, j = 1, \ldots, m$, let $Q_\theta(i, j, \cdot)$ be a distribution function (d.f.) (not necessarily a proper one) such that :

$$Q_\theta(i,j,x) = 0, \quad x \leq 0, \text{ and } \sum_{j=1}^{m} Q_\theta(i,j,\infty) = 1, \quad i = 1, \ldots, m. \tag{8.2}$$

Consider the transition probabilities

$$p_\theta(i,j) = Q_\theta(i,j,\infty), \tag{8.3}$$

and define the d.f. $F_\theta(i,j,\cdot)$ by:

$$F_\theta(i,j,\cdot) = \frac{Q_\theta(i,j,\cdot)}{p_\theta(i,j)}, \; p_\theta(i,j) > 0, \text{ and } F_\theta(i,j,\cdot) = \text{ arbitrary, otherwise.} \tag{8.4}$$

Then $F_\theta(i,j,\cdot)$ is a proper probability distribution, and $F_\theta(i,j,x)$ represents the probability that the next transition is to state j, within x time units, after entering state i; x is the sojourn time. Next, let $\{(Z_n, R_n), n \geq 0\}$ be a 2-dimensional Markov process defined on $(\Omega, \mathcal{A}, P_\theta)$ such that, for each $\theta \in \Theta$ and $j = 1, \ldots, m$, the (initial) distribution of Z_0, the joint (initial) distribution of (Z_0, R_0), and the (stationary) transition distribution of (Z_n, R_n) are as follows:

$$P_\theta(Z_0 = j) = p_\theta(j) \quad (p_\theta(j) \geq 0, \sum_{j=1}^{m} p_\theta(j) = 1), \tag{8.5}$$

$$P_\theta(Z_0 = i, R_0 \leq x) = \pi_\theta(i,x), \quad x \geq 0, \tag{8.6}$$

$$P_\theta\left[Z_n = j, R_n \leq x | (Z_m, R_m), m = 0, \ldots, n-1\right] = F_\theta(Z_{n-1}, j, x), \; x \in \Re. \tag{8.7}$$

Also, in terms of the process $\{R_n, n \geq 0\}$, define the r.v.'s $N(t), t \geq 0$, by:

$$N(t) = \inf\left\{s \geq 1; R_1 + \ldots + R_s > t\right\}, \quad t \geq 0. \tag{8.8}$$

Then $N(t), t \geq 0$, are random times taking values in $\{1, 2, \ldots\}$, since $P_\theta[N(t) < \infty$, for all $t \geq 0] = 1$ (see Pyke (1961), page 1236). Then $\{X(t), t \geq 0\}$ may be represented as follows :

$$X(t) = Z_{N(t)-1} \text{ a.s. } [P_\theta], t \geq 0, \theta \in \Theta. \tag{8.9}$$

The intuitive interpretation of $X(t)$, in terms of Z_n, R_n and $N(t)$, is this: Z_n denotes the state occupied by the S.M.P. after the n-th transition, and R_n denotes the time the process has spent at state Z_{n-1}. Finally, $N(t)$ denotes the number of transitions which have occurred up to time t. It would be convenient if $F_\theta(i,j,\cdot)$, defined in (8.4), were independent of j. This, actually, can be assumed without loss of generality (see Pyke and Schaufele (1964), page 1751). Thus, from this point on,

$$F_\theta(i,j,\cdot) = F_\theta(i,\cdot) \text{ for all } j, \tag{8.10}$$

and therefore

$$P_\theta\left[Z_n = j, R_n \le x | (Z_m, R_m), m = 0, \dots, n-1\right] = F_\theta(Z_{n-1,}x), \quad x \in \Re. \tag{8.11}$$

It is assumed throughout that, for each $\theta \in \Theta$, there is only one positive recurrent class C which is independent of θ. It follows (see Theorem 5.1 in Pyke (1961)) that

$$\eta_\theta(i) = \int_0^\infty x dF_\theta(i, x) < \infty, \quad i \in C, \quad \theta \in \Theta. \tag{8.12}$$

Also, as a consequence of (8.12), there are no absorbing states ($p_\theta(i,i) < 1, i = 1, \dots, m, \theta \in \Theta$). Then the position may be adopted that transition from a state to itself plays no significant role, and therefore, without loss of generality (see Pyke (1961)), it is assumed that $p_\theta(i,i) = 0$, $i = 1, \dots, m$, $\theta \in \Theta$. Finally, the process $\{(Z_n, R_n), n \ge 0)\}$ is (strictly) stationary and ergodic, and, with $p_\theta(i)$ and $\eta_\theta(i)$ as in (8.5) and (8.12), respectively, we have

$$\frac{N(t)}{t} \to c(\theta) \text{ a.s. } [P_\theta], \text{ where } 0 < c(\theta) = \left[\sum_{i=1}^m p_\theta(i)\eta_\theta(i)\right]^{-1} < \infty. \tag{8.13}$$

This section is concluded with the recording of two auxiliary results established in Akritas and Roussas (1980) (see Theorems 1.3.1 and 1.3.2).

Proposition 1 *Let $Q_\theta(i,j,\cdot)$ and $\pi_\theta(i,\cdot)$ be as in (8.2) and (8.6), respectively, and use the same notation for the induced measures. It is assumed that, for each θ, θ^* in Θ and for all $i, j = 1, \dots, m$, it holds $Q_{\theta^*}(i,j,\cdot) \approx Q_\theta(i,j,\cdot)$ and $\pi_{\theta^*}(i,\cdot) \approx \pi_\theta(i,\cdot)$. Then:*

(i) For all $T > 0$, $P_{T,\theta^*} \approx P_{T,\theta}$, and indeed,

(ii)

$$\begin{aligned}\frac{dP_{T,\theta^*}}{dP_{T,\theta}} &= \frac{d\pi_{\theta^*}(Z_0,R_0)}{d\pi_\theta(Z_0,R_0)} \cdot \prod_{j=1}^{K} \frac{p_{\theta^*}(Z_j,Z_{j+1})}{p_\theta(Z_j,Z_{j+1})} \cdot \frac{dF_{\theta^*}(Z_j,R_{j+1})}{dF_\theta(Z_j,R_{j+1})} \\ &\cdot \frac{1 - F_{\theta^*}(Z_{N(t)-1}, U_T)}{1 - F_\theta(Z_{N(t)-1}, U_T)},\end{aligned} \tag{8.14}$$

where $U_T = T - R_1 - \dots - R_{N(T)}$, $K = N(T) - 2$, and $p_\theta(i,j)$ is given in (8.3).

Remark 1 *The assumption $Q_{\theta^*}(i,j,\cdot) \approx Q_\theta(i,j,\cdot)$ implies $F_{\theta^*}(i,j,\cdot) \approx F_\theta(i,j,\cdot)$, by means of (8.4), where $F_\theta(i,j,\cdot)$ also stands for the measure induced by the d.f.. Furthermore, recall that $F_\theta(i,j,\cdot) = F_\theta(i,\cdot)$ for all j, by means of (8.10).*

2. ASSUMPTIONS AND FORMULATION OF MAIN RESULTS

The assumptions stated below are taken from Akritas and Roussas (1980). These assumptions ensure the validity of the main results in this paper, Theorems 1-3. The first two of these theorems are also discussed to a certain extent in the reference just cited. Here, one of the conditions used there is dropped, and detailed background information is provided for the validity of all three theorems. Theorem 3 does not appear in Akritas and Roussas (1980), and the justification of Theorems 1 and 2 is rather sketchy there.

Assumptions.

(A1) Let $Q_\theta(i,j,\cdot)$ and $\pi_\theta(i,\cdot)$ be the measures induced by $Q_\theta(i,j,\cdot)$ and $\pi_\theta(i,\cdot)$ as they are given in (8.2) and (8.6). Then, for $i,j = 1,\ldots,m$ and any θ,θ^* in Θ, $Q_{\theta^*}(i,j,\cdot) \approx Q_\theta(i,j,\cdot)$ and $\pi_{\theta^*}(i,\cdot) \approx \pi_\theta(i,\cdot)$. Assumption (A1) allows the formation of the likelihood as given in (8.1). In reference to it, set:

$$\varphi_j(\theta,\theta^*) = \varphi(Z_j, Z_{j+1}; \theta,\theta^*) = [p_{\theta^*}(Z_j,Z_{j+1})/p_\theta(Z_j,Z_{j+1})]^{1/2}, \tag{8.15}$$

$$\psi_j(\theta,\theta^*) = \psi(Z_j, R_{j+1}; \theta,\theta^*) = [dF_{\theta^*}(Z_j,R_{j+1})/dF_\theta(Z_j,R_{j+1})]^{1/2}. \tag{8.16}$$

Regarding the random functions $\varphi_j(\theta,\theta^*)$ and $\psi_j(\theta,\theta^*)$, the following assumption is made.

(A2)(i) For each $\theta \in \Theta$, the random functions $\varphi_j(\theta,\theta^*)$ and $\psi_j(\theta,\theta^*)$ are differentiable in quadratic mean (q.m.) with respect to θ^* at θ when the probability measure P_θ is employed.

Let $\dot\varphi_j(\theta)$ and $\dot\psi_j(\theta)$ be the q.m. derivatives of $\varphi_j(\theta,\theta^*)$ and $\psi_j(\theta,\theta^*)$, respectively. Then:

(ii) $\dot\varphi_j(\theta)$ and $\dot\psi_j(\theta)$ are $\mathcal{F}_{j+1} \times \mathcal{C}$-measurable, where

$$\mathcal{F}_j = \sigma(Z_0,\ldots,Z_j,R_1,\ldots,R_{j+1}), \tag{8.17}$$

and $\mathcal{C}$ is the σ-field of Borel subsets of Θ.

Define $\Gamma_i(\theta),\ i = 1,2$ by

$$\Gamma_1(\theta) = 4\mathcal{E}_\theta\left[\dot{\varphi}_j(\theta)\dot{\varphi}_j'(\theta)\right], \quad \Gamma_2(\theta) = 4\mathcal{E}_\theta\left[\dot{\psi}_j(\theta)\dot{\psi}_j'(\theta)\right], \tag{8.18}$$

and set

$$\Gamma\theta) = \Gamma_1(\theta) + \Gamma_2(\theta) \quad \text{(see also relation (8.32))}. \tag{8.19}$$

Then :

(iii) The covariance matrices $\Gamma_i(\theta),\ i = 1,2$ are positive definite for each $\theta \in \Theta$.

(iv) For every $t \in \Re^k$, $\frac{1}{n}\sum_{j=0}^{n} \mathcal{E}_\theta\left\{[t'\dot{\varphi}_j(\theta)]^2 \,|\mathcal{F}_j\right\} \to \Gamma_1(\theta)$ in P_θ-probability, and

$$\frac{1}{n}\sum_{j=0}^{n} \mathcal{E}_\theta\left\{\left[t'\dot{\psi}_j(\theta)\right]^2 |\mathcal{F}_{j-1}\right\} \to \Gamma_2(\theta) \text{ in } P_\theta - \text{probability.}$$

Let $\pi_\theta(i,x)$, $F_\theta(i,x)$, $\varphi(Z_0, Z_1; \theta, \theta^*)$ and $\psi(Z_0, R_1; \theta, \theta^*)$ be as in (8.6), (8.10) (in conjunction with Remark 1), (8.15), and (8.16), respectively. Then :

(A3) For each $\theta \in \Theta$:

(i) $\frac{d\pi_{\theta^*}(Z_0,R_0)}{d\pi_\theta(Z_0,R_0)} \to 1$ in P_θ-probability, as $\theta^* \to \theta$.

(ii) $\varphi(Z_0, Z_1; \theta, \theta^*)$ is $\mathcal{F}_1 \times \mathcal{C}$-measurable and $\psi(Z_0, R_1; \theta, \theta^*)$ is $\mathcal{F}_0 \times \mathcal{C}$-measurable, where $\mathcal{F}_j$ is given in (8.17).

(iii) For each $i = 1, \ldots, m$ and uniformly in $x \geq 0$, $F_{\theta^*}(i,x) - F_\theta(i,x) \to 0$, as $\theta^* \to \theta$.

Remark 2 *Assumption (A3)(iii) implies that, for each $i = 1, \ldots, m$ and uniformly in $x \geq 0$, $[1 - F_{\theta^*}(i,x)]/[1 - F_\theta(i,x)] \to 1$, as $\theta^* \to \theta$. This is so because, for $F_\theta(i,x) \neq 1$, and as $\theta^* \to \theta$:*

$$\frac{1 - F_{\theta^*}(i,x)}{1 - F_\theta(i,x)} \to 1 \textit{ if and only if } \frac{1 - F_{\theta^*}(i,x)}{1 - F_\theta(i,x)} - 1 \to 0,$$

$$\textit{or} \quad \frac{F_{\theta^*}(i,x) - F_\theta(i,x)}{F_\theta(i,x) - 1} \to 0, \qquad \textit{or} \quad F_{\theta^*}(i,x) - F_\theta(i,x) \to 0.$$

With $c(\theta)$ given in (8.13), set

$$\theta_T = \theta + h_T \left[c(\theta)T\right]^{-1/2}, \tag{8.20}$$

and denote by $\Lambda_{N(T)}(\theta)$ the logarithm of the likelihood function $dP_{T,\theta_T}/dP_{T,\theta}$, where the index $N(T)$ rather than T is used in order to emphasize its dependence on the random time $N(T)$. That is,

$$\Lambda_{N(T)}(\theta) = \log(dP_{T,\theta_T}/dP_{T,\theta}). \tag{8.21}$$

Then, on the basis of (8.1) and by means of (8.15) and (8.16), $\Lambda_{N(T)}(\theta)$ may be rewritten as follows:

$$\Lambda_{N(T)}(\theta) = \log \frac{d\pi_{\theta_T}(Z_0, R_0)}{d\pi_\theta(Z_0, R_0)} + \log \frac{1 - F_{\theta_T}(Z_{N(T)-1}, U_T)}{1 - F_\theta(Z_{N(T)-1}, U_T)} + \Lambda_{1N(T)}(\theta) + \Lambda_{2N(T)}(\theta),$$

where

$$\Lambda_{1N(T)}(\theta) = 2 \sum_{j=0}^{N(T)-2} \log \varphi_j(\theta, \theta_T), \quad \Lambda_{2N(T)}(\theta) = 2 \sum_{j=0}^{N(T)-2} \log \psi_j(\theta, \theta_T), \tag{8.22}$$

and U_T is defined right after (8.14).

Regarding the asymptotic behavior of $\Lambda_{N(T)}(\theta)$, all one has to consider are $\Lambda_{iN(T)}(\theta)$, $i = 1, 2$, on account of assumption (A3)(i),(iii). Finally, by means of the q.m. derivatives $\dot{\varphi}_j(\theta)$ and $\dot{\psi}_j(\theta)$, define the k-dimensional random vector $\Delta_{N(T)}(\theta)$ by:

$$\Delta_{N(T)}(\theta) = \Delta_{1N(T)}(\theta) + \Delta_{2N(T)}(\theta), \tag{8.23}$$

where

$$\Delta_{1N(T)}(\theta) = 2[N(T)]^{-1/2} \sum_{j=0}^{N(T)-2} \dot{\varphi}_j(\theta),$$

$$\Delta_{2N(T)}(\theta) = 2[N(T)]^{-1/2} \sum_{j=0}^{N(T)-2} \dot{\psi}_j(\theta). \tag{8.24}$$

Remark 3 *In relations (8.22)-(8.23), the index $N(T)$ rather than T was used for the reason explained above; namely, in order to indicate dependence on a random number of random vectors. At this point, let*

$$\tilde{\mathcal{F}}_{N(T)} = \sigma\left(Z_0, \ldots, Z_{N(T)}, R_1, \ldots, R_{N(T)+1}\right), \quad \tilde{P}_{T,\theta} = P_\theta | \tilde{\mathcal{F}}_{N(T)}. \tag{8.25}$$

Then it is seen that $\Lambda_{N(T)}(\theta)$ and $\Delta_{N(T)}(\theta)$ are $\tilde{\mathcal{F}}_{N(T)}$-measurable (for details, see, for example, Proposition 3.1 in Roussas and Bhattacharya (1998a)). The first two main results are the following :

Theorem 1 *Let $\Gamma(\theta)$, $\Lambda_{N(T)}(\theta)$ and $\Delta_{N(T)}(\theta)$ be defined by (8.19), (8.21)-(8.22), and (8.23)-(8.24), respectively, and in (8.20), suppose that $h_T \to h \in \Re^k$, as $T \to \infty$. Then, for each $\theta \in \Theta$, and as $T \to \infty$, the following hold, under assumptions (A1)-(A3), and in the notation introduced in (8.25):*

(i) $\Lambda_{N(T)}(\theta) - h'_T\Delta_{N(T)}(\theta) \to -\frac{1}{2}h'\Gamma(\theta)h$ *in $\tilde{P}_{T,\theta}$-probability,*

(ii) $\mathcal{L}\left[\Delta_{N(T)}(\theta)|\tilde{P}_{T,\theta}\right] \Longrightarrow N(0,\Gamma(\theta))$,

(iii) $\mathcal{L}\left[\Lambda_{N(T)}(\theta)|\tilde{P}_{T,\theta}\right] \Longrightarrow N(-\frac{1}{2}h'\Gamma(\theta)h, h'\Gamma(\theta)h)$.

Theorem 2 *With the same notation as that in Theorem 1, and under assumptions (A1)-(A3), it holds :*

(i) $\Lambda_{N(T)}(\theta) - h'_T\Delta_{N(T)}(\theta) \to -\frac{1}{2}h'\Gamma(\theta)h$ *in $\tilde{P}_{T,\theta_T}$-probability,*

(ii) $\mathcal{L}\left[\Delta_{N(T)}(\theta)|\tilde{P}_{T,\theta_T}\right] \Longrightarrow N(\Gamma(\theta)h,\Gamma(\theta))$,

(iii) $\mathcal{L}\left[\Lambda_{N(T)}(\theta)|\tilde{P}_{T,\theta_T}\right] \Longrightarrow N(\frac{1}{2}h'\Gamma(\theta)h, h'\Gamma(\theta)h)$.

For the formulation of the third result, Theorem 3, some additional notation is needed. To this effect, an arbitrary parameter point θ is fixed hereafter and is denoted by θ_0. Let $\Delta^*_{N(T)}(\theta)$ be a certain truncated version of $\Delta_{N(T)}(\theta)$ (see relation (3.4) in Roussas and Bhattacharya (1998b)) such that :

$$\int_\Omega \exp\left[h'\Delta^*_{N(T)}(\theta)\right] d\tilde{P}_{T,\theta_0} = \exp\left[\tilde{B}_T(h)\right] < \infty, \tag{8.26}$$

and define the probability measure $\tilde{R}_{T,h}$ as follows:

$$\tilde{R}_{T,h}(A) = \exp\left[-\tilde{B}_T(h)\right] \int_A \exp\left[h'\Delta^*_{N(T)}(\theta_0)\right] d\tilde{P}_{T,\theta_0},\ A \in \tilde{\mathcal{F}}_{N(T)}. \tag{8.27}$$

Then

$$\frac{d\tilde{R}_{T,h}}{d\tilde{P}_{T,\theta_0}} = \exp\left[-\tilde{B}_T(h)\right] \exp\left[h'\Delta^*_{N(T)}(\theta_0)\right], \tag{8.28}$$

and this probability density function is exponential in $h \in \Re^k$. Then the result referred to above is as follows.

Theorem 3 *Let θ_0 be an arbitrarily chosen and fixed point in Θ, and consider the families of families of probability measures $\left\{\tilde{P}_{T,\theta_0+h[c(\theta_0)T]^{-1/2}};\ h \in \Re^k\right\}$, $T \geq 0$. Also, consider the families of exponential families of probability measures $\left\{\tilde{R}_{T,h};\ h \in \Re^k\right\}$, $T \geq 0$, defined by (8.26) and (8.27) in terms of a suitable truncated version $\Delta^*_{N(T)}$ of $\Delta_{N(T)}$ (see Remark 4 below). Then, under assumptions (A1)-(A3),*

$$\sup\left\{\|\tilde{R}_{T,h} - \tilde{P}_{T,\theta_0+h[c(\theta_0)T]^{-1/2}}\|;\ h \in B,\ \text{any bounded subset of } \Re^k\right\} \to 0. \tag{8.29}$$

The complete justification of Theorems 1 and 2 depends heavily on results in Roussas and Bhattacharya (1998a). The third result, Theorem 3, is justified on the basis of material discussed in Roussas and Bhattacharya (1998b). The relevant proofs are given in Section 4, after some background information has been presented in Section 3.

Remark 4 *As already mentioned, a suitable truncated version of $\Delta_{N(T)}(\theta)$, in the framework of assumptions (B1)-(B5) below, is given in relation (3.4) of Roussas and Bhattacharya (1998b). The ideas involved are roughly as follows : Let $\{M_r\}$ be a certain sequence of positive integers increasingly tending to ∞, and let $\{\alpha_r\}$ be a sequence of positive numbers non-decreasingly tending to ∞. The truncated version $\Delta^{\alpha_r}_{N(T)}(\theta)$ of $\Delta_{N(T)}(\theta)$ is defined by :*

$$\Delta^{\alpha_r}_{N(T)}(\theta) = \begin{cases} \Delta_{N(T)}(\theta) & \text{if } \|\Delta_{N(T)}(\theta)\| \leq \alpha_r \\ 0 & \text{otherwise} \end{cases} \quad (r = 1, 2, \ldots).$$

Then, $\Delta^*_{N(T_n)}(\theta)$ is defined thus:

$$\Delta^*_{N(T_n)}(\theta) = \begin{cases} \Delta^{\alpha_r}_{N(T_n)}(\theta) & \text{if } M_r \leq T_n < M_{r+1} \\ 0 & \text{otherwise} \end{cases} \quad (r = 1, 2, \ldots).$$

(N_r in Roussas and Bhattacharya (1998b) is replaced by M_r here to avoid possible confusion).

3. SOME AUXILIARY RESULTS ON RANDOM TIMES

In Section 1, it was seen that the S.M.P. $\{X(t),\, t \geq 0\}$ may be looked upon as a (strictly) stationary ergodic Markov chain indexed by a random stopping time (see relation (8.9)). Furthermore, the study of the log-likelihood ratio based on the S.M.P. is, essentially, the summation of two log-likelihood ratios based on two randomly stopped Markov processes (see relations (8.20)-(8.22)). These observations suggest that results obtained in Roussas and Bhattacharya (1998a,b) should be useful in the present context, provided, of course, the relevant assumptions are fulfilled. For the sake of comparison, the set-up and the assumptions used in Roussas and Bhattacharya (1998a,b) are recorded below.

Set-up. For each θ in an open set Θ in $\Re^k$, $k \geq 1$, $\{Y_n, n = 0, 1, \ldots\}$ is a stochastic process with the Y_n's defined on the probability space $(\Omega, \mathcal{A}, P_\theta)$ and taking values in $\Re^d$, $d \geq 1$. $\mathcal{A}_n$ is the σ-field induced by $Y_0, \ldots, Y_n$, and $P_{n,\theta}$ is the restriction of P_θ to $\mathcal{A}_n$.

3.1 ASSUMPTIONS

(B1) For each $n \geq 0$ integer, $P_{n,\theta^*} \approx P_{n,\theta}$, for all $\theta, \theta^* \in \Theta$. Set

$$\begin{aligned} q_j\left(\mathbf{Y}_j; \theta, \theta^*\right) &= dP_{j,\theta^*}/dP_{j,\theta}, \\ q_j(Y_j|\mathbf{Y}_{j-1}; \theta, \theta^*) &= q_j(\mathbf{Y}_j; \theta, \theta^*)/q_{j-1}(\mathbf{Y}_{j-1}; \theta, \theta^*), \quad \text{and} \\ \lambda_j(\theta, \theta^*) &= \lambda_j(Y_j|\mathbf{Y}_{j-1}; \theta, \theta^*) = [q_j(Y_j|\mathbf{Y}_{j-1}; \theta, \theta^*)]^{1/2}. \end{aligned}$$

Then :

(B2)(i) For each $\theta \in \Theta$, the random functions $\lambda_j(\theta; \cdot)$ are differentiable in q.m. $[P_\theta]$ uniformly in $j \geq 1$, and let $\dot{\lambda}_j(\theta)$ be the q.m. derivatives involved.

(ii) For $j \geq 1$, $\dot{\lambda}_j(\theta)$ is $\mathcal{A}_j \times \mathcal{C}$-measurable.

(B3)(i) For each $\theta \in \Theta$ and each $t \in \Re^k$, $\left\{t'\dot{\lambda}_j(\theta)\right\}^2$, $j \geq 1$, are uniformly integrable with respect to P_θ. For each $\theta \in \Theta$ and $n \geq 1$, let the $k \times k$ covariance matrix $\bar{\Gamma}_n(\theta)$ be defined by :

$$\bar{\Gamma}_n(\theta) = \frac{1}{n}\sum_{j=1}^{n} \Gamma_j(\theta), \text{ where } \Gamma_j(\theta) = 4\mathcal{E}_\theta\left[\dot{\lambda}_j(\theta)\dot{\lambda}_j'(\theta)\right]. \text{ Then :}$$

(ii) $\bar{\Gamma}_n(\theta) \to \bar{\Gamma}(\theta)$ (in any one of the standard norms) and $\bar{\Gamma}(\theta)$ is positive definite for all $\theta \in \Theta$.

(iii) For each $\theta \in \Theta$ and each $t \in \Re^k$, the WLLN's holds under P_θ for the sequence of r.v.'s $\left\{t'\dot{\lambda}_j(\theta)\right\}^2$, $j \geq 1$; or $\frac{1}{n}\sum_{j=1}^{n}\left[t'\dot{\lambda}_j(\theta)\right]^2 \to \frac{1}{4}t'$ $\bar{\Gamma}(\theta)t$ in P_θ-probability, by means of assumption (B3)(ii).

(iv) For each $\theta \in \Theta$ and each $t \in \Re^k$,

$$\frac{1}{n}\sum_{j=1}^{n}[\mathcal{E}_\theta\{\left[t'\dot{\lambda}_j(\theta)\right]^2|\mathcal{A}_{j-1}\} - \left[t'\dot{\lambda}_j(\theta)\right]^2] \to 0 \text{ in } P_\theta\text{-probability; or}$$

$$\frac{1}{n}\sum_{j=1}^{n}\mathcal{E}_\theta\{\left[t'\dot{\lambda}_j(\theta)\right]^2|\mathcal{A}_{j-1}\} \to \frac{1}{4}t'\bar{\Gamma}(\theta)t \text{ in } P_\theta\text{-probability,}$$

on account of assumption (B3)(iii).

(B4) For each $\theta \in \Theta$, $q_0(Y_0; \theta, \theta^*)$ is $\mathcal{A}_0 \times \mathcal{C}$-measurable and continuous in P_θ-probability, as $\theta^* \to \theta$.

(B5) Let ν_n be stopping times defined on $\{Y_n, n \geq 0\}$, and let $0 < \tau_n$ be reals $\uparrow \infty$. Then, for each $\theta \in \Theta$, $\frac{\nu_n}{\tau_n} \to 1$ in P_θ-probability.

Let $L_n(\theta;\theta^*) = dP_{n,\theta^*}/dP_{n,\theta}$ and set $\mathcal{A}_{\nu_n} = \sigma(Y_0, \ldots, Y_{\nu_n})$. Also, denote by $\tilde{P}_{n,\theta} = P_\theta|\mathcal{A}_{\nu_n}$. Then, it is seen that $\tilde{P}_{n,\theta^*} \approx \tilde{P}_{n,\theta}$ for any two θ, θ^* in Θ, and that $d\tilde{P}_{n,\theta^*}/d\tilde{P}_{n,\theta} = L_{\nu_n}(\theta, \theta^*)$, what one obtains from $L_n(\theta, \theta^*)$ by replacing n by ν_n. Furthermore, $\Lambda_{\nu_n}(\theta, \theta^*) = \log L_{\nu_n}(\theta, \theta^*)$ and $\Delta_{\nu_n}(\theta)$ defined below, are $\mathcal{A}_{\nu_n}$-measurable; here

$$\Delta_{\nu_n}(\theta) = 2\nu_n^{-1/2}(\theta)\sum_{j=1}^{\nu_n}\dot{\lambda}_j(\theta)$$

(see Proposition 3.1 in Roussas and Bhattacharya (1998a)).

Next, set $\theta_{\tau_n} = \theta + h_n\tau_n^{-1/2}$, $h_n \to h \in \Re^k$, and denote $\Lambda_{\nu_n}(\theta, \theta_{\tau_n})$ by $\Lambda_{\nu_n}(\theta)$. Then with the notation introduced, one has

Proposition 2 *Suppose (B1)-(B5) hold. Then, for every $\theta \in \Theta$ and any $h_n \to h \in \Re^k$, one has :*

(i) $\Lambda_{\nu_n}(\theta) - h_n'\Delta_{\nu_n}(\theta) \to -\frac{1}{2}h'\bar{\Gamma}(\theta)h$ *in* $\tilde{P}_{n,\theta}$*-probability,*

(ii) $\mathcal{L}[\Delta_{\nu_n}(\theta)|\tilde{P}_{n,\theta}] \Rightarrow N(0, \bar{\Gamma}(\theta))$,

(iii) $\mathcal{L}[\Lambda_{\nu_n}(\theta)|\tilde{P}_{n,\theta}] \Rightarrow N(-\frac{1}{2}h'\bar{\Gamma}(\theta)h, h'\bar{\Gamma}(\theta)h)$.

Remark 5 *It should be mentioned at this point that the justification of the convergences in (i)-(iii) is rather involved. The fact that the number of r.v.'s (or random vectors) involved is random necessitates the checking of a condition given in Anscombe (1952); see also the introductory part in the Appendix in Roussas and Bhattacharya (1998a)). Furthermore,* *in the process of the proofs, one utilizes sums of α_n r.v.'s, where α_n is the largest integer $\leq \tau_n$.*

It is also shown that

Lemma 1 *Under (B1)-(B5), the sequences of probability measures $\left\{\tilde{P}_{n,\theta}\right\}$ and $\left\{\tilde{P}_{n,\theta_{\tau_n}}\right\}$ are contiguous, where now $\theta_{\tau_n} = \theta + h_n \tau_n^{-1/2}$ and $\{h_n\}$ is bounded.*

A combination of Proposition 2 and Lemma 1 also yields

Proposition 3 *Under the set-up in Proposition 2, one has :*

(i) $\Lambda_{\nu_n}(\theta) - h'_n \Delta_{\nu_n}(\theta) \to -\frac{1}{2} h' \bar{\Gamma}(\theta) h$ *in* $\tilde{P}_{n,\theta_{\tau_n}}$*-probability,*

(ii) $\mathcal{L}[\Delta_{\nu_n}(\theta) | \tilde{P}_{n,\theta_{\tau_n}}] \Rightarrow N(\bar{\Gamma}(\theta)h, \bar{\Gamma}(\theta))$,

(iii) $\mathcal{L}[\Lambda_{\nu_n}(\theta) | \tilde{P}_{n,\theta_{\tau_n}}] \Rightarrow N(\frac{1}{2} h' \bar{\Gamma}(\theta) h, h' \bar{\Gamma}(\theta) h)$.

The proof of these results is found in Roussas and Bhattacharya (1998a).

Remark 6 *Once again, the justification of the convergences in (i)-(iii) necessitates the repetition of the procedures described in Remark 5 under the probability measures $\tilde{P}_{n,\theta_{\tau_n}}$ employed here.*

The local asymptotic behavior of the log-likelihood function, when based on stopping times, has been studied recently by Luschgy (1995) in the framework of continuous time semimartingales.

4. PROOF OF MAIN RESULTS AND SOME CLOSING COMMENTS

We may now proceed with the proof of Theorems 1 and 2. To this effect, the following statement must be first verified; namely,

Statement 1 *Assumptions (A1)-(A3) (to be referred to as the (A)-assumptions) imply assumptions (B1)-(B5) (to be referred to as the (B)-assumptions).*

Justification of Statement 1. In view of assumptions (A3)(i),(iii), it suffices to focus on $\Lambda_{iN(T)}$, $i = 1, 2$, only. In the proofs of the theorems, the limits are actually taken through monotone sequences $(0 <)\ T_n \to \infty$, as $n \to \infty$. So, in the justifications to follow, we restrict ourselves to such sequences. By (8.13), $N(T_n)/c(\theta)T_n \to 1$ a.s. $[P_\theta]$, so that assumption (B5) is satisfied with $\nu_n = N(T_n)$ and $\tau_n = c(\theta)T_n$. Thus, it suffices to show that the (A)-assumptions imply the remaining (B)-assumptions with $\lambda_j(\theta)$ replaced by $\varphi_j(\theta) + \psi_j(\theta)$, $\dot{\lambda}(\theta)$ by $\dot{\varphi}_j(\theta) + \dot{\psi}_j(\theta)$, and $\mathcal{A}_j$ by $\mathcal{F}_{j+1}$. In the first place, assumption (A1) implies $P_{T_{n,\theta^*}} \approx P_{T_{n,\theta}}$ on account of Proposition 1 (i), which is what assumption (B1) states. Assumption (B2)(i) is implied by assumption (A2)(i) through the c_2-inequality, since

$$\mathcal{E}_\theta\left\{\frac{1}{\lambda}[\lambda_j(\theta,\theta+\lambda h)-1]-\lambda h'\dot{\lambda}_j(\theta)\right\}^2$$

$$\leq 2\mathcal{E}_\theta\left\{\frac{1}{\lambda}[\varphi_j(\theta,\theta+\lambda h)-1]-\lambda h'\dot{\varphi}_j(\theta)\right\}^2$$

$$+2\mathcal{E}_\theta\left\{\frac{1}{\lambda}[\psi_j(\theta,\theta+\lambda h)-1]-\lambda h'\dot{\psi}_j(\theta)\right\}^2 \to 0 \text{ as } \lambda \to 0$$

uniformly on bounded sets of $h \in \Re^k$.
Assumption (B2)(ii) follows from assumption (A2)(ii). With regards to assumption (B3), first observe that

$$\mathcal{E}_\theta[t'\dot{\lambda}_j(\theta)]^2 = \mathcal{E}_\theta[t'\dot{\varphi}_j(\theta)+t'\dot{\psi}_j(\theta)]^2 = \mathcal{E}_\theta[t'\dot{\varphi}_j(\theta)]^2+\mathcal{E}_\theta[t'\dot{\psi}_j(\theta)]^2,$$

since

$$\mathcal{E}_\theta\{[t'\dot{\varphi}_j(\theta)][t'\dot{\psi}_j(\theta)]\} = \mathcal{E}_\theta\left[\mathcal{E}_\theta\{[t'\dot{\varphi}_j(\theta)][t'\dot{\psi}_j(\theta)]|\mathcal{F}_j\}\right]$$

$$= \mathcal{E}_\theta\left\{t'\dot{\psi}_j(\theta)\mathcal{E}_\theta[t'\dot{\varphi}_j(\theta)|\mathcal{F}_j]\right\} \text{ (a.s. } [P_\theta] \text{ as } t'\dot{\psi}_j(\theta) \text{ is } \mathcal{F}_j\text{-measurable)}$$

$$= 0, \quad \text{since} \quad \mathcal{E}_\theta[t'\dot{\varphi}_j(\theta)|\mathcal{F}_j] = 0 \quad \text{a.s. } [P_\theta]; \tag{8.30}$$

this is seen as in Lemma 5.5(i), page 60, in Roussas (1972). Thus, for every $t \in \Re^k$,

$$\mathcal{E}_\theta[t'\dot{\lambda}_j(\theta)]^2 = \mathcal{E}_\theta[t'\dot{\varphi}_j(\theta)]^2+\mathcal{E}_\theta[t'\dot{\psi}_j(\theta)]^2, \tag{8.31}$$

which, by means of (8.18), implies:

$$\Gamma(\theta) = \mathcal{E}_\theta[t'\dot{\lambda}_j(\theta)]^2 = \Gamma_1(\theta)+\Gamma_2(\theta). \tag{8.32}$$

Now, assumption (B3)(i) follows by stationarity; (B3)(ii) is implied again by stationarity and assumption (A2)(iii). Assumption (B3)(iii) follows by stationarity, and relations (8.18), (8.31) and (8.32). On account of (8.30),

$$\mathcal{E}_\theta\{[t'\dot{\lambda}_j(\theta)]^2|\mathcal{F}_j\} = \mathcal{E}_\theta\{[t'\dot{\varphi}_j(\theta)]^2|\mathcal{F}_j\}+[t'\dot{\psi}_j(\theta)]^2 \quad \text{a.s. } [P_\theta],$$

so that assumption (B3)(iv) is implied by assumption (A2)(iv). Finally, assumption (B4) is implied by assumption (A3)(ii). □

Remark 7 *From the justification of Statement 1 it is clear that the (B)-assumptions holds separately for each $\Lambda_{iN(T_n)}$, $i = 1, 2$, and all quantities associated with them.*

In the following proofs, all limits are taken as $n \to \infty$.

Proof of Theorem 1. (i) By Remark 7, the (A)-assumptions imply the (B)-assumptions as they apply to each $\Lambda_{iN(T_n)}$, $i = 1, 2$. Then, by Proposition 2(i), and in the notation introduced in (8.25),

$$\Lambda_{iN(T_n)}(\theta) - h'_{T_n}\Delta_{iN(T_n)}(\theta) \to -\frac{1}{2}h'\Gamma_i(\theta)h \text{ in } \tilde{P}_{T_n,\theta}\text{-probability, } i = 1, 2. \tag{8.33}$$

In (8.33), let $i = 1, 2$, add-up the expressions, and use relations (8.18)-(8.19), (8.21)-(8.22), and (8.23)-(8.24) to obtain:

$$\Lambda_{N(T_n)}(\theta) - h'_{T_n}\Delta_{N(T_n)}(\theta) \to -\frac{1}{2}h'\Gamma(\theta)h \text{ in } \tilde{P}_{T_n,\theta}\text{-probability.} \tag{8.34}$$

From the definition of $\Delta_{N(T_n)}(\theta)$ by (8.23)-(8.24), we have

$$\Delta_{N(T_n)}(\theta) = 2[N(T_n)]^{-\frac{1}{2}} \sum_{j=0}^{N(T_n)-2} [\dot{\varphi}_j(\theta) + \dot{\psi}_j(\theta)], \tag{8.35}$$

and, by (8.30),

$$\begin{aligned}\Gamma(\theta) &= \mathcal{E}_\theta\left\{[\dot{\varphi}_j(\theta) + \dot{\psi}_j(\theta)][\dot{\varphi}_j(\theta) + \dot{\psi}_j(\theta)]'\right\} = \mathcal{E}_\theta[\dot{\varphi}_j(\theta)\dot{\varphi}'_j(\theta)] \\ &\quad + \mathcal{E}_\theta[\dot{\psi}_j(\theta)\dot{\psi}'_j(\theta)] = \Gamma_1(\theta) + \Gamma_2(\theta).\end{aligned} \tag{8.36}$$

Relations (8.34)-(8.36) complete the proof of part (i).

(ii) With $\Delta_{N(T_n)}$ given by (8.35), we wish to show that

$$\mathcal{L}[\Delta_{N(T_n)}|\tilde{P}_{T_n,\theta}] \Rightarrow N(0, \Gamma(\theta)).$$

This follows by Proposition 2(ii), on account of Statement 1.

(iii) This is an immediate consequence of parts (i) and (ii). □

Proof of Theorem 2. First, consider the sequences of probability measures $\left\{\tilde{P}_{T_n,\theta}\right\}$ and $\left\{\tilde{P}_{T_n,\theta_{T_n}}\right\}$, where $\tilde{P}_{T,\theta}$ is defined in (8.25), and θ_T is given in (8.20). By Statement 1, Lemma 1 applies and gives that

$\left\{\tilde{P}_{T_n,\theta}\right\}$ and $\left\{\tilde{P}_{T_n,\theta_{T_n}}\right\}$ are contiguous. Then parts (i)-(iii) follow, respectively, from parts (i)-(iii) of Proposition 3. □

Proof of Theorem 3. One would have to show that

$$\sup\left\{\|\tilde{R}_{T_n,h} - \tilde{P}_{T_n,\theta_0+h[c(\theta)T_n]^{-1/2}}\|;\, h \in B, \text{ any bounded subset of } \Re^k\right\} \to 0. \tag{8.37}$$

Instead of (8.37), it suffices to show that

$$\|\tilde{R}_{T_n,h_n} - \tilde{P}_{T_n,\theta_0+h_n[c(\theta)T_n]^{-1/2}}\| \to 0 \tag{8.38}$$

for every bounded sequence $\{h_n\}$ in $\Re^k$. This is done by contradiction. If (8.38) does not hold as described, there must exist $\{h_m\} \subseteq \{h_n\}$ with $h_m \to h \in \Re^k$, such that

$$\|\tilde{R}_{T_m,h_m} - \tilde{P}_{T_m,\theta_0+h_m[c(\theta)T_m]^{-1/2}}\| \to 0. \tag{8.39}$$

It is shown that (8.39) cannot happen by showing that the L_1-norm below converges to 0; namely,

$$\mathcal{E}_{\theta_0}|L^*_{T_m}(h_m) - L_{T_m}(h_m)| \to 0, \tag{8.40}$$

where

$$L_{T_m}(h_m) = d\tilde{P}_{T_m,\theta_0+h_m[c(\theta)T_m]^{-1/2}}/d\tilde{P}_{T_m,\theta_0},\quad L^*_{T_m}(h_m) = d\tilde{R}_{T_m}/d\tilde{P}_{T_m,\theta_0}.$$

The proof of (8.40) is concluded by showing the following results :

$$L_{T_m}(h_m) - L^*_{T_m}(h_m) \to 0 \text{ in } \tilde{P}_{T_m,\theta_0}\text{-probability, and}$$

the three sequences $\{L_{T_m}(h_m)\}$, $\{L^*_{T_m}(h_m) - L_{T_m}(h_m)\}$ and $\{L^*_{T_m}(h_m)\}$ are sequences of $\tilde{P}_{T_m,\theta_0}$-uniformly integrable r.v.'s. The relevant details are extensive and can be found in Section 5 of Roussas and Bhattacharya (1998b). □

4.1 SOME CLOSING COMMENTS

As it has already been mentioned in the introductory section, the results obtained here, and especially the exponential approximation, can be used for statistical inference purposes. Such inference may include situations as the ones outlined below. The construction of asymptotically uniformly most powerful tests for testing a simple hypothesis against a one-sided alternative for a real-valued parameter, or for testing a one-sided hypothesis against a one-sided alternative. The construction of asymptotically uniformly most powerful unbiased tests for testing a simple hypothesis against a two-sided alternative. The study of asymptotic efficiency of estimates. Or, for the multi-parameter case, the construction of tests for testing a simple hypothesis against composite alternatives; such tests may be shown to enjoy several optimal asymptotic properties. Some such results are already available in the literature. Thus, Malinovskii (1992) considered a semi-Markov process whose probability law depends on a real-valued parameter $\theta \geq 0$, and studied the problem of testing the simple hypothesis $H_0 : \theta = 0$ against the one-sided contiguous alternatives $H_1 : \theta = hT^{-1/2}$, $0 < h \leq c$. Let Z_T be a statistic based on observations on the process up to time T, let $\beta_T^Z(h)$ be the power of a test based on Z_T, and let $\beta_T^*(h)$ be the envelope function. Then the author proposes a test which is second-order efficient, in the sense that $\sup\left\{\beta_T^*(h) - \beta_T^Z(h); 0 \leq h \leq c\right\} = o(T^{-1/2})$ as $T \to \infty$. The log-likelihood ratio is instrumental in the relevant derivations. For a review of estimation procedures, both parametric and non-parametric, in semi-Markov models the reader is referred to Jain (1990). We intend to discuss specific statistical applications in a future paper.

References

[1] Akritas, M.G. and Roussas, G.G. (1980). Asymptotic inference in continuous time semi-Markov processes, *Scandinavian Journal of Statistics*, **7**, 73-79.

[2] Anscombe, F.J. (1952). Large sample theory of sequential estimation. *Proceedings of the Cambridge Philosophical Society,* **48**, 600-607.

[3] Greenwood, P.E. and Wefelmeyer, W. (1996). Empirical estimators for semi-Markov processes, *Mathematical Methods in Statistics*, **5**, 299-315.

[4] Jain, P.K. (1990). Statistical estimation for semi-Markov models, *Metron*, **48**, 223-242.

[5] Janssen, J. (1986). *Semi-Markov Models: Theory and Applications,* Plennum Press, New York.

[6] Luschgy, H. (1995). Local asymptotic quadraticity of stochastic process models based on stopping times, *Stochastic Processes and Their Applications,* **57**, 305-317.

[7] Malinovskii, V.K. (1992). Asymptotic optimality of criteria in the problem of testing hypotheses for a recurrent semi-Markov process, *Journal of Soviet Mathematics,* **59**(4), 955-959.

[8] Pyke, R. (1961). Markov renewal processes: definition and preliminary properties, *Annals of Mathematical Statistics,* **32**, 1231-1242.

[9] Pyke, R. and Schaufele, R. (1964). Limit theorems for Markov renewal processes, *Annals of Mathematical Statistics,* **35**, 1746-1764.

[10] Roussas, G.G. (1972). *Contiguity of Probability Measures: Some applications in statistics,* Cambridge University Press, Cambridge.

[11] Roussas, G.G. and Bhattacharya, D. (1998a). Asymptotic behavior of the log-likelihood function in stochastic processes when based on a random number of random variables, this volume.

[12] Roussas, G.G. and Bhattacharya, D. (1998b). *Exponential approximation of distributions*, Technical Report No. 347, Division of Statistics, University of California, Davis.

Chapter 9

MARKOV RENEWAL PROCESSES AND EXPONENTIAL FAMILIES

Valeri T. Stefanov
The University of Western Australia, Australia

Abstract Exponential families of Markov renewal processes are considered. Many Markov renewal models, that are important from a practical point of view, belong to these families. General properties together with limit results and applications are discussed.

Keywords: Markov renewal process, exponential family, stopping time, limit theory, cluster, run, composite event.

1. INTRODUCTION

Let $(Z(t), A(t))_{t\geq 0}$ be a Markov renewal process, where $Z(t)$ is the embedded m-state Markov chain, and $A(t)$ is the additive component. We assume that for each i, $i = 1, \cdots, m$, the conditional density of $A(t+1) - A(t)$ given $Z(t) = i$, with respect to a σ-finite measure σ_i, is given by

$$\exp(\vartheta_i x + f_i(\vartheta_i)), \tag{9.1}$$

where ϑ_i is a real parameter, $\vartheta_i \in \Psi_i \subset R$. In other words the sojourn time distributions belong to one-dimensional exponential families. We assume that Ψ_i is an open subset of the natural parameter space of the exponential family. Without loss of generality we shall further assume that $Z(0) = 1$ with probability 1. Note that most of the popular distributions belong to one-dimensional exponential families. We confine ourselves to these distributions although our methodology also covers the case of sojourn distributions which are marginals in finite-dimensional noncurved exponential families. Note that, in view of results of Stefanov (1991), the so-called phase-type distributions are such marginals.

The Radon-Nikodym derivative of the measure generated by the trajectories of the process up to time t, with respect to a σ-finite measure σ_t, has the following form :

$$\exp\left(\sum_{i,j=1}^{m} N_{i,j}(t)(\ln p_{i,j} + f_i(\vartheta_i)) + \sum_{i=1}^{m} S_i(t)\vartheta_i\right) \tag{9.2}$$

where $\pi = (p_{i,j})_{i,j=1}^{m}$ is the transition probability matrix of the embedded Markov chain $Z(t)$, $\vartheta = (\theta_1, \cdots, \vartheta_m)$, $N_{i,j}(t)$ is the number of the one-step transitions from state i to state j in the time interval $[0, t]$, and $S_i(t)$ is the sojourn time at state i in the time interval $[0, t]$. This family is a curved exponential family. Roughly speaking a curved exponential family is an exponential family whose minimal sufficient statistic has a higher dimension than that of the parameter space. If both dimensions are equal then the family is a noncurved one. For rigorous definitions see Barndorff-Nielsen (1980) and Brown (1986). Noncurved exponential families have various nice analytical properties while curved ones are less tractable.

The sequential analogues of the above derivative, corresponding to random stopping rules, usually belong to curved exponential families. For a few *nice* stopping rules these belong to noncurved exponential families (Stefanov (1995)). The latter stopping rules embrace well-known sequences of regeneration times, in particular, the consecutive entries to a state. A brief discussion of this result is provided in Section 2. below. This property, when amalgamated with other results, provide a useful technology for deriving explicit limit results in the functional limit theory for Markov renewal processes. This is discussed in Section 3. and a simple example provided. Many applied problems from areas, such as genetics, reliability, computer science, and communication theory, are embeddable into waiting time problems for suitably selected Markov renewal processes. Such waiting time problems are very often associated with the aforementioned (*nice*) stopping rules. In this paper we confine ourselves to two such problems: the distribution theory associated with formation of clusters of points generated by renewal processes (Section 4.), and its discrete-time analogue associated with formation of composite events (Section 5.). A detailed discussion of the topics in Sections 2., 3. and 4. is found in Stefanov (1995) and Stefanov (1999). The detailed treatment of composite events, provided in Section 5., is not found elsewhere. Sections 2. and 3. are consistent as far as notation is concerned. A few symbols from Sections 4. and 5. identify quantities that have nothing to do with those identified by them in earlier sections. This slight inconsistency is conspicuous and will not confuse the readers.

2. NONCURVED EXPONENTIAL FAMILIES ASSOCIATED WITH MARKOV RENEWAL PROCESSES

In view of the fundamental identity in sequential analysis (cf. Jacod and Shiryaev (1987, p. 153)) for each finite stopping time τ the sequential version of the Radon-Nikodym derivative (9.2), say $L(\pi, \vartheta, \tau)$, is given by

$$\exp\left(\sum_{i,j=1}^{m} N_{i,j}(\tau)(\ln p_{i,j} + f_i(\vartheta_i)) + \sum_{i=1}^{m} S_i(\tau)\vartheta_i\right) \tag{9.3}$$

The following proposition describes a class of *nice* stopping times. For precise definitions of the terms associated with exponential families one may refer to Barndorff-Nielsen (1980) and Brown (1986).

Proposition 1 *(Stefanov (1995)). Let I be the state space of the embedded Markov chain $Z(t)$. Then for each $i \in I$ and each $J \subseteq I$ such that $\sum_{j\in J} p_{j,i} > 0$, the stopping times $\tau_J^i(s)$ given by*

$$\tau_J^i(s) = \inf\left(t : \sum_{j\in J} N_{j,i}(t) = s\right), \qquad s = 1, 2, \cdots$$

have the property that their corresponding Radon-Nikodym derivatives $L(\pi, \vartheta, \tau_J^i(s))$, belong to noncurved exponential families of order equal to the dimension of the parameter (π, ϑ).

In particular the consecutive entries to state 1, that is $\tau_I^1(s)$ (for brevity we will denote these by $\tau(s)$) are amongst the aforementioned stopping times. Proposition 1 also says that, after a suitable reparametrization, $L(\pi, \vartheta, \tau(s))$ has the following minimal canonical representation

$$\exp\left(\sum_{i=1}^{n} \theta_i X_i(\tau(s)) + \varphi(\theta)s\right), \tag{9.4}$$

where n is the dimension of the parameter (π, ϑ) and the X_i's are linearly independent components of the $N_{i,j}$'s and S_i's. Such a noncurved exponential representation can be derived explicitly in each particular case and we demonstrate this on a simple example.

Example 1 *Consider a Markov renewal process whose embedded Markov chain has two states and the sojourn times are gamma distributed with* *parameters* λ_i *and* β_i *(*$i = 1, 2.$*) If the* β_i*'s are fixed and the* λ_i*'s are parameters these distributions are one-dimensional exponential families and (9.1) takes the following form:*

$$\exp(-\lambda_i x + \beta_i \ln(\lambda_i)), \quad i = 1, 2.$$

Note that in view of Stefanov (1995) (cf. (19) on p. 1085) we have

$$N_{1,2}(\tau(s)) = N_{2,1}(\tau(s)), \qquad N_{1,1}(\tau(s)) = s - N_{2,1}(\tau(s)).$$

Therefore, one of the minimal canonical representations (9.4) is such that $n = 4, X_1(\tau(s)) = N_{2,1}(\tau(s)), X_2(\tau(s)) = N_{2,2}(\tau(s)), X_3(\tau(s)) = S_1(\tau(s)), X_4(\tau(s)) = S_2(\tau(s))$ *and*

$$\theta_1 = \ln\left(\frac{p_{1,2}p_{2,1}(-\lambda_2)^{\beta_2}}{p_{1,1}}\right), \ \theta_2 = \ln\left(p_{2,2}(-\lambda_2)^{\beta_2}\right), \ \theta_3 = -\lambda_1, \ \theta_4 = -\lambda_2,$$

$$\varphi(\theta) = \ln\left[\frac{1 - e^{-\theta_1}((-\theta_4)^{\beta_2} - e^{\theta_2})}{(-\theta_3)^{\beta_1} e^{-\theta_1}((-\theta_4)^{\beta_2} - e^{\theta_2})}\right].$$

3. APPLICATION TO LIMIT THEORY

In this section we discuss briefly how the results from the preceding section can be applied to derive explicit expressions for the limiting parameters in the functional central limit theorem for Markov renewal processes. Details are found in Stefanov (1995). Also, calculable expressions for the variance parameter in the functional central limit theorem for some Markov processes are found in Bhattacharya (1982) and Bhattacharya and Waymire (1990, pages 313-315). The process $(X(\tau(s)))_{s\geq 0}$ (Section 2.) has stationary independent increments. All known functional and nonfunctional limit results for processes with stationary independent increments are valid for this process. Furthermore, well-known analytic properties of noncurved exponential families (cf. Barndorff-Nielsen (1978; p.114) or Brown (1986)) yield

$$Cov\,(X_i(\tau(s)), X_j(\tau(s))) = -s\frac{\partial^2\varphi(\theta)}{\partial\theta_i\partial\theta_j}.$$

Also for each $s > 0$ and each $\alpha > 0$,

$$E\,(X_i(\tau(s)) - X_i(\tau(s-1)))^{\alpha} < +\infty\,, \quad i = 1, 2, \cdots, n,$$

$$\tau(s)/s \to E\tau(1) \overset{def}{=} \delta(\theta) > 0 \ a.s. \ as \ s \uparrow +\infty. \tag{9.5}$$

The latter follows from the fact that $\tau(s)$ is a linear function of the components of $X(\tau(s))$ and thus $(\tau(s))_{s\geq 0}$ has stationary independent increments.

Define for each $k, k > 1$, the processes $Y_k(t)$ and $\tilde{Y}_k(t)$ as follows:

$$Y_k(t) = \frac{f(X(kt)) + kt\delta^{-1}(\theta)f(\nabla\varphi(\theta))}{\sqrt{k}}$$

$$\tilde{Y}_k(t) = \frac{f(X(\tau([kt])) + \tau([kt])\delta^{-1}(\theta)f(\nabla\varphi(\theta))}{\sqrt{k}}$$

where $\nabla\varphi(\theta)) = (\partial\varphi(\theta)/\partial\theta_i, i = 1, 2, \ldots, n)$, $f : R^n \to R$ is a linear function, i.e.,

$$f(x) = b_1x_1 + b_2x_2 + \cdots, +b_nx_n \ ,$$

and $\delta(\theta)$ is defined in (9.5). Recall that $\tau([kt])$ is a linear function of the components of the minimal sufficient statistic $X_1(\tau([kt])), \cdots, X_n(\tau([kt]))$, say

$$\tau([kt]) = \Sigma_{i=1}^n a_i X_i(\tau([kt]) + a_{n+1}(kt)$$

where $a_{n+1}(kt)$ depends on kt in general. Of course, $\tilde{Y}_k(t)$ can be represented also as follows :

$$\tilde{Y}_k(t) = \frac{\Sigma_{i=1}^n \left(b_i + a_i\delta^{-1}(\theta)f(\nabla\varphi(\theta))\right) X_i(\tau([kt])) + a_{n+1}(kt)\delta^{-1}(\theta)f(\nabla\varphi(\theta))}{\sqrt{k}}.$$

Bearing in mind that the expected value of $\tau(s)$ equals $s\delta^{-1}(\theta)$ it is straightforward to see that the expected value of $\tilde{Y}_k(t)$ is equal to 0. Of course, from the functional central limit theorem for processes with stationary independent increments we have that, as $k \to +\infty$, $\tilde{Y}_k$ converges in distribution to a Brownian motion starting at the origin with zero drift and diffusion coefficient σ^2, such that

$$\sigma^2 = -\Sigma_{i,j=1}^n c_i c_j \frac{\partial^2\varphi(\theta)}{\partial\theta_i\partial\theta_j},$$

where $c_i = b_i + a_i\delta^{-1}(\theta)f(\nabla\varphi(\theta))$. General embedding techniques are applicable here for deriving limit results for Y_k from such for $\tilde{Y}_k$ (cf. Serfozo (1975) and the discussion at the beginning of section 7 of Stefanov (1995)). All regularity conditions on the associated processes, as required by these techniques, are satisfied and this has been proved in Stefanov (1995, section 7). Finally, the following limit result holds.

Proposition 2 *The process $(Y_k(t))_{t\geq 0}$ converges in distribution to a Brownian motion starting at the origin with zero drift and diffusion coefficient $\sigma^2\delta^{-1}(\theta)$, where σ^2 is given above.*

Example 2 *(continuation of Example 1). Note that $\tau(s) = S_1(\tau(s)) + S_2(\tau(s))$. Thus*

$$\delta(\theta) = -\partial\varphi(\theta)/\partial\theta_3 - \partial\varphi(\theta)/\partial\theta_4,$$

where $\varphi(\theta)$ has been given above. Consider, for example, the sequence of processes $(S_1^{(k)}(t))_{t\geq 0}$, $k = 1, 2, \cdots$, where $S_1^{(k)}(t) := S_1(kt)$. Note that $b_1 = b_2 = b_4 = 0$, $b_3 = 1$ and $a_1 = a_2 = 0$, $a_3 = a_4 = 1$. Therefore

$$\frac{S_1^{(k)}(t) + kt\left(-\frac{\partial\varphi(\theta)}{\partial\theta_3} - \frac{\partial\varphi(\theta)}{\partial\theta_4}\right)^{-1}\frac{\partial\varphi(\theta)}{\partial\theta_3}}{\sqrt{k}}$$

converges to a Brownian motion with zero drift and diffusion coefficient $\sigma^2\delta^{-1}(\theta)$ where

$$\sigma^2 = -\left(1+\delta^{-1}(\theta)(\frac{\partial\varphi(\theta)}{\partial\theta_3})\right)^2\left(\frac{\partial^2\varphi(\theta)}{\partial\theta_3^2}\right) - \left(\delta^{-1}(\theta)(\frac{\partial\varphi(\theta)}{\partial\theta_4}\right)^2\left(\frac{\partial^2\varphi(\theta)}{\partial\theta_4^2}\right)$$
$$-2\left(1+\delta^{-1}(\theta)(\frac{\partial\varphi(\theta)}{\partial\theta_3})\right)\left(\delta^{-1}(\theta)(\frac{\partial\varphi(\theta)}{\partial\theta_4}\right)\frac{\partial^2\varphi(\theta)}{\partial\theta_3\partial\theta_4}.$$

Further elaborations of this technology on examples drawn from ion channel models are found in Stefanov and Yeo (1997).

4. CLUSTERS OF POINTS

In this section we discuss the distribution theory associated with formation of clusters of points. The points are generated by a renewal process whose interarrival times are distributed according to a member from a general exponential family. A collection of m points with the property that each two consecutive points are separated by no more than δ units of time, where δ is a positive number fixed in advance, will be called a δ-cluster of m points. This concept of clustering has been suggested by Leslie (1969) who modelled firing of neurons. Also, our clusters are 'clumps of a fixed length' in the terminology of spatial statistics (cf. Hall (1988)). We are interested in the joint distribution of the waiting time till the occurrence of a δ-cluster of m points and the associated counts of occurrences of δ-clusters of l, or at least l, points, $l = 1, 2, \ldots, m-1$. It turns out that these random quantities can be embedded into the canonical statistic of a noncurved exponential family associated with a Markov renewal process. The embedded Markov chain of the latter has $m+1$ states and a special transition matrix. Also the associated sojourn distributions are members of exponential families that are related to the interarrival distribution of the original renewal process. A suitable application of the exponential family technology leads to explicit closed expressions of the joint moment generating function of the aforementioned quantities. More details follow. A complete treatment is found in Stefanov (1999).

Consider a Markov renewal process $(Z(t), A(t))$, $t \geq 1$, whose embedded Markov chain $Z(t)$, $t = 0, 1, \dots$, has $(m+1)$ states denoted by $\{0, 1, \dots, m\}$. Assume its one-step transition probability matrix is given by

$$\begin{bmatrix} 0 & 1 & 0 & \dots & \dots & \dots & 0 \\ 0 & q_1 & p_1 & 0 & \dots & \dots & 0 \\ 0 & q_2 & 0 & p_2 & 0 & \dots & 0 \\ \vdots & \vdots & \vdots & \vdots & \vdots & \vdots & \vdots \\ 0 & q_{m-1} & \dots & \dots & \dots & 0 & p_{m-1} \\ 0 & \dots & \dots & \dots & \dots & 0 & 1 \end{bmatrix}$$

Denote by $F_{i,j}$ the holding time distribution at state j given that the previous state is i. The $F_{i,j}$'s are selected amongst three different distribution functions, say F_1, F_2 and F_3, as follows:

$$F_{0,1} = F_1, \quad F_{i,i+1} = F_2, \quad F_{i,1} = F_3, \quad i = 1, 2, \dots, m-1,$$

where the F_j, $j = 1, 2, 3$, belong to one-dimensional exponential families whose densities with respect to σ-finite measures, say σ_j's, are given by $f_j(x) = \exp(\vartheta_j x + \varphi_j(\vartheta_j))$. Let τ_a be the time to absorption in state m. The motivation for introducing this model comes from the following

Example 3 *Consider a renewal process whose interarrival times, $Y_1, Y_2, \dots$, follow the distribution of a random variable Y. Assume that this distribution belongs to a general one-dimensional exponential family, with a support lying in $[0, +\infty)$, and density, with respect to some σ-finite measure, say σ, given by*

$$g(y) = \exp(\vartheta y + \varphi(\vartheta)). \tag{9.6}$$

Define the densities f_j, $j = 1, 2, 3$, as follows

$$f_1(x) = g(x), \quad f_2(x) = \frac{g(x) I_{[0,\delta]}(x)}{P_\vartheta(Y \leq \delta)}, \quad f_3(x) = \frac{g(x) I_{(\delta,+\infty)}(x)}{P_\vartheta(Y > \delta)},$$

where $\delta > 0$ and $I_A(\cdot)$ is the indicator function of the set A. Note that f_2 and f_3 also belong to exponential families as they are derived from an exponential family via censoring its support and normalization. The relevant canonical representations of these, with respect to σ-finite measures which are derived from σ via the same censoring mechanism, are

$$f_2(x) = \exp\{\vartheta x + \varphi_2(\vartheta)\}, \quad f_3(x) = \exp\{\vartheta x + \varphi_3(\vartheta)\},$$

with

$$\varphi_2(\vartheta) = \varphi(\vartheta) - \ln P_\vartheta(Y \le \delta), \quad \varphi_3(\vartheta) = \varphi(\vartheta) - \ln P_\vartheta(Y > \delta),$$

where φ is the function appearing in (9.6) above. Now it is easy to see that if we select the p_i's, ($q_i = 1 - p_i$), appearing in the above transition probability matrix as follows

$$p_i = P_\vartheta(Y \le \delta), \qquad i = 1, 2, \dots, m-1,$$

then the stopping time τ_a represents the waiting time till obtaining a δ-cluster of m points in this renewal process.

Example 4 *In the particular case of a Poisson process with parameter λ the relevant quantities take the following form:*

$$\varphi(\lambda) = \varphi_1(\lambda) = \ln\lambda, \quad \varphi_2(\lambda) = \ln\frac{\lambda}{1 - e^{-\lambda\delta}}, \quad \varphi_3(\lambda) = \ln\frac{\lambda}{e^{-\lambda\delta}}.$$

The following results are found in Stefanov (1999).

Proposition 3 *The Radon-Nikodym derivative of the measure generated by the trajectories of the aforementioned Markov renewal process on the time interval $[0, \tau_a]$, with respect to a σ-finite measure, belongs to a noncurved exponential family of order $m + 2$.*

$(N, T) = (N_{1,1}(\tau_a), N_{1,2}(\tau_a), N_{2,3}(\tau_a), \dots, N_{m-2,m-1}(\tau_a), T_1(\tau_a), T_2(\tau_a), T_3(\tau_a))$ is a minimal sufficient statistic for this family, where

$$\begin{aligned} T_1(\tau_a) &= \sum_{i=1}^{\tau_a} (A_i - A_{i-1}) I_{\{Z_{i-1}=0, Z_i=1\}} \\ T_2(\tau_a) &= \sum_{i=1}^{\tau_a} (A_i - A_{i-1}) I_{\bigcup_{j=1}^{m-1}\{Z_{i-1}=j, Z_i=j+1\}} \\ T_3(\tau_a) &= \sum_{i=1}^{\tau_a} (A_i - A_{i-1}) I_{\bigcup_{j=1}^{m-1}\{Z_{i-1}=j, Z_i=1\}}. \end{aligned}$$

Proposition 4 *The moment generating function of (N, T), say $M(t_1, t_2, \dots, t_{m-1}, s_1, s_2, s_3)$, is given by*

$$\frac{C_1 C_2^{m-1} C_{3,s} p_1 p_2 \dots p_{m-1} e^{t_2 + \dots + t_{m-1}}}{C_{1,s} C_{2,s}^{m-1} \left[C_{3,s} - C_3 \sum_{i=1}^{m-1} \left(p_1 p_2 \dots p_{i-1} q_i e^{t_2 + \dots + t_i} (C_2 / C_{2,s})^{i-1} \right) \right]},$$

with the convention $t_2 + \dots + t_i = t_1$ if $i = 1$, where

$$C_i = \exp(\varphi_i(\vartheta_i))), \quad C_{i,s} = \exp(\varphi_i(\vartheta_i + s_i))), \quad i = 1, 2, 3.$$

The components of (N,T), or linear combinations of these, contain the aforementioned random quantities that are associated with the occurrence of δ-clusters.

Example 5 *(continuation of example 4) Let τ be the waiting time till a δ-cluster of m points occurr and N_i be the associated count of δ-clusters of at least i points. Then applying the general result from Proposition 4 we derive the following explicit expression for the joint moment generating function, say $\psi(t_1, t_2, \ldots, t_{m-1})$, of $(\tau, N_2, \ldots, N_{m-1})$*

$$\frac{\lambda^m e^{t_2+t_3+\ldots+t_{m-1}}}{(\lambda+t_1)\left(\frac{\lambda+t_1}{1-e^{-(\lambda+t_1)\delta}}\right)^{m-1} - e^{-(\lambda+t_1)\delta}\sum_{i=1}^{m-1}\left[\lambda^i e^{t_2+\ldots+t_i}\left(\frac{\lambda+t_1}{1-e^{-(\lambda+t_1)\delta}}\right)^{m-i}\right]}$$

with the convention $t_2 + t_3 + \ldots + t_i = 0$ if $i = 1$.

We restricted our attention to clusters formed by points generated by renewal processes. However, our technology is amenable to more general models. For example the points may be generated by Markov renewal processes and δ-clusters of points of the same type, or of a mixed type, can be introduced. Then the clustering processes can be embedded into suitable, albeit more complicated, Markov renewal processes. Simple cases of such models would lead to explicit closed expressions although the algebra involved may be somewhat heavier.

5. COMPOSITE EVENTS

In this section we provide a detailed discussion of the discrete-time analogue of the clustering problem from the preceding section. Here we observe binary trials instead of a renewal process. The two possible outcomes in a binary trial are conventionally called success (denoted by 1) and failure (denoted by 0). We are interested in the waiting time, and associated random quantities, till the first occurrence of m successes with the property that each two consecutive successes are not separated by more than k failures. For brevity we will call such a collection of successes a k-gap run of m successes. Also such *composite* runs are called composite events (cf. Leslie (1967)). Some of the associated random quantities that are of interest are the counts of occurrences of k-gap runs of l, or at least l, successes, $l = 1, 2, \ldots, m-1$, till the occurrence of a k-gap run of m successes. We assume that the counting rule is the natural extension of Mood's (1940) counting of ordinary runs. That is a k-gap run of l successes is not counted if it turns out to be a subset of a longer k-gap run. We find an explicit expression for the joint moment generating function of the random quantities mentioned above when either independent, or Markov-dependent, binary trials are performed.

Such results have applications to areas such as physiology, communication theory, reliability theory and ecology (cf. Koutras (1996)). Special cases have been treated by Leslie (1967) and Koutras (1996). For simpler events Stefanov and Pakes (1997) suggested a unified approach for deriving joint generating functions. This approach can be extended to handle the case of composite events.

Note that if the aforementioned waiting time and the associated counts of occurrences of k-gap runs of shorter length than m were of sole interest then we could follow the same embedding techniques applied in the preceding section. The one-step transition probability matrix of the embedded Markov chain has the same structure as that in Section 4. and the holding time distributions are discrete and also belong to exponential families. More specifically, the latter are geometric and censored geometric distributions (for Bernoulli trials) or slight generalizations of these (for Markov-dependent binary trials). Therefore, proceeding along the same lines as those in Section 4. we can find a discrete-time analogue of Proposition 4, and subsequently an explicit expression for the generating function of the aforementioned quantities in discrete-time analogues of Example 4.

However, if we would like to capture more random quantities associated with the occurrence of a k-gap run of m successes then we can embed the problem into a suitable Markov chain. The one-step transition probability matrix of the latter is not as simple as that of the embedded Markov chain of the aforementioned Markov renewal process. We follow this approach and provide all details below. Also the technicalities associated with the application of the exponential family technology, both for Markov renewal processes and Markov chains, are due to the (embedded) Markov chain whereas the holding times do not usually introduce additional difficulties. Therefore, we demonstarate below how this technology works in a nontrivial case.

Consider a homogeneous Markov chain, $Z(t)$, $t = 0, 1, \dots$, with $(m-1)(k+1)+2$ states and one-step transition probabilities $p_{i,j}$, $i, j = 0, 1, \dots, (m-1)(k+1)+1$, where m and k are arbitrary, but fixed, natural numbers. Assume the following are the nonzero entries of the one-step transition probability matrix:

$$
\begin{aligned}
p_{i,i+1} \quad & \text{for} \quad i = 0, 1, \dots, (m-1)(k+1), \\
p_{i(k+1),0} \quad & \text{for} \quad i = 0, 1, \dots, m-1, \\
p_{i(k+1)+1-j,i(k+1)+1} \quad & \text{for} \quad j = 1, 2, \dots, k+1, \quad \text{and} \quad i = 1, 2, \dots, (m-1)
\end{aligned}
$$

and $p_{(m-1)(k+1)+1,(m-1)(k+1)+1} = 1$, that is the last state is an absorbing one. Note that each row, except for the last one, of the transition matrix has exactly two nonzero entries. Also deleting the first and last rows as well as the first two columns of the transition matrix results in a block-diagonal structure. There are $m-1$ blocks and each block is a $(k+1) \times (k+1)$ matrix whose entries along the main diagonal and last column are the only nonzero entries.

Denote by $N_{i,j}(t)$ the number of one-step transitions from state i to state j up to time t. The only nonzero $N_{i,j}$'s are those associated with the nonzero $p_{i,j}$'s. For clarity and brevity we shall use the following notation

$$\begin{aligned} p_i &= p_{i,i+1}, \\ q_i &= 1 - p_i, \quad i = 0, 1, \ldots, (m-1)(k+1). \end{aligned}$$

Also denote by L_i's the $N_{i,j}$'s which are associated with the p_i's and by R_i's those associated with the q_i's; for example $N_{i,i+1}(t) = L_i(t)$. Let τ be the time to absoption in state $(m-1)(k+1)+1$, that is

$$\tau = \inf\{t : Z(t) = (m-1)(k+1)+1\}.$$

In special cases this absorption time may be interpreted as the waiting time till obtaining a k-gap run of m successes in some binary trials. The meaning of a k-gap run has been explained above. The following two examples concern the case of independent and Markov-dependent binary trials.

Example 6 *Assume*

$$\begin{aligned} p_i &= p, && \text{if } \; i = j(k+1),\; j = 0, 1, \ldots, m-1, \\ p_i &= q && \text{otherwise,} \end{aligned}$$

where $q = 1 - p$. It is easy to see that τ has the interpretation mentioned above if Bernoulli trials, with success probability p, are performed.

Example 7 *Assume*

$$\begin{aligned} p_i &= v_{01}, && \text{if } \; i = j(k+1),\; j = 0, 1, \ldots, m-1, \\ p_i &= v_{10}, && \text{if } \; i = j(k+1)+1,\; j = 0, 1, \ldots, m-2, \\ p_i &= v_{00} && \text{otherwise,} \end{aligned}$$

where v_{ij}, $i, j = 0, 1$, are the one step transition probabilities in a two-state Markov chain with state 0 identified as failure and state 1 identified as success and initial state 0. Then τ has the above interpretation if such Markov-dependent binary trials are performed.

The Radon-Nikodym derivative of the measure generated by the trajectories of the chain up to time t with respect to a σ-finite measure is

$$\exp\left\{\sum_{i,j=0}^{(m-1)(k+1)}[N_{i,j}(t)\ln p_{i,j}]\right\}=\exp\left\{\sum_{i=0}^{(m-1)(k+1)}[L_i(t)\ln p_i+R_i(t)\ln q_i]\right\}. \tag{9.7}$$

In view of the fundamental identity in sequential analysis the sequential version of the above derivative corresponding to the measures generated up to time τ is

$$\exp\left\{\sum_{i=0}^{(m-1)(k+1)}[L_i(\tau)\ln p_i+R_i(\tau)\ln q_i]\right\}. \tag{9.8}$$

Whereas the family of densities given by (9.7) is a curved-exponential family, the one given by (9.8) is a noncurved exponential family (cf. Stefanov (1991) and Remark 1 in Stefanov and Pakes (1997)). The order of the noncurved exponential family given by (9.8) is $(m-1)(k+1)+1$, that is there are $(m-1)(k+1)+1$ linear constraints on the L_i's and R_i's. For brevity L_i and R_i will stand for $L_i(\tau)$ and $R_i(\tau)$ respectively. The following linear constraints hold (cf. the explanation below)

$$\begin{aligned}
R_1 &= L_0-L_1,\\
R_{i+1} &= L_i-L_{i+1}, \qquad i=1,2,\ldots,k,\\
R_{k+2} &= L_{k+1}-L_{k+2}+(L_0-L_k),\\
R_{i+1} &= L_i-L_{i+1}, \qquad i=k+2,k+3,\ldots,2(k+1)-1,\\
R_{2(k+1)+1} &= L_{2(k+1)}-L_{2(k+1)+1}\\
&+ (L_0-L_k)+(L_{k+1}-L_{2(k+1)-1}),\\
R_{i+1} &= L_i-L_{i+1},\\
&\quad i=2(k+1)+1,2(k+1)+2,\ldots,3(k+1)-1,\\
R_{3(k+1)+1} &= L_{3(k+1)}-L_{3(k+1)+1}+\sum_{i=0}^{2}(L_{i(k+1)}-L_{(i+1)(k+1)-1}),\\
\vdots &= \vdots\\
R_{(m-2)(k+1)} &= L_{(m-2)(k+1)-1}-L_{(m-2)(k+1)},\\
R_{(m-2)(k+1)+1} &= L_{(m-2)(k+1)}-L_{(m-2)(k+1)+1}\\
&\quad +\sum_{i=0}^{m-3}(L_{i(k+1)}-L_{(i+1)(k+1)-1}),\\
R_{i+1} &= L_i-L_{i+1},\\
&\quad i=(m-2)(k+1)+1,\ldots,(m-1)(k+1)-1,
\end{aligned}$$

$$L_{(m-1)(k+1)} = 1 - \sum_{i=0}^{m-2} (L_{i(k+1)} - L_{(i+1)(k+1)-1}).$$

Actually these identities are derived via counting the number of entries to, and exits from, a state up to time τ (cf. (3) in Stefanov (1991)). Note that for each state i, $i = 1, 2, \ldots, (m-1)(k+1)$, these two quantities are equal. More precisely L_{i-1} is the number of entries to state i, and $L_i + R_i$ is the number of exits from state i. Thus we obtain the first $k+1$ identities above. The next identity, that is

$$L_{k+1} = L_{k+2} + R_{k+2} - (L_0 - L_k),$$

is also obtained via comparing the number of entries, and exits from, state $k+2$, but incorporating the information from the previous $k+1$ identities; note that

$$L_0 - L_k = R_1 + R_2 + \ldots + R_k.$$

The remaining identities are derived likewise. These linear constraints suggest that each R_i, $i = 1, 2, \ldots (m-1)(k+1)$, has a linear expression in terms of the L_i's, $i = 1, 2, \ldots (m-1)(k+1) - 1$. Note that in view of the last two identities

$$R_{(m-1)(k+1)} = \sum_{i=0}^{m-3} (L_{i(k+1)} - L_{(i+1)(k+1)-1}) + L_{(m-2)(k+1)} - 1.$$

Also $L_{(m-1)(k+1)}$ is a linear function of the L_i's. Replace the R_i's and $L_{(m-1)(k+1)}$ by these expressions and get the following minimal canonical representation of (9.8):

$$\exp\left\{ \theta_0 R_0 + \sum_{i=1}^{(m-1)(k+1)} [\theta_i L_{i-1}] + \varphi(\theta) \right\},$$

where $\varphi(\theta) = \ln(p_{(m-1)(k+1)}/q_{(m-1)(k+1)})$, whose explicit expression in terms of the θ's will be found below, and

$$\begin{aligned}
\theta_0 &= \ln q_0, \\
\theta_1 &= \ln\left(\frac{p_0 q_1 q_{(m-1)(k+1)} \prod_{j=1}^{m-2} q_{j(k+1)+1}}{p_{(m-1)(k+1)}} \right), \\
\theta_{i(k+1)} &= \ln\left(\frac{p_{i(k+1)-1} q_{i(k+1)}}{q_{i(k+1)-1}} \frac{p_{(m-1)(k+1)}}{q_{(m-1)(k+1)} \prod_{j=i}^{m-2} q_{j(k+1)+1}} \right),
\end{aligned}$$

$$\theta_{i(k+1)+1} = \ln\left(\frac{p_{i(k+1)}q_{i(k+1)+1}}{q_{i(k+1)}}\frac{q_{(m-1)(k+1)}\prod_{j=i+1}^{m-2} q_{j(k+1)+1}}{p_{(m-1)(k+1)}}\right),$$
$$i = 1, 2, \ldots, m-2,$$
$$\theta_i = \ln\left(\frac{p_{i-1}q_i}{q_{i-1}}\right) \qquad \text{otherwise,} \tag{9.9}$$

with the convention $\prod_{j=1}^{0}(\cdot) = 1$. This and the convention $\sum_{j=i}^{i-1}(\cdot) = 0$ will apply throughout, unless otherwise explicitly stated. For brevity we will use the following notation

$$A_1 = \exp(\theta_1)\left(1 + \sum_{i=2, i\neq k+1}^{k+2} \exp\left(\sum_{j=2}^{i}\theta_j\right)\right),$$

$$B_1 = 1 - \exp(\theta_0) - \exp\left(\sum_{j=1}^{k+1}\theta_j\right),$$

$$A_r = 1 + \sum_{i=2, i\neq k+1}^{k+2} \exp\left(\sum_{j=2}^{i}\theta_{(r-1)(k+1)+j}\right), \qquad r = 2, 3, \ldots, m-2,$$

$$B_r = -\exp\left(\sum_{j=2}^{k+1}\theta_{(r-1)(k+1)+j}\right), \qquad r = 2, 3, \ldots, m-2,$$

$$A_{m-1} = 1 + \sum_{i=2}^{k+1} \exp\left(\sum_{j=2}^{i}\theta_{(m-2)(k+1)+j}\right),$$

$$B_{m-1} = -\exp\left(\sum_{j=2}^{k+1}\theta_{(m-2)(k+1)+j}\right).$$

Note that in the case $m = 2$ we have $A_1 \neq A_{m-1}$ and $B_1 \neq B_{m-1}$. In this particular case the expressions for B_1 and A_{m-1} will be valid. We will show that

$$\exp(\varphi(\theta)) = \frac{\sum_{i=1}^{m-1}(A_0A_1\ldots A_{i-1}B_i)}{A_1A_2\ldots A_{m-1}}, \tag{9.10}$$

where $A_0 = 1$. Note that from the identities for θ_0 and θ_1 in (9.8) one can express q_1 in terms of θ_0, θ_1 and $p_{(m-1)(k+1)}, q_{(m-1)(k+1)}, q_{i(k+1)+1}$, $i = 1, 2, \ldots, m-2$, that is

$$q_1 = \frac{p_{(m-1)(k+1)}e^{\theta_1}}{q_{(m-1)(k+1)}\prod_{i=1}^{m-2} q_{i(k+1)+1}(1 - e^{\theta_0})}.$$

From this and the identity for θ_2 in (9.8) one gets an expression for q_2 in terms of $\theta_0, \theta_1, \theta_2$ and $p_{(m-1)(k+1)}, q_{(m-1)(k+1)}, q_{i(k+1)+1}, i = 1, 2, \dots, m-2$. Proceeding likewise one arrives at a simple equation for q_{k+2} involving $\theta_0, \dots, \theta_{k+2}$ and $p_{(m-1)(k+1)}, q_{(m-1)(k+1)}, q_{i(k+1)+1}, i = 2, 3, \dots, m-2$. This yields

$$q_{k+2} = \frac{p_{(m-1)(k+1)} A_1}{q_{(m-1)(k+1)} B_1 \prod_{i=2}^{m-2} q_{i(k+1)+1}}.$$

Proceeding along the same lines one arrives at a simple equation for $q_{2(k+1)+1}$ involving $\theta_0, \dots, \theta_{2(k+1)+1}$ and $p_{(m-1)(k+1)}, q_{(m-1)(k+1)}, q_{i(k+1)+1}$, $i = 2, \dots, m-2$, (note that i runs from 2 on) and subsequently to the following:

$$q_{2(k+1)+1} = \frac{p_{(m-1)(k+1)} A_1 A_2}{q_{(m-1)(k+1)} (B_1 + A_1 B_2) \prod_{i=3}^{m-2} q_{i(k+1)+1}}$$

(note i runs from 3 onward in the product expression above). Likewise proceeding along these lines one arrives at the following expression:

$$q_{(m-2)(k+1)+1} = \frac{p_{(m-1)(k+1)} A_1 A_2 \dots A_{m-2}}{q_{(m-1)(k+1)} (B_1 + A_1 B_2 + A_1 A_2 B_3 + \dots + A_1 A_2 \dots A_{m-3} B_{m-2})}.$$

This expression involves $p_{(m-1)(k+1)}, q_{(m-1)(k+1)}$ and θ_i's for $i = 0, 1, \dots, (m-2)(k+1)+1$, only. From this expression and the remaining identities in (9.8) one gets an expression for $p_{(m-1)(k+1)}$ in terms of θ_i's for $i = 0, 1, \dots, (m-1)(k+1)$. The latter implies formula (9.10) for $p_{(m-1)(k+1)}/q_{(m-1)(k+1)}$ $(= \exp(\varphi(\theta)))$.

Let $t = (t_0, t_1, \dots, t_{(m-1)(k+1)})$ where the t_i's are real numbers. Substitute first the θ_j's by $(\theta_j + t_j)$'s in the expressions for A_i and B_i and second, substitute the θ_j's in the new equations by their expressions in terms of the p_i's and q_i's (see (9.8)). The derived expressions we denote by $A_i^{(t)}$ and $B_i^{(t)}$ respectively. Then

$$\begin{aligned} A_1^{(t)} &= \frac{q_{(m-1)(k+1)}}{p_{(m-1)(k+1)}} \prod_{i=1}^{m-2} q_{i(k+1)+1} (e^{t_1} p_0 q_1 + e^{t_1+t_2} p_0 p_1 q_2 \\ &+ \dots + e^{t_1+t_2+\dots+t_k} p_0 p_1 \dots p_{k-1} q_k + e^{t_1+t_2+\dots+t_{k+2}} p_0 p_1 \dots p_{k+1}), \\ B_1^{(t)} &= 1 - e^{t_0} q_0 - e^{t_1+t_2+\dots+t_{k+1}} p_0 p_1 \dots p_k q_{k+1}, \end{aligned}$$

$$
\begin{aligned}
A^{(t)}_{r+1} &= 1+\frac{1}{q_{r(k+1)+1}}(e^{t_{r(k+1)+2}}p_{r(k+1)+1}q_{r(k+1)+2}\\
&+ e^{t_{r(k+1)+2}+t_{r(k+1)+3}}p_{r(k+1)+1}p_{r(k+1)+2}q_{r(k+1)+3}+\dots\\
&+ e^{t_{r(k+1)+2}+\dots+t_{(r+1)(k+1)-1}}p_{r(k+1)+1}\cdots p_{(r+1)(k+1)-2}q_{(r+1)(k+1)-1}\\
&+ e^{t_{r(k+1)+2}+t_{r(k+1)+3}+\dots+t_{(r+1)(k+1)+1}}p_{r(k+1)+1}p_{r(k+1)+2}\cdots p_{(r+1)(k+1)}),\\
B^{(t)}_{r+1} &= -e^{t_{r(k+1)+2}+t_{r(k+1)+3}+\dots+t_{(r+1)(k+1)}}\\
&\times \frac{p_{r(k+1)+1}p_{r(k+1)+2}\cdots p_{(r+1)(k+1)-1}q_{(r+1)(k+1)}p_{(m-1)(k+1)}}{q_{r(k+1)+1}q_{(m-1)(k+1)}\prod_{i=r+1}^{m-2}q_{i(k+1)+1}},\\
A^{(t)}_{m-1} &= 1+\frac{1}{q_{(m-2)(k+1)+1}}(e^{t_{(m-2)(k+1)+2}}p_{(m-2)(k+1)+1}q_{(m-2)(k+1)+2}\\
&+ e^{t_{(m-2)(k+1)+2}+t_{(m-2)(k+1)+3}}p_{(m-2)(k+1)+1}p_{(m-2)(k+1)+2}q_{(m-2)(k+1)+3}+\dots\\
&+ e^{t_{(m-2)(k+1)+2}+\dots+t_{(m-1)(k+1)-1}}p_{(m-2)(k+1)+1}\cdots p_{(m-1)(k+1)-2}q_{(m-1)(k+1)-1}\\
&+ e^{t_{(m-2)(k+1)+2}+\dots+t_{(m-1)(k+1)}}p_{(m-2)(k+1)+1}\cdots p_{(m-1)(k+1)-1}p_{(m-1)(k+1)}),\\
B^{(t)}_{m-1} &= -e^{t_{(m-2)(k+1)+2}+\dots+t_{(m-1)(k+1)}}\frac{p_{(m-2)(k+1)+1}\cdots p_{(m-1)(k+1)-1}p_{(m-1)(k+1)}}{q_{(m-2)(k+1)+1}},
\end{aligned}
$$

where $r = 1, 2, \dots m-3$. Therefore in view of (9.10) the expression for $\exp(\varphi(\theta+t))$ in terms of the p_i's and q_i's is

$$
\exp(\varphi(\theta+t)) = \frac{\sum_{i=1}^{m-1}(A_0^{(t)}A_1^{(t)}\dots A_{i-1}^{(t)}B_i^{(t)})}{A_1^{(t)}A_2^{(t)}\dots A_{m-1}^{(t)}}, \tag{9.11}
$$

where $A_0^{(t)} = 1$. Recall that $\exp(\varphi(\theta))$, expressed in terms of the p_i's and q_i's, is equal to $p_{(m-1)(k+1)}/q_{(m-1)(k+1)}$. Thus, from (9.11) and well-known analytic properties of noncurved exponential families (see Barndorff-Nielsen (1978, p. 114)) we get the following

Proposition 5 *For the moment generating function of the vector* (R_0, L) *(*$= (R_0, L_0, L_1, \dots, L_{(m-1)(k+1)-1})$*) we have*

$$
M_{R_0,L}(t) = \frac{p_{(m-1)(k+1)}}{q_{(m-1)(k+1)}}\frac{A_1^{(t)}A_2^{(t)}\dots A_{m-1}^{(t)}}{\sum_{i=1}^{m-1}(A_0^{(t)}A_1^{(t)}\dots A_{i-1}^{(t)}B_i^{(t)})}. \tag{9.12}
$$

The L_i's, R_0 and linear combinations of these embrace various random quantities that might be of interest. Now we elaborate on a collection of such quantities. Note that in view of the identities for the R_i's, given earlier, we have

$$
R_1+R_2+\dots+R_{(m-1)(k+1)} = L_0-1+\sum_{i=0}^{m-2}(m-1-i)(L_{i(k+1)}-L_{(i+1)(k+1)-1}).
$$

Recall that

$$L_{(m-1)(k+1)} = 1 - \sum_{i=0}^{m-2} (L_{i(k+1)} - L_{(i+1)(k+1)-1}).$$

From these and the fact that

$$\tau = \sum_{i=0}^{(m-1)(k+1)} (L_i + R_i)$$

we get

$$\begin{aligned}
\tau &= R_0 + 2L_0 + \sum_{i=1}^{(m-1)(k+1)-1} L_i \\
&\quad + \sum_{i=0}^{m-2} (m-2-i)(L_{i(k+1)} - L_{(i+1)(k+1)-1}) \qquad (9.13) \\
&= R_0 + mL_0 + \sum_{i=1}^{m-2} (m-1-i) L_{i(k+1)} \\
&\quad - \sum_{i=1}^{m-2} (m-2-i) L_{i(k+1)-1} + \sum_{\text{all other i's}} L_i.
\end{aligned}$$

In particular, the joint distribution of the following quantities τ, L_0, L_{k+1}, $\ldots$, $L_{(m-2)(k+1)}$ is of interest. The interpretation of the $L_{i(k+1)}$'s, $i = 0, 1, \ldots, m-2$, in the context of Examples 6 and 7, is that $L_{i(k+1)}$ is the number of occurrences of k-gap success runs of length at least $i+1$ till the occurrence of the first k-gap success run of length m. Since $(\tau, L_0, L_{k+1}, .., L_{(m-2)(k+1)})$ is derived from $(R_0, L_0, L_1, .., L_{(m-1)(k+1)-1})$ via a linear transformation we get its moment generating function, say $\psi(s, s_0, s_1, \ldots, s_{m-2})$, from the expression given in (9.12) after substituting the t_i's as follows (cf. (9.12) above):

$$\begin{aligned}
t_1 &= ms + s_0, \\
t_{i(k+1)} &= -(m-2-i)s, \qquad i = 1, 2, \ldots, m-2, \\
t_{i(k+1)+1} &= (m-1-i)s + s_i, \qquad i = 1, 2, \ldots, m-2, \\
t_j &= s, \qquad \text{otherwise.}
\end{aligned}$$

We will derive closed expressions for this moment generating function in the cases of the models introduced in Example 6 and 7 above.

Example 8 *(continuation of example 6). Note that the relevant expressions for the $A_i^{(t)}$'s and $B_i^{(t)}$'s, after the above substitution of the t_i's, are as follows:*

$$\begin{aligned}
A_1^{(t)} &= \frac{qp^{m-1}e^{ms+s_0}\left[1-(1-e^{s_1})(qe^s)^k - e^{s_1}(qe^s)^{k+1}\right]}{1-qe^s},\\
B_1^{(t)} &= 1-qe^s - e^{s_0}pe^s(qe^s)^{k+1},\\
A_{r+1}^{(t)} &= \frac{1-(1-e^{s_{r+1}})(qe^s)^k - e^{s_{r+1}}(qe^s)^{k+1}}{1-qe^s},\\
B_{r+1}^{(t)} &= -q^k p^{r+2-m} e^{(k+r+2-m)s},\\
A_{m-1}^{(t)} &= \frac{1-(qe^s)^{k+1}}{1-qe^s},\\
B_{m-1}^{(t)} &= -(qe^s)^k,
\end{aligned}$$

where $r = 1, 2, \ldots, m-3$.
Note that with the convention $s_{m-1} = 0$ the expressions for the $A_{m-1}^{(t)}$ and $B_{m-1}^{(t)}$ take the same form as that for $A_{r+1}^{(t)}$ and $B_{r+1}^{(t)}$, respectively. Finally, in view of (9.12), we get the following expression for the moment generating function of $(\tau, L_0, L_{k+1}, \ldots, L_{(m-2)(k+1)})$, $\psi(s, s_0, \ldots, s_{m-2})$, in the case of independent Bernoulli trials with probability of success p, $(q = 1-p)$:

$$\frac{\frac{p^m e^{ms+s_0}}{(1-qe^s)^{m-1}}\prod_{j=1}^{m-1}\left(1-(1-e^{s_j})(qe^s)^k - e^{s_j}(qe^s)^{k+1}\right)}{1-qe^s-e^{s_0}pe^s(qe^s)^{k+1}\sum_{i=1}^{m-1}\left[\left(\frac{pe^s}{1-qe^s}\right)^{i-1}\prod_{j=1}^{i-1}\left(1-(1-e^{s_j})(qe^s)^k - e^{s_j}(qe^s)^{k+1}\right)\right]},$$

where $s_{m-1} = 0$. In the particular case when $s_j = 0$, $j = 0, 1, \ldots, m-2$, the expression reduces to the moment generating function of τ, as found earlier by Leslie (1967).

Example 9 *(continuation of example 7). The relevant expressions for the $A_i^{(t)}$'s and $B_i^{(t)}$'s are as follows:*

$$\begin{aligned}
A_1^{(t)} &= \left[v_{11}(1-v_{00}e^s) + v_{10}v_{01}e^s\left(1-(1-e^{s_1})(v_{00}e^s)^{k-1} - e^{s_1}(v_{00}e^s)^k\right)\right]\\
&\quad\times\frac{v_{00}v_{11}^{m-1}e^{ms+s_0}}{v_{11}(1-v_{00}e^s)},
\end{aligned}$$

$$\begin{aligned}
B_1^{(t)} &= 1 - v_{00}e^s - e^{s_0}v_{10}v_{01}e^{2s}(v_{00}e^s)^k,\\
A_{r+1}^{(t)} &= \frac{v_{11}(1-v_{00}e^s) + v_{10}v_{01}e^s\left[1-(1-e^{s_{r+1}})(v_{00}e^s)^{k-1} - e^{s_{r+1}}(v_{00}e^s)^k\right]}{v_{11}(1-v_{00}e^s)},\\
B_{r+1}^{(t)} &= -v_{10}v_{01}v_{00}^{k-1}v_{11}^{r+1-m}e^{(k+r+2-m)s},\\
A_{m-1}^{(t)} &= \frac{v_{11}(1-v_{00}e^s) + v_{10}v_{01}e^s\left[1-(v_{00}e^s)^k\right]}{v_{11}(1-v_{00}e^s)},\\
B_{m-1}^{(t)} &= -v_{10}v_{01}v_{00}^{k-1}v_{11}^{-1}e^{ks}.
\end{aligned}$$

Finally we get the following expression for the moment generating function of $(\tau, L_0, L_{k+1}, \ldots, L_{(m-2)(k+1)})$, *in the case of Markov-dependent binary trials with initial state* 0 *(recall the convention* $s_{m-1} = 0$*):*

$$\frac{v_{01}e^{ms+s_0}D_{m-1}}{(1-v_{00}e^s)^{m-1}\left(1 - v_{00}e^s - e^{s_0}v_{10}v_{01}e^{2s}(v_{00}e^s)^k\sum_{i=1}^{m-1}\left[\left(\frac{e^s}{1-v_{00}e^s}\right)^{i-1}D_{i-1}\right]\right)},$$

where

$$D_i = \prod_{j=1}^{i}\left(v_{11}(1-v_{00}e^s) + v_{10}v_{01}e^s\left[1-(1-e^{s_j})(v_{00}e^s)^{k-1} - e^{s_j}(v_{00}e^s)^k\right]\right).$$

If the initial state is 1 *then we have* $p_0 = v_{11}$ *and* $q_0 = v_{10}$. *This change affects the expressions for the* $A_1^{(t)}$ *and* $B_1^{(t)}$ *only. The new* $A_1^{(t)}$ *is equal to the old one multiplied by* v_{11}/v_{01} *and the new* $B_1^{(t)}$ *is equal to*

$$1 - v_{10}e^s - e^{s_0}v_{10}v_{01}e^{2s}(v_{00}e^s)^k\frac{v_{11}}{v_{01}}.$$

Therefore, the expression for the above mentioned moment generating function, in the case of Markov-dependent binary trials with initial state 1 *is*

$$\frac{v_{11}e^{ms+s_0}D_{m-1}}{(1-v_{00}e^s)^{m-1}\left(1 - v_{10}e^s - e^{s_0}v_{10}v_{11}e^{2s}(v_{00}e^s)^k\sum_{i=1}^{m-1}\left[\left(\frac{e^s}{1-v_{00}e^s}\right)^{i-1}D_{i-1}\right]\right)}.$$

where the D_i*'s have been introduced above.*

References

[1] Barndorff-Nielsen, O. (1978). *Information and Exponential Families*, Wiley, Chichester.

[2] Barndorff-Nielsen, O. (1980). Conditionality resolutions, *Biometrika*, **67**, 293–310.

[3] Bhattacharya, R.N. (1982). On the functional central limit theorem and the law of the iterated logarithm for Markov processes, *Zeischrift fur Wahrscheinlichkeitstheorie und verwande Gebiete*, **60**, 185–201.

[4] Bhattacharya, R.N. and Waymire, E.C. (1990). *Stochastic Processes with Applications*, Wiley, New York.

[5] Brown, L. (1986). *Fundamentals of Statistical Exponential Families*, Institute of Mathematical Statistics Vol. 9 , Hayward.

[6] Hall, P. (1988). *Introduction to the Theory of Coverage Processes*, Wiley, New York.

[7] Jacod, J. and Shiryaev, A.N. (1987). *Limit Theorems for Stochastic Processes*, Springer-Verlag, Berlin.

[8] Koutras, M. V. (1996). On a waiting time distribution in a sequence of Bernoulli trials, *Annals of the Institute of Statistical Mathematics*, **48**, 789–806.

[9] Leslie, R. T. (1967). Recurrent composite events, *Journal of Applied Probability*, **4**, 34–64.

[10] Leslie, R. T. (1969). Recurrence times of clusters of Poisson points, *Journal of Applied Probability*, **6**, 372–388.

[11] Mood, A. M. (1940). The distribution theory of runs, *Annals of Mathematical Statistics*, **11**, 367–392.

[12] Serfozo, R. F. (1975). Functional limit theorems for stochastic processes based on embedded processes, *Advances in Applied Probability*, **7**, 123–139.

[13] Stefanov, V.T. (1991). Noncurved exponential families associated with observations over finite state Markov chains, *Scandinavian Journal of Statistics*, **18**, 353–356.

[14] Stefanov, V.T. (1995). Explicit limit results for minimal sufficient statistics and maximum likelihood estimators in some Markov processes: Exponential families approach, *Annals of Statistics*, **23**, 1073–1101.

[15] Stefanov, V.T. (1999). On the occurrence of composite events and clusters of points, *Journal of Applied Probability*, **36**, (to appear).

[16] Stefanov, V.T. and Pakes, A. G. (1997). Explicit distributional results in pattern formation, *Annals of Applied Probability*, **7**, 666–678.

[17] Stefanov, V.T. and Yeo, G. F. (1997). Explicit asymptotic results for open times in ion channel models, *Advances in Applied Probability*, **29**, 947–964.

Chapter 10

ON HOMOGENEITY OF TWO SEMI-MARKOV SAMPLES

Larisa Afanasyeva
Moscow State University, Russia

Peter Radchenko
Moscow State University, Russia

Abstract In this paper the Mann-Whitney test, which works for samples of independent observations, is adapted and extended to the semi-Markovian case. The central limit theorem for the U-statistic is proved and the dominant term of its dispersion is computed. The results are used for estimating homogeneity of stock market data.

Keywords: The sign test, U-statistic, semi-Markovian process.

1. INTRODUCTION

The sign test, introduced by Wilcoxon (1945) for samples of the same size, was extended by Mann and Whitney (1947) for samples of different sizes. The observations inside the samples were considered to be independent. These results may have a following financial application. Suppose that we have records of stock prices over two different years, and records of the same year are statistically homogeneous and independent. The Mann-Whitney test provides a tool for estimating homogeneity of all the data (over both years).

However, it seems more natural to regard data of the same year as dependent in probability sense. Consider the following evolution of stock price S_n : $S_n = (1 + \xi_n)S_{n-1}$. The sign of ξ_n indicates whether the stock price has risen or fallen. We assume that ξ_n depends on previous observations, but only through the sign of ξ_{n-1}. In other words, behavior of a stock price depends only on whether there has been a fall or a rise of the previous price. We test homogeneity of two independent samples, $(\xi_1,\ldots,\xi_n)$ and $(\tilde{\xi}_1,\ldots,\tilde{\xi}_m)$, which represent two different years.

In this paper a more general approach is taken by dividing the state space $\Re$ into R regions. Thus, ξ_n depends on what region the value of ξ_{n-1} was in. When $R = 2$, we have as usual: $\Re = (-\infty, 0) \cup [0, +\infty)$. In section 2 a formalization of the model is provided, the fundamental theorem is formulated, and ideas of the proof are given. In section 3 the dominant term of dispersion of the U-statistic is calculated. Proof of the theorem is given in section 4. In section 5 the parameters are evaluated and the statistical test is built.

2. THE MODEL

Consider a stationary sequence $\{\xi_i\}$. To explain dependence among the variables we introduce the following partition of the state space $\Re$:

$$\Re = \bigsqcup_{k=1}^{R} A_k, \quad A_k \text{ such that } x_i \in A_i,\ x_j \in A_j,\ i < j \text{ implies } x_i < x_j.$$

Let $e(x)$ be a function of R values

$$e(x) = \begin{cases} 1, & x \in A_1 \\ 2, & x \in A_2 \\ & \ldots \\ R, & x \in A_R. \end{cases} \tag{10.1}$$

Let us denote $e_i = e(\xi_i)$, $\tilde{e}_j = e(\tilde{\xi}_j)$.
Assume that $\{\xi_i\}$ satisfies the following conditions

$$P\{\xi_i < x | \xi_1, \ldots, \xi_{i-1}\} = P\{\xi_i < x | e_{i-1}\} \tag{10.2}$$

$$P\{\xi_i < x | e_i, e_{i-1}\} = P\{\xi_i < x | e_i\}. \tag{10.3}$$

These conditions mean, in particular, that random variables $\xi_1, \ldots, \xi_k$ are conditionally independent if the values of $e(\xi_1), \ldots, e(\xi_k)$ are fixed. Condition (10.2) implies that $\{\xi_i\}$ forms a semi-Markovian process. Condition (10.3) is not really a necessary one. It limits the number of parameters for evaluation and facilitates the calculations.

Consider two independent samples $(\xi_1,\ldots,\xi_n)$ and $(\tilde{\xi}_1,\ldots,\tilde{\xi}_m)$ of stationary i.d. random variables with distribution functions F and G respectively. Assume that both sequences satisfy conditions (10.2) and (10.3). Let H_0 denote the hypothesis of their statistical homogeneity, i.e. H_0 : samples $(\xi_1,\ldots,\xi_n)$ and $(\tilde{\xi}_1,\ldots,\tilde{\xi}_m)$ are distributed identically.

To verify H_0 we introduce new random variables

$$z_{ij} = \begin{cases} 1, & \xi_i < \tilde{\xi}_j \\ 0, & \xi_i > \tilde{\xi}_j \end{cases} \qquad i = 1, \ldots, n \quad j = 1, \ldots, m \tag{10.4}$$

and form the U-statistic

$$U = \sum_{i=1}^{n} \sum_{j=1}^{m} z_{ij}. \tag{10.5}$$

If we put elements of both samples in ascending order, then U represents the number of times that elements of the first sample precede elements of the second sample.

Theorem 1 *Assume that $\{\xi_i\}$ and $\{\tilde{\xi}_j\}$ satisfy conditions (10.2) and (10.3). If H_0 is valid, then*

$$\frac{U - EU}{\sqrt{DU}} \longrightarrow N(0,1) \tag{10.6}$$

when $m, n \to \infty$ and $m \sim cn$, where c is an arbitrary positive constant.

In order to prove theorem 1 we introduce the following notations

$$\alpha_i = E(z_{ij}|\xi_i), \qquad \beta_j = E(z_{ij}|\tilde{\xi}_j) \tag{10.7}$$

$$V = \sum_{i=1}^{n} \sum_{j=1}^{m} (\alpha_i + \beta_j) \tag{10.8}$$

$$\Delta = \sum_{i=1}^{n} \sum_{j=1}^{m} (z_{ij} - \alpha_i - \beta_j) \tag{10.9}$$

Thus, $U = V + \Delta$. To establish asymptotic normality of the U-statistic it is sufficient to prove the following

$$\frac{\Delta - E\Delta}{\sqrt{DU}} \xrightarrow{P} 0 \tag{10.10}$$

$$\frac{V - EV}{\sqrt{DU}} \longrightarrow N(0,1) \tag{10.11}$$

when $n, m \to \infty$, $\frac{n}{m} \to c$.

In section 3 we calculate the dominant term of DU. In section 4 we establish that $D\Delta = o(DU)$, hence proposition (10.10) follows from Chebyshev inequality. We show that $DU \sim DV$ and we prove proposition (10.11) utilizing the central limit theorem for Markov chain functionals (Chung, 1960).

3. CALCULATION OF DU

Condition (10.2) implies that sequences $\{e_i\}$ and $\{\tilde{e}_j\}$ form Markov chains. Let $P = (p_{st})$ denote the transition matrix and $\{\pi_s, s = 1, \ldots, R\}$ be the stationary distribution. Define $p_{st}(j)$ as the j-step probabilities. Since H_0 is valid, these characteristics are identical for both sequences. It is easy to see that $\pi_s = P(e_i = s) = \int_{A_s} dF(x)$.

Clearly,

$$(EU)^2 = \frac{n^2m^2}{4}. \tag{10.12}$$

Observe that

$$EU^2 = 4\sum_{i=1}^{n-1}\sum_{k=i+1}^{n}\sum_{j=1}^{m-1}\sum_{l=j+1}^{m} Ez_{ij}z_{kl} + 2\sum_{i=1}^{n}\sum_{j=1}^{m-1}\sum_{l=j+1}^{m} Ez_{ij}z_{il}+$$

$$2\sum_{i=1}^{n-1}\sum_{k=i+1}^{n}\sum_{j=1}^{m} Ez_{ij}z_{kj} + \sum_{i=1}^{n}\sum_{j=1}^{m} Ez_{ij}^2 = J_1 + J_2 + J_3 + J_4 \tag{10.13}$$

To calculate J_1 we write

$$\begin{aligned}
Ez_{ij}z_{kl} &= P\{\xi_i < \tilde{\xi}_j, \xi_k < \tilde{\xi}_l\} \\
&= \sum_{s,t,p,q\in\{1..R\}} P\{\xi_i < \tilde{\xi}_j | e_i = s, \tilde{e}_j = p\} P\{\xi_k < \tilde{\xi}_l | e_k = t, \tilde{e}_l = q\} \\
&\quad \times \pi_s \pi_t p_{sp}(k-i) p_{tq}(l-j) \\
&= \sum_{s,t\in\{1..R\}} \frac{1}{4}\pi_s^2 p_{st}(k-i) p_{st}(l-j) \\
&+ \sum_{s=1}^{R}\sum_{t=1}^{R-1}\sum_{q=t+1}^{R} \frac{1}{2}\pi_s^2 p_{st}(k-i) p_{sq}(l-j) \\
&+ \sum_{s=1}^{R-1}\sum_{t=1}^{R}\sum_{p=s+1}^{R} \frac{1}{2}\pi_s\pi_p p_{st}(k-i) p_{pt}(l-j) \\
&+ \sum_{s=1}^{R-1}\sum_{t=1}^{R-1}\sum_{p=s+1}^{R}\sum_{q=t+1}^{R} \pi_s\pi_p p_{st}(k-i) p_{pq}(l-j).
\end{aligned}$$

Let Π denote the matrix which rows are vectors $(\pi_1, \ldots, \pi_R)$. Applying the following formula from Kemeny and Snell (1960)

$$\sum_{k=1}^{n} p_{ij}(k) = n\pi_j + c_{ij}, \quad C = (E - (P - \Pi))^{-1} - E \tag{10.14}$$

we have

$$J_1=\sum_{s=1}^{R}\sum_{t=1}^{R}(\frac{n^2m^2}{4}\pi_t^2+\frac{n^2m}{4}\pi_t(2c_{st}-\pi_t)+\frac{nm^2}{4}\pi_t(2c_{st}-\pi_t))\pi_s^2+$$
$$+\sum_{s=1}^{R}\sum_{t=1}^{R-1}\sum_{q=t+1}^{R}(\frac{n^2m^2}{2}\pi_t\pi_q+\frac{n^2m}{2}\pi_t(2c_{sq}-\pi_q)+\frac{nm^2}{2}\pi_q(2c_{st}-$$
$$-\pi_t))\pi_s^2+\sum_{s=1}^{R-1}\sum_{t=1}^{R}\sum_{p=s+1}^{R}(\frac{n^2m^2}{2}\pi_t^2+\frac{n^2m}{2}\pi_t(2c_{pt}-\pi_t)+$$
$$+\frac{nm^2}{2}\pi_t(2c_{st}-\pi_t))\pi_s\pi_p+\sum_{s=1}^{R-1}\sum_{t=1}^{R-1}\sum_{p=s+1}^{R}\sum_{q=t+1}^{R}(n^2m^2\pi_t\pi_q+$$
$$+n^2m\pi_t(2c_{pq}-\pi_q)+nm^2\pi_q(2c_{st}-\pi_t))\pi_s\pi_p+O(n^2)$$

Hence, we obtain the coefficient multiplying n^4 in J_1

$$\tfrac{1}{4}(\textstyle\sum_{s=1}^{R-1}\sum_{t=1}^{R-1}\sum_{p=s+1}^{R}\sum_{q=t+1}^{R}4\pi_s\pi_t\pi_p\pi_q+$$
$$+\textstyle\sum_{s=1}^{R-1}\sum_{t=1}^{R}\sum_{p=s+1}^{R}2\pi_s\pi_t^2\pi_p+\sum_{s=1}^{R}\sum_{t=1}^{R-1}\sum_{q=t+1}^{R}2\pi_s^2\pi_t\pi_q+$$
$$\textstyle\sum_{s=1}^{R}\sum_{t=1}^{R}\pi_s^2\pi_t^2)=\tfrac{1}{4}(\sum_{s=1}^{R}\sum_{t=1}^{R}\sum_{p=1}^{R}\sum_{q=1}^{R}\pi_s\pi_t\pi_p\pi_q)=\tfrac{1}{4}$$

By analogy, we find that the sum of $3rd$ order terms that do not include c_{ij} equals to $-\frac{nm(n+m)}{4}$. Finally, we get the sum of $3rd$ order terms that include c_{ij}

$$\tfrac{nm}{2}\Big[\textstyle\sum_{s=1}^{R}\sum_{t=1}^{R}(n+m)c_{st}\pi_s^2\pi_t+\sum_{s=1}^{R}\sum_{t=1}^{R-1}\sum_{q=t+1}^{R}(2nc_{sq}\pi_s^2\pi_t^2+$$
$$+2mc_{st}\pi_s^2\pi_q)+\textstyle\sum_{s=1}^{R-1}\sum_{t=1}^{R}\sum_{p=s+1}^{R}(2nc_{pt}\pi_s\pi_t\pi_p+2mc_{st}\pi_s\pi_t\pi_p)+$$
$$+\textstyle\sum_{s=1}^{R-1}\sum_{t=1}^{R-1}\sum_{p=s+1}^{R}\sum_{q=t+1}^{R}(4nc_{pq}\pi_s\pi_t\pi_p+4mc_{st}\pi_s\pi_p\pi_q)\Big].$$

We now calculate J_2. For $l > j$, $Ez_{ij}z_{il}$ can be rewritten as

$$Ez_{ij}z_{il} = P\{\xi_i < \tilde{\xi}_j, \xi_i < \tilde{\xi}_l\} =$$

$$= \sum_{s=1}^{R} P\{\xi_i < \tilde{\xi}_j, \xi_i < \tilde{\xi}_l | e_i = s, \tilde{e}_j = s, \tilde{e}_l = s\}\pi_s^2 p_{ss}(l-j) +$$

$$+ \sum_{s=1}^{R-1} \sum_{t=s+1}^{R} P\{\xi_i < \tilde{\xi}_j | e_i = s, \tilde{e}_j = s\}\pi_s^2 p_{st}(l-j) +$$

$$+ \sum_{s=1}^{R-1} \sum_{t=s+1}^{R} P\{\xi_i < \tilde{\xi}_l | e_i = s, \tilde{e}_l = s\}\pi_s \pi_t p_{ts}(l-j) +$$

$$+ \sum_{s=1}^{R-1} \sum_{t=s+1}^{R} \sum_{p=s+1}^{R} \pi_s \pi_t p_{tp}(l-j).$$

In view of property (10.14) we obtain

$$J_2 = nm^2 \left(\sum_{s=1}^{R} \frac{\pi_s^3}{3} + \sum_{s=1}^{R-1} \sum_{t=s+1}^{R} \pi_s^2 \pi_t + \sum_{s=1}^{R-1} \sum_{t=s+1}^{R} \sum_{p=s+1}^{R} \pi_s \pi_t \pi_p + O(n^2)\right)$$

$$= \frac{nm^2}{3} \sum_{s=1}^{R} \sum_{t=1}^{R} \sum_{p=1}^{R} \pi_s \pi_t \pi_p + O(n^2) = \frac{nm^2}{3} + O(n^2). \qquad (10.15)$$

By analogy,

$$J_3 = \frac{n^2 m}{3} + O(n^2) \qquad \text{and} \qquad J_4 = O(n^2). \qquad (10.16)$$

Thus

$$DU \sim \tfrac{nm(n+m)}{12} + \tfrac{nm}{2}\Big[\textstyle\sum_{s=1}^{R} \sum_{t=1}^{R} (n+m) c_{st} \pi_s^2 \pi_t +$$

$$\textstyle\sum_{s=1}^{R} \sum_{t=1}^{R-1} \sum_{q=t+1}^{R} (2nc_{sq} \pi_s^2 \pi_t^2 + 2mc_{st} \pi_s^2 \pi_q) +$$

$$\textstyle\sum_{s=1}^{R-1} \sum_{t=1}^{R} \sum_{p=s+1}^{R} (2nc_{pt} \pi_s \pi_t \pi_p + 2mc_{st} \pi_s \pi_t \pi_p) +$$

$$+ \textstyle\sum_{s=1}^{R-1} \sum_{t=1}^{R-1} \sum_{p=s+1}^{R} \sum_{q=t+1}^{R} (4nc_{pq} \pi_s \pi_t \pi_p + 4mc_{st} \pi_s \pi_p \pi_q)\Big].$$

Let us simplify the expression in square brackets

$$\Big[\dots\Big] = n\textstyle\sum_{s=1}^{R}\sum_{t=1}^{R} c_{st}\pi_s(2\pi_1 + \dots + 2\pi_{s-1} + \pi_s)(2\pi_1 + \dots$$

$$+2\pi_{t-1} + \pi_t) + m\textstyle\sum_{s=1}^{R}\sum_{t=1}^{R} c_{st}\pi_s(\pi_s + 2\pi_{s+1} + \dots$$

$$+2\pi_R)(\pi_t + 2\pi_{t+1} + \dots + 2\pi_R) = nS_1 + mS_2.$$

We use the following properties of c_{ij} (Kemeny and Snell, 1960)

$$\sum_{j=1}^{R} c_{ij} = 0, \qquad \sum_{i=1}^{R} c_{ij}\pi_i = 0 \tag{10.17}$$

and write S_2 as

$$\begin{aligned} S_2 &= \sum_{s=1}^{R}\sum_{t=1}^{R}\pi_s\pi_t[(4c_{11}\pi_1 + \dots + 4c_{s-11}\pi_{s-1} + 2c_{s1}\pi_s) + \dots \\ &+ (4c_{1t-1}\pi_1 + \dots + 4c_{s-1t-1}\pi_{s-1} + 2c_{st-1}\pi_s) + (2c_{1t}\pi_1 + \dots \\ &+ 2c_{s-1t}\pi_{s-1} + c_{st}\pi_s)] = \sum_{s=1}^{R}\sum_{t=1}^{R}\pi_s\pi_t[-(2c_{s1}\pi_s + 4c_{s+11}\pi_{s+1} + \dots \\ &+ 4c_{R1}\pi_R) - \dots - (c_{st}\pi_s + 2c_{s+1t}\pi_{s+1} + \dots + 2c_{Rt}\pi_R)] \\ &= \sum_{s=1}^{R}\sum_{t=1}^{R}\pi_s\pi_t[(c_{st}\pi_s + 2c_{s+1t}\pi_{s+1} + \dots + 2c_{Rt}\pi_R) + \dots \\ &+ (2c_{sR}\pi_s + 4c_{s+1R}\pi_{s+1} + \dots + c_{RR}\pi_R)] \\ &= \sum_{s=1}^{R}\sum_{t=1}^{R} c_{st}\pi_s(2\pi_1 + \dots + 2\pi_{s-1} + \pi_s)(2\pi_1 + \dots + 2\pi_{t-1} + \pi_t). \end{aligned}$$

Hence

$$DU \sim nm(n+m)\left[\frac{1}{12} + 2\sum_{s=1}^{R}\sum_{t=1}^{R} c_{st}\pi_s\Big(\sum_{i=1}^{s-1}\pi_i + \frac{\pi_s}{2}\Big)\Big(\sum_{j=1}^{t-1}\pi_j + \frac{\pi_t}{2}\Big)\right]. \tag{10.18}$$

For the 2-dimensional case we can write explicit expressions for parameters c_{ij} and dispersion DU

$$c_{11} = \frac{\pi_2 - p_{12}}{p_{12} + p_{21}} \quad c_{12} = \frac{p_{12} - \pi_2}{p_{12} + p_{21}} \quad c_{21} = \frac{p_{21} - \pi_1}{p_{12} + p_{21}} \quad c_{22} = \frac{\pi_1 - p_{21}}{p_{12} + p_{21}}.$$

$$DU \sim \frac{nm(n+m)}{12}\left[1 + \frac{6\rho\pi_1\pi_2}{1-\rho}\right], \text{ where } \rho = 1 - p_{12} - p_{21}.$$

4. PROOF OF THE THEOREM

First consider proposition (10.10). To establish it we compute the dominant term of DV. Random variables α_i and β_j have uniform distribution on [0,1]. Indeed,

$$\alpha_i = E(z_{ij}|\xi_i) = P\{\tilde{\xi}_j > \xi_i|\xi_i\} = 1 - F(\xi_i), \qquad \beta_j = F(\tilde{\xi}_j).$$

Observe that

$$\begin{aligned} DV &= D\sum_{i=1}^{n}\sum_{j=1}^{m}(\alpha_i + \beta_j) = m^2 D\sum_{i=1}^{n}\alpha_i + n^2 D\sum_{j=1}^{m}\beta_j \\ &= m^2 E(\sum_{i=1}^{n}\alpha_i)^2 + n^2 E(\sum_{j=1}^{m}\beta_j)^2 - \frac{n^2m^2}{2}. \end{aligned} \tag{10.19}$$

It is easily seen that

$$E\left(\sum_{i=1}^{n}\alpha_i\right)^2 = \sum_{i=1}^{n}E\alpha_i^2 + 2\sum_{i=1}^{n-1}\sum_{k=i+1}^{n}E\alpha_i\alpha_k = \frac{n}{3} + 2\sum_{i=1}^{n-1}\sum_{k=i+1}^{n}E[\alpha_i E(\alpha_k|\xi_i)].$$

Note that for $k > i$

$$E(\alpha_k|\xi_i) = E(\alpha_k|e_i),$$

$$E(\alpha_k|e_i = s) = \sum_{t=1}^{R-1}\sum_{q=t+1}^{R}\pi_q p_{st}(k-i) + \sum_{t=1}^{R}\frac{\pi_t}{2}p_{st}(k-i).$$

Consequently,

$$E[\alpha_i E(\alpha_k|\xi_i)] = \sum_{s=1}^{R}\pi_s\sum_{t=1}^{R}(\sum_{q>t}\pi_q + \frac{\pi_t}{2})(\sum_{p>s}\pi_p + \frac{\pi_s}{2})p_{st}(k-i).$$

Using properties (10.14), we obtain:

$$\begin{aligned}
E(\sum_{i=1}^{n}\alpha_i)^2 &= \frac{n}{3}+\sum_{s=1}^{R}\sum_{t=1}^{R}(\frac{n^2}{4}\pi_t+\frac{n}{4}(2c_{st}-\pi_t))\pi_t\pi_s^2\\
&+ \sum_{s=1}^{R}\sum_{t=1}^{R-1}\sum_{q=t+1}^{R}(\frac{n^2}{2}\pi_t+\frac{n}{2}(2c_{st}-\pi_t))\pi_q\pi_s^2\\
&+ \sum_{s=1}^{R-1}\sum_{t=1}^{R}\sum_{p=s+1}^{R}(\frac{n^2}{2}\pi_t+\frac{n}{2}(2c_{st}-\pi_t))\pi_t\pi_p\pi_s\\
&+ \sum_{s=1}^{R-1}\sum_{t=1}^{R-1}\sum_{p=s+1}^{R}\sum_{q=t+1}^{R}(n^2\pi_t+n(2c_{st}-\pi_t))\pi_q\pi_p\pi_s+O(n).
\end{aligned}$$

We deduce the expression for $E\Big(\sum_{j=1}^{m}\beta_j\Big)^2$, by analogy. It follows from (10.12), (10.13), (10.14), (10.15),(10.16), and (10.19), that the dominant term of DV coincides with the dominant term of DU. Hence,

$$DU \sim DV. \tag{10.20}$$

Clearly,

$$D\Delta = DU + DV - 2cov(\sum_{i=1}^{n}\sum_{j=1}^{m} z_{ij}, \sum_{i=1}^{n}\sum_{j=1}^{m}(\alpha_i+\beta_j)) \tag{10.21}$$

We claim that

$$Cov(\sum_{i=1}^{n}\sum_{j=1}^{m} z_{ij}, \sum_{i=1}^{n}\sum_{j=1}^{m}(\alpha_i+\beta_j)) = DV. \tag{10.22}$$

Indeed,

$$\begin{aligned}
Cov(\sum_{i=1}^{n}\sum_{j=1}^{m} z_{ij}, \sum_{i=1}^{n}\sum_{j=1}^{m}(\alpha_i+\beta_j)) &= m\sum_{i=1}^{n}\sum_{j=1}^{m}\sum_{k=1}^{n}Ez_{ij}\alpha_k\\
&+ n\sum_{i=1}^{n}\sum_{j=1}^{m}\sum_{l=1}^{m}Ez_{ij}\beta_l-\frac{n^2m^2}{2}.
\end{aligned}$$

It is obvious that for $k < i$,

$$\begin{aligned}
Ez_{ij}\alpha_k &= E[\alpha_k E(z_{ij}|\xi_k)] = E[\alpha_k E(z_{ij}|e_k)]\\
E(z_{ij}|e_k) &= E(E(z_{ij}|e_i)|e_k) = E(\alpha_i|e_k)\\
E[\alpha_k E(z_{ij}|e_k)] &= E[\alpha_k E(\alpha_i|e_k)] = E(\alpha_i\alpha_k).
\end{aligned}$$

For $k > i$ we write

$$\begin{aligned} Ez_{ij}\alpha_k &= E[E(z_{ij}\alpha_k|\xi_i)] = E[E(z_{ij}|\xi_i)E(\alpha_k|\xi_i)] \\ &= E[\alpha_i E(\alpha_k|\xi_i)] = E(\alpha_i\alpha_k). \end{aligned}$$

For $k = i$ we have

$$E[\alpha_i E(z_{ij}|\xi_i) = E\alpha_i^2.$$

For β we reason by analogy. Hence

$$Cov(\sum_{i=1}^{n}\sum_{j=1}^{m} z_{ij}, \sum_{i=1}^{n}\sum_{j=1}^{m}(\alpha_i + \beta_j) = m^2 E(\sum_{i=1}^{n}\alpha_i)^2 + n^2 E(\sum_{j=1}^{m}\beta_j)^2 - \frac{n^2 m^2}{2}.$$

This expression coincides with expression (10.19) for DV. Thus, statement (10.22) is proved. Combining (10.20),(10.21), and (10.22), we conclude that $D\Delta = o(DU)$, and proposition (10.10) holds by Chebyshev inequality.

Finally consider proposition (10.11). To prove it we introduce the following notations. Consider state 1 of the Markov chain $\tilde{e}_j$. Let us denote

- $\tau_i(1)$, the ith passage time.
- $T_i = \tau_{i+1}(1) - \tau_i(1)$, the ith recurrence time.
- $Y_i(1) = \sum_{n=\tau_i}^{\tau_{i+1}-1} \beta_n$.

Since $Y_i(1)$ are distributed identically, we omit the index in the following definition:

$$M_1 = \pi_1 EY(1) \tag{10.23}$$

Denote

$$Z_i(1) = Y_i(1) - M_1 T_i(1) = \sum_{n=\tau_i}^{\tau_{i+1}-1} (\beta_n - M_1). \tag{10.24}$$

Random variables $Z_i(1)$ are also distributed identically. Denote

$$B_1 = \pi_1 EZ^2(1). \tag{10.25}$$

We now apply the central limit theorem for Markov chain functionals (Chung, 1960). It states that

$$\frac{\sum_{j=1}^{m}\beta_j - M_1 m}{\sqrt{B_1 m}} \longrightarrow N(0,1) \tag{10.26}$$

To calculate M_1 and B_1 we need the following formulas from Chung (1960)

$$EY(1) = \sum_{j=1}^{R} {}_1p^*_{1j} E(F(\tilde{\xi}_k)|\tilde{e}_k = j),$$

where $${}_kp^*_{ij} = \sum_{n=1}^{\infty} {}_kp^{(n)}_{ij}, \quad {}_kp^{(n)}_{ij} = P\{\tilde{e}_n = j; \tilde{e}_l \neq k, 0 < l < n|\tilde{e}_0 = i\}.$$

It is known that (Chung, 1960)

$${}_ip^*_{ij} = \tfrac{\pi_j}{\pi_i}, \quad {}_kp^*_{ij} = \tfrac{\pi_j}{\pi_k}(c_{kk} - c_{ik} + 1) + (c_{ij} - c_{kj}).$$

Therefore

$$M_1 = \sum_{j=1}^{R} \pi_j E(F(\tilde{\xi}_k)|\tilde{e}_k = j) = EF(\tilde{\xi}_k) = \frac{1}{2}.$$

It holds that (Chung, 1960)

$$\begin{aligned} EZ^2(1) &= \sum_j {}_1p_{1j} E([F(\tilde{\xi}_k) - \frac{1}{2}]^2|\tilde{e}_k = j) \\ &+ 2\sum_j {}_1p_{1j} E(F(\tilde{\xi}_l) - \frac{1}{2}|\tilde{e}_l = j) \sum_{k\neq 1} {}_1p^*_{jk} E(F(\tilde{\xi}_l) - \frac{1}{2}|\tilde{e}_l = k). \end{aligned}$$

Hence, B_1 can be written as

$$\begin{aligned} B_1 &= \pi_1 EZ^2(1) = \sum_j \pi_j E([F(\tilde{\xi}_k) - \frac{1}{2}]^2|\tilde{e}_k = j) \\ &+ 2\sum_j \pi_j E([F(\tilde{\xi}_l) - \frac{1}{2}]|e_l = j)\Big[\sum_{k\neq 1} \frac{\pi_k}{\pi_1}(c_{11} - c_{j1} + 1) \times \\ &\times E(F(\tilde{\xi}_l) - \frac{1}{2}|\tilde{e}_l = k) + \sum_{k\neq i}(c_{jk} - c_{1k}) E(F(\tilde{\xi}_l) - \frac{1}{2}|\tilde{e}_l = k)\Big]. \end{aligned}$$

We simplify the expression in brackets using the following relation

$$\sum_k \pi_k E(F(\tilde{\xi}_l) - \frac{1}{2}|e_l = k) = 0 \qquad (10.27)$$

We obtain

$$\Big[\dots\Big] = \sum_{k=1}^{R}(c_{jk} - c_{1k}) E(F(\tilde{\xi}_l) - \tfrac{1}{2}|\tilde{e}_l = k) - E(F(\tilde{\xi}_l) - \tfrac{1}{2}|\tilde{e}_l = 1).$$

It follows from (10.14) and (10.27) that

$$B_1 = \frac{1}{12} + 2\sum_j \sum_k c_{jk}\pi_j E(F(\tilde{\xi}_l)|\tilde{e}_l = j)E(F(\tilde{\xi}_l)|\tilde{e}_l = k).$$

If we denote $F^k(x) = P(\xi_i < x|e_i = k)$, then

$$\begin{aligned} E(F(\tilde{\xi}_l)|\tilde{e}_l = k) &= \int F(x)dF^k(x) \\ &= \pi_1 \int F^1(x)dF^k(x) + \ldots + \pi_R \int F^R(x)dF^k(x) \\ &= \pi_1 + \pi_2 + \cdots + \pi_{k-1} + \frac{\pi_k}{2}. \end{aligned}$$

Hence,

$$B_1 = \frac{1}{12} + 2\sum_j \sum_k c_{jk}\pi_j(\pi_1 + \cdots + \pi_{j-1} + \frac{\pi_j}{2})(\pi_1 + \cdots + \pi_{k-1} + \frac{\pi_k}{2}).$$

It is easily seen that $DU \sim nm(n+m)B_1$. Therefore

$$\frac{\sum_{j=1}^m \beta_j - \frac{m}{2}}{\sqrt{B_1 m}} \longrightarrow N(0,1), \quad m \to \infty. \tag{10.28}$$

By analogy

$$\frac{\sum_{i=1}^n \alpha_i - \frac{n}{2}}{\sqrt{B_1 n}} \longrightarrow N(0,1), \quad n \to \infty. \tag{10.29}$$

Thus

$$\frac{V - EV}{\sqrt{DU}} \sim \frac{m(\sum_{i=1}^n \alpha_i - \frac{n}{2})}{\sqrt{nm(n+m)B_1}} + \frac{n(\sum_{j=1}^m \beta_j - \frac{m}{2})}{\sqrt{nm(n+m)B_1}} =$$

$$= \frac{\sum_{i=1}^n \alpha_i - \frac{n}{2}}{\sqrt{B_1 n}} \frac{1}{\sqrt{\frac{n}{m}+1}} + \frac{\sum_{j=1}^m \beta_j - \frac{m}{2}}{\sqrt{B_1 m}} \frac{1}{\sqrt{\frac{m}{n}+1}} \longrightarrow N(0,1)$$

when $n, m \to \infty$, $\frac{n}{m} \to const.$

Hence, proposition (10.11) and the asymptotic normality are proved. □

5. THE TEST

In (10.18) we have unknown parameters, as opposed to the case of independent observations where the test is distribution-free. All these parameters are expressed in terms of elements of transition matrix P. In fact we have to evaluate $R \times (R-1)$ parameters. In the 2-dimensional case these parameters are p_{12} and p_{21}. We use both samples for their evaluation and form the following statistics:

$$\hat{p}_{12} = \frac{\sum_{i=1}^{n-1} I(e_i = 1, e_{i+1} = 2) + \sum_{j=1}^{m-1} I(\tilde{e}_j = 1, \tilde{e}_{j+1} = 2)}{\sum_{i=1}^{n-1} I(e_i = 1) + \sum_{j=1}^{n-1} I(\tilde{e}_j = 1)} \tag{10.30}$$

$$\hat{p}_{21} = \frac{\sum_{i=1}^{n-1} I(e_i = 2, e_{i+1} = 1) + \sum_{j=1}^{m-1} I(\tilde{e}_j = 2, \tilde{e}_{j+1} = 1)}{\sum_{i=1}^{n-1} I(e_i = 2) + \sum_{j=1}^{n-1} I(\tilde{e}_j = 2)} \tag{10.31}$$

In R-dimensional case we evaluate parameters by analogy. After evaluation we obtain the dominant term of DU using (10.18) and the U-statistic using (10.5). Theorem 1 implies that $\tilde{U} = \frac{U - \frac{mn}{2}}{\sqrt{DU}}$ has standard normal distribution for enough large m and n. If we denote by $\lambda_{\frac{\alpha}{2}}$ the normal quantile then the critical domain of the test is defined as

$$\left| \frac{U - \frac{mn}{2}}{\sqrt{DU}} \right| > \lambda_{\frac{\alpha}{2}}. \tag{10.32}$$

If the hypothesis of statistical homogeneity is valid, then U belongs to the critical domain with probability α.

References

[1] Wilcoxon, F. (1945). Individual comparisons by ranking methods, *Biometrics*, Vol. 1, 80–83.

[2] Mann, H.B. and Whitney, D.R. (1947). On a test of whether one of two random variables is stochastically larger than the other, *The Annals of Mathematical Statistics*, Vol. 18, 50–60.

[3] Wilks, S.S. (1963). *Mathematical Statistics*, New York, London, Wiley.

[4] Chung, K.L. (1960). *Markov Chains With Stationary Transition Probabilities*, Springer-Verlag, Berlin Gottingen Heidelberg.

[5] Kemeny, J.G., Snell, J.L. (1960). *Finite Markov chains*, Prinston, N. J., Van Nostrand.

Chapter 11

PRODUCT-TYPE ESTIMATOR OF CONVOLUTIONS

Ilya Gertsbakh
Ben Gurion University of the Negev, Israel

I. Spungin
Ben Gurion University of the Negev, Israel

Abstract An unbiased estimator of $P(\tau_1+\ldots+\tau_k \le T)$ is suggested, where τ_i are independent random variables (r.v.) with density $f_i(t)$ and distribution function $F_i(t)$. This estimator is constructed sequentially. First, a r.v. X_1 with density $f_1(x)/F_1(T)$ is generated. Its support is $[0,T]$. After observing $X_1 = x_1$, the second r.v. X_2 is generated from the density $f_2(x)/F(T-x_1)$. Its support is $[0, T-x_1]$, etc. The desired estimator has the form $B(T) = \prod \psi_i(X_i)$. We investigate the properties of this estimator and show how to use it to simulate the distribution function of the time to absorption for a Semi-Markov process.

Keywords: Monte Carlo simulation, convolutions, product-type unbiased estimator, time to absorption, semi-Markov process.

1. THE PRODUCT-TYPE ESTIMATOR OF A CONVOLUTION

Let $\tau_1,\ldots,\tau_k$ be i.r.v.'s with CDF $F_i(t)$ and density function (d.f.) $f_i(t)$. Consider a random vector

$$\mathbf{X} = (X_1, X_2, \ldots, X_{k-1})$$

with the joint density $f_{\mathbf{X}}(v_1,\ldots,v_{k-1})$, defined as follows :

$$f_{\mathbf{X}}(v_1,\ldots,v_{k-1}) =$$

$$f_{X_1}(v_1) f_{X_2|X_1}(v_2|v_1) \ldots f_{X_{k-1}|X_1\ldots X_{k-2}}(v_{k-1}|v_1,\ldots,v_{k-2}),$$

where

$$f_{X_1}(v_1) = \frac{f_1(v_1)}{F_1(T)}, \quad v_1 \in [0, T],$$

and

$$f_{X_j|X_1,\ldots,X_{j-1}}(v_j|v_1,\ldots,v_{j-1}) = \frac{f_j(v_j)}{F_j(T - v_1 - \ldots - v_{j-1})},$$

$v_j \in [0, T - v_1 - \ldots - v_{j-1}]$, for $j = 2, \ldots, k-1$. Define the following r.v. :

$$B_k(T) = F_1(T) \cdot F_2(T - X_1) \cdot \ldots \cdot F_k(T - X_1 - \ldots - X_{k-1}), \text{ for } k > 1. \tag{11.1}$$

Claim 1.
$B_k(T)$ is an unbiased estimator of $P(T) = P(\tau_1 + \ldots + \tau_k \le T)$.

Proof. Note that outside the region $A(T) = \{\mathbf{v} = (v_1, v_2, \ldots, v_{k-1}) : v_i \ge 0, 0 \le v_1 + \ldots + v_{k-1} \le T\}$, $f_{\mathbf{X}}(\cdot) = 0$. Inside $A(T)$,

$$f_{\mathbf{X}}(\cdot) = \prod_{i=1}^{k-1} \frac{f_i(v_i)}{F_i(T - v_1 - \ldots - v_{i-1})}.$$

Then by the definition of mean,

$$\begin{aligned}
E(B_k(T)) &= \int (k-1) \int_{\mathbf{v} \in A(T)} \prod_{i=1}^{k} F_i(T - v_1 - \ldots - v_{i-1}) f_{\mathbf{X}}(\mathbf{v}) d\mathbf{v} \\
&= \int (k-1) \int_{\mathbf{v} \in A(T)} F_k(T - v_1 - \ldots - v_{k-1}) \prod_{i=1}^{k} f_i(v_i) d\mathbf{v} \\
&= \int (k) \int_{(v_i : v_i \ge 0 \& 0 \le v_1 + \ldots + v_k \le T)} \prod_{i=1}^{k} f_i(v_i) d\mathbf{v}.
\end{aligned} \tag{11.2}$$

The method of obtaining the estimator $B_k(T)$ in a sequential fashion, as described above, can be viewed as a special case of a rather general procedure for weighted simulation of a random process described by Kovalenko and Kuznetzov (1988), Chapter 4.

2. GENERATING AN ESTIMATE OF $B_K(T)$

1. Generate $\mathbf{X} = (X_1, \ldots, X_{k-1})$ recursively : generate $X_1 = x_1$ from the population with d.f. $f_{X_1}(\cdot)$. Use x_1 to generate the value x_2 of r.v. X_2 from the conditional d.f. $f_{X_2|X_1}(\cdot)$, etc.

2. Substitute $\mathbf{X} = \mathbf{x} = (x_1, x_2, \ldots, x_{k-1})$ into (11.1).

For $\tau_i \sim Exp(\Lambda_i)$ the generation procedure is as follows. Put $\xi \sim \mathcal{U}(0,1)$ - the uniform distribution. Generate $\xi_1 = \alpha_1$; Put $x_1 = -\Lambda_1^{-1}\log(1 - \alpha_1(1-\exp(-\Lambda_1 T)))$; Generate $\xi_i = \alpha_i$, $i = 2, \ldots, k-1$. Denote $\xi_i := \alpha_i$; put

$$y_i = -\Lambda_i^{-1}\log(1 - \alpha_i(1 - \exp(-\Lambda_i(T - x_1 - \ldots - x_{i-1})))).$$

The vector $\mathbf{x} = (x_1, \ldots, x_{k-1})$ is a realization of $\mathbf{X} = (X_1, X_2, \ldots, X_{k-1})$.

3. CRUDE MONTE CARLO VS PRODUCT-TYPE ESTIMATOR

Crude Monte Carlo (CMC) estimator of $P(T)$ is the following one. Let I be the indicator of the event $(\tau_1 + \ldots + \tau_k \leq T)$. Obviously, $E[I] = P(T)$. Simulate N times the r.v. I and denote by I_r the r-th replica of r.v. I. Then the CMC estimator of $P(T)$ is defined as

$$\hat{P}(T) = \frac{\sum_{r=1}^{N} I_r}{N}. \tag{11.3}$$

Suppose that we are able to produce an unbiased estimator of $P(T)$ which has a product - type form

$$P^*(T) = \prod_{i=1}^{k} J_i,$$

where $J_i, i = 1, \ldots, k$ are nonnegative r.v.'s separated from 1 :

$$J_i \leq \epsilon_i < 1 \tag{11.4}$$

Claim 2.

(i) $Var[P^*(T)] \leq \epsilon_1 \ldots \epsilon_k P(T)(1 - P(T))$.

(ii) For $N = 1$, $Var[P^*(T)]/Var[\hat{P}(T)] \leq \prod_{i=1}^{k} \epsilon_i$.

Proof. By (11.4),

$$\begin{aligned} Var[P^*(T)] &= E[J_1^2 \ldots J_k^2] - E^2[P^*] \\ &\leq \prod_{i=1}^{k} \epsilon_i \cdot E[J_1 \ldots J_k] - E^2[J_1 \ldots J_k]. \end{aligned} \tag{11.5}$$

Multiply the second term in (11.5) by $\prod_{i=1}^{k} \epsilon_i$. (i) follows after simple algebra. (ii) follows immediately from (11.3) because $Var[\hat{P}(T)] = P(T)(1 - P(T))$.

4. COMPARISON OF THE PRODUCT-TYPE ESTIMATOR WITH BROWN'S ESTIMATOR

Brown et al (1981) suggested a Monte Carlo estimator of the renewal function which is based on the following estimator of convolutions. Let $\tau_i \sim F_i(t)$ be independent random variables. Define

$$Z_k(T) = F_k\left(T - \sum_{i=1}^{k-1} \tau_i\right),$$

if $(\tau_1 + \ldots + \tau_{k-1} \leq T)$, and zero otherwise.

Using the definition of mean value, it is easy to prove that $E[Z_k(T)] = P(\tau_1 + \ldots + \tau_{k-1} \leq T)$.

Claim 3.

$$\frac{Var[B_k(T)]}{Var[Z_k(T)]} \leq F_1(T) \ldots F_{k-1}(T).$$

Proof. Since $B_k(T)$ and $Z_k(T)$ are both unbiased,

$$Var[Z_k(T)] - Var[B_k(T)] = E[Z_k^2(T)] - E[B_k^2(T)] =$$

$$\int (k-1) \int F_k^2(T - v_1 - \ldots - v_{k-1}) f_1(v_1) \ldots f_{k-1}(v_{k-1}) d\mathbf{v} -$$

$$\int (k-1) \int F_1^2(T) \ldots F_k^2(T - v_1 - \ldots - v_{k-1})$$

$$\prod_{i=1}^{k-1} (f_i(v_i)/F_i(T - v_1 - \ldots - v_{i-1})) d\mathbf{v}.$$

Now replace in the second integral of the last expression $F_i(\cdot)$ by $F_i(T)$. After simple algebra, one can find out that

$$Var[Z_k(T)] - Var[B_k(T)] \geq Var[Z_k(T)](1 - \prod_{j=1}^{k-1} F_j(T)).$$

Now Claim 3 follows from this last expression. ■

5. CONVOLUTIONS IN SIMULATING THE TIME TO ABSORPTION IN A SEMI-MARKOV PROCESS

Suppose that we are given a finite-state semi-Markov process (SMP) $\xi_t, t \geq 0$, see Pyke (1961).

Let $E_0, E_1, \ldots, E_n$ be the states of the SMP $\xi_0 = E_0$. Suppose that the unique absorbing state E_n is reachable from all other states. The SMP is defined through the matrix

$$\mathbf{Q}(t) = ||Q_{ij}(t)||,$$

see the above cited paper of R.Pyke. The Markov chain $\{I_m, m \geq 1\}$ imbedded into the SMP has the transition probabilities

$$P(I_{m+1} = E_j | I_m = E_i) = Q_{ij}(\infty) = p_{ij}, \text{ and}$$

$$F_{ij}(t) = Q_{ij}(t)/Q_{ij}(\infty)$$

is the CDF of the sitting time $X_{i,j}$ in state E_i *given* that the next transition is into the state E_j.

Suppose now that we are interested in obtaining an unbiased estimator for the CDF of the transition time $\tau_{0,n}$ from state E_0 into the absorbing state E_n. Let Ω be the set of all trajectories $\{\omega\}$ leading from E_0 to E_n. Each ω is a sequence of states $E_0, E_1(\omega), \ldots, E_{r-1}(\omega), E_r(\omega) = E_n$.

By the formula of the total probability,

$$P(\tau_{0,n} \leq T) = \sum_{\omega \in \Omega} p(\omega) P(\tau_{0,n} \leq T | \omega),$$

where $p(\omega)$ is the measure of $\omega = (E_0, E_1(\omega), \ldots, E_r(\omega) = E_n)$, and

$$P(\tau_{0,n} \leq T|\omega) = P(X_{E_0,E_1(\omega)} + X_{E_1(\omega),E_2(\omega)} + \ldots + X_{E_{(r-1)}(\omega),E_r(\omega)} \leq T).$$

For a *given* ω, the last expression has the form $P(\tau_1 + \tau_2 + \ldots + \tau_m \leq T)$, where τ_i are conditionally (given ω) independent r.v.'s with known CDF's. Therefore, our previously developed technique for obtaining unbiased estimators for convolutions can be used to obtain an unbiased estimator for $P(\tau_{0,n} \leq T)$. In particular, the following procedure could be suggested.

(i) Generate a trajectory ω_j leading from the initial state E_0 to the absorbing state E_n.

(ii) Calculate for this trajectory an estimate $\hat{P}_j(T|\omega_j)$ of the probability that on given ω_j the absorption time does not exceed T.

Repeat (i),(ii) M times and calculate

$$\hat{P}(T) = M^{-1} \sum \hat{P}_j(T|\omega_j). \qquad \blacksquare$$

References

[1] M. Brown et al. (1981). Monte Carlo Simulation of the Renewal Function, *J. Appl. Probability*, **18**, 426-434.

[2] I.N. Kovalenko and N.Yu. Kuznetsov. (1988). *Calculation Methods for Highly Reliable Systems.* Moscow, Sovietskoe Radio (in Russian).

[3] R.Pyke. (1961). Markov Renewal Processes: Definitions and Preliminary Properties, *Ann. Math. Statist.*, **33**, 1231-1242.

Chapter 12

NON-PARAMETRIC FAILURE RATE ESTIMATION OF SEMI-MARKOV SYSTEMS

Brahim Ouhbi
Ecole Nationale Supérieure d'Arts et Métiers de Meknès, Morocco

Nikolaos Limnios
Université de Technologie de Compiègne, France

Abstract In this paper we propose an estimator of the failure rate of a semi-Markov system and derive some of its theoretical asymptotic properties. To illustrate, a numerical example is given.

Keywords: Estimation, semi-Markov process, reliability, failure rate.

1. INTRODUCTION

Semi-Markov models have found important applications in reliability analysis, cf. Limnios (1997), Limnios and Oprişan (1997), etc. Estimators of reliability and availability of semi-Markov systems and their properties are given in Ouhbi and Limnios (1996, 1998) and a method for constructing confidence intervals for the availability of semi-Markov systems is given in Ouhbi and Limnios (1997).

In many fields of engineering science it is of great interest to estimate the failure rate of semi-Markov systems and to our knowlege, there has been no work carried out on this problem.

In the present paper, we define the failure rate of semi-Markov systems and introduce its estimator. We derive for the first time, uniform strong consistency and a central limit theorem for the previous estimator. We end this work by an application of our results to estimate the failure rate of a semi-Markov system.

2. PRELIMINARIES

Let us consider a Markov renewal process, $(J,S) = (J_n, S_n)_{n\geq 0}$, defined on a probability complete space, where $(J_n)_{n\geq 0}$ is a Markov chain with values in $E = \{1, ..., s\}$, which is the state space of the process, and $(S_n)_{n\geq 0}$ are the jump times which will be in $\overline{\mathbb{R}}_+ = [0, \infty]$. J_0, J_1, ..., J_n,... are the consecutive states to be visited by the MRP and $X_1, X_2, ...$ defined by $X_n = S_n - S_{n-1}$, for $n \geq 1$, are the sojourn times in these states taking values in $\overline{\mathbb{R}}_+$.

The semi-Markov process (SMP) $(Z_t)_{t\geq 0}$ is defined by $Z_t = J_{N_t}$.

A MRP or a SMP can be completely determined if we know its initial law and its semi-Markov kernel (which we shall estimate) defined respectively by: $P(J_0 = k) = p(k)$ and

$$P(J_{n+1} = k, X_{n+1} \leq x | J_0, J_1, ..., J_n, X_1, ..., X_n) = Q_{J_n k}(x), \quad (a.s.) \qquad (12.1)$$

for all $x \in \overline{\mathbb{R}}_+$ and $1 \leq k \leq s$.

The probabilities $p_{ij} = Q_{ij}(\infty)$. $(= \lim_{t\to+\infty} Q_{ij}(t))$ are the transition probabilities of the imbedded Markov chain $(J_n)_{n\geq 0}$.

Let us also consider the distribution function associated to sojourn time in state i before going to state j defined by:

$$F_{ij}(.) = \begin{cases} p_{ij}^{-1} \times Q_{ij}(.) & \text{if} \quad p_{ij} > 0 \\ 0 & \text{otherwise,} \end{cases}$$

and $H_i(.) = 1 - \sum_{j=1}^{s} Q_{ij}(.)$ is the survival function in state i. Suppose that $F_{ij}(.)$ is absolutely continuous with respect to the Lebesgue's measure and let $f_{ij}(.)$ be its density and $h_j(t) = \frac{dH_j}{dt}(t)$.

From (12.1) we take:

$$P(J_n = j / J_0, J_1, ..., J_{n-1} = i) = p_{ij}, \quad \text{for all } n > 0.$$

$$P(X_n \leq x / J_0, ..., J_{n-1} = i, J_n = j) = F_{ij}(x), \quad \text{for all } n \geq 0 \quad \text{and} \quad x \geq 0$$

$$P(X_{n_1} \leq x_1, X_{n_2} \leq x_2, ...X_{n_k} \leq x_k / J_n, n \geq 0) = \prod_{i=1}^{k} F_{J_{n_i-1} J_{n_i}}(x_i), \quad (a.s.$$

for $0 \leq n_1 \leq n_2 \leq ... \leq n_k$ and $x_i \geq 0$ for $i = 1, ..., k$.

We define, for all i and j in E, the transition rate, $\lambda_{ij}(t)$, of a semi-Markov kernel by:

$$\lambda_{ij}(t) = \begin{cases} \frac{\frac{dQ_{ij}}{dt}(t)}{1 - \sum_{k=1}^{s} Q_{ik}(t)} & \text{if} \quad p_{ij} > 0 \text{ and } \sum_{j=1}^{s} Q_{ik}(t) < 1 \\ 0 & \text{otherwise.} \end{cases}$$

Let us define also the cumulative hazard rate from state i to state j at time t by $\Lambda_{ij}(t) = \int_0^t \lambda_{ij}(u)du$ and the total cumulative hazard rate of state i at time t by $\Lambda_i(t) = \Sigma_{j=1}^{s}\Lambda_{ij}(t)$.

For all $i, j \in E$ and $t \in \overline{\mathbb{R}}_+$ we have:

$$Q_{ij}(t) = \int_0^t \exp(-\Lambda_i(u))\lambda_{ij}(u)du. \tag{12.2}$$

Hence, for every continuity point $t \geq 0$ of f_{ij}, we have:

$$Q'_{ij}(t) = p_{ij}f_{ij}(t) = \exp(-\Lambda_i(t))\lambda_{ij}(t). \tag{12.3}$$

Consider a censored history, at fixed time T, of a MRP which is described by:

$$\mathcal{H}(T) = (J_0, J_1, ..., J_{N_T}, X_1, X_2, ..., X_{N_T}, U_T), \tag{12.4}$$

where $U_T = T - S_{N_T}$ is the backward recurrence time and N_T is the number of jumps of the process before T.

In Ouhbi and Limnios (1996a), we have approximated the transition rate $\lambda_{ij}(t)$ by the piecewise constant function $\lambda^*_{ij}(t)$ defined by: $\lambda^*_{ij}(t) = \lambda_{ij}(v_k) = \lambda_{ijk}$ for $t \in (v_k, v_{k+1}] = I_k$, where $(v_k)_{0 \leq k \leq M}$ is a regular subdivision of [0,T], ($v_0 = 0$ and $v_M = T$), with step $\Delta_T = \frac{T}{M}$ and $M = [T^{1+\alpha}]$, $0 < \alpha < 1$ and $[x]$ is the integer part of x this implies that Δ_T is asymptotically equivalent to $T^{-\alpha}$, when T tends to infinity.

So,

$$\lambda^*_{ij}(t) = \sum_{k=0}^{M-1} \lambda_{ijk} 1_{(v_k, v_{k+1}]}(t),$$

where 1_A(x) is equal to one if x $\in A$ and zero otherwise. The maximum likelihood estimator (MLE) of λ_{ijk} is given by, (cf. Ouhbi and Limnios (1996a)):

$$\hat{\lambda}_{ijk} = \begin{cases} d_{ijk}/\nu_{ik} & \text{if} \quad p(i,j) > 0 \\ 0 & \text{otherwise} \end{cases}$$

with the assumption: $\frac{0}{0} = 0$. The estimator of $\lambda_{ij}(t)$ is then given by

$$\hat{\lambda}_{ij}(t, T) = \sum_{k=0}^{M-1} \hat{\lambda}_{ijk} 1_{(v_k, v_{k+1}]}(t), \tag{12.5}$$

where ν_{ik} is the sojourn time in state i on the interval time I_k, given for $N_T \geq 1$ by

$$\begin{aligned} \nu_{ik} &= \sum_{l=0}^{N_T-1} (X_{l+1} \wedge v_{k+1} - v_k) 1_{\{J_l = i, X_{l+1} \geq v_k\}} \\ &\quad + (U_T \wedge v_{k+1} - v_k) 1_{\{J_{N_T} = i, U_T \geq v_k\}}, \end{aligned}$$

and $d_{ijk} = \sum_{l=0}^{N_T-1} 1_{\{J_l=i, J_{l+1}=j, X_{l+1}\in I_k\}}$ is the number of transitions to state j from state i and for which the observed sojourn time in state i belongs to I_k.

3. RELIABILITY AND FAILURE RATE FUNCTIONS OF A SEMI-MARKOV SYSTEM

In reliability analysis, the state space, E, is naturally partitioned into two sets, U and D, where U is the set of working (or 'up') states, and D is the set of repair (or 'down') states, i.e., $E = U \cup D$ and $U \cap D = \emptyset$. Finite semi-Markov reliability models whose states space is partitioned in the above manner will be considered here. The transition from one state to another means, physically speaking, the failure or the repair of one of the components of the system. In the set of up states, U, the system is operational. No service is delivered when the system is in the set of down states, D. However, a repair will return the system from D to U. To model this situation, it will be assumed that the MRP is irreducible. The quantities examined here are the system's availability and reliability defined respectively by

$$A(t) = P[J_{N_t} \in U]$$

and

$$R(t) = P[J_{N_s} \in U, \text{for all } s \leq t].$$

In the sequel, for a matrix X, we will denote X^U its restriction to U and $\Delta X(t,T) = \hat{X}(t,T) - X(t)$ and for two functions X and Y, $[X * Y](t)$ is the Stieltjes convolution at point t of X and Y.

The reliability R(t) of the system is given by:

$$R(t) = \sum_{i\in U} p(i) R_i(t)$$

Let $R_i(t)$ to be the conditional probability that the first failure does not occur up to time t, given that the system started in state $i \in U$ at time zero.

So

$$R(t) = p^U.P^U(t).\mathbf{1},$$

where $P^U(t) = (I - Q^U(t))^{(-1)} * (I - \text{diag}(Q(t)\mathbf{1})^U)$ and $\mathbf{1} = (1, 1, ..., 1)^t$.

The failure rate function of the above described semi-Markov system is:

$$\begin{aligned}
\lambda_{SM}(t) &= \lim_{h \longrightarrow 0} P[Z_{t+h} \in D | Z_u \in U; \forall u \leq t]/h \\
&= \frac{R'(t)}{R(t)} \\
&= \frac{p^U \cdot \psi^U * [I - \text{diag}(Q(t)1)^U]' \cdot 1}{p^U \cdot \psi^U * (I - \text{diag}(Q(t)1)^U) \cdot 1} \\
&= \frac{p^U \cdot \psi^U * \text{diag}[H^U]'^{(t)} \cdot 1}{p^U \cdot \psi^U * (I - \text{diag}(Q(t)1)^U) \cdot 1}, \qquad (12.6)
\end{aligned}$$

where $\text{diag}(H^U(t))$ is a diagonal matrix with the ith component $H_i^U(t) = 1 - \sum_{j=1}^{s} Q_{ij}(t)$ for all $i \in U$. Thus

$$[H^U]_i'(t) = -\sum_{j=1}^{s} p_{ij} \cdot \frac{dF_{ij}}{dt}(t) = -\sum_{j=1}^{s} p_{ij} \cdot f_{ij}(t).$$

4. ESTIMATOR OF THE FAILURE RATE OF A SEMI-MARKOV SYSTEM

Let $\hat{Q}$ be the MLE of the transition probability of the semi-Markov kernel Q. Then we propose the following estimator for $P^U(.)$:

$$\hat{P}^U(t,T) = (I - \hat{Q}^U(t,T))^{(-1)} * (I - \text{diag}(\hat{Q}(t,T)1)^U)$$

The reliability estimator of the system is:

$$\hat{R}(t,T) = \hat{p}_U . \hat{P}^U(t,T).1 \qquad (12.7)$$

If we replace in (12.5) each quantity by its estimator we obtain an estimator of the failure rate of the semi-Markov system which is given by

$$\hat{\lambda}_{SM}(t) = \frac{\hat{p}^U \cdot \hat{\psi}^U * \text{diag}[\hat{H}^U]'(t) \cdot 1}{\hat{p}^U \cdot \hat{\psi}^U * (I - \text{diag}(\hat{Q}(t,T)1)^U) \cdot 1} \qquad (12.8)$$

Theorem 1
If the semi-Markov kernel is C^1, then the estimator of the failure rate (12.8), $\hat{\lambda}_{SM}(\cdot)$, is uniformly strongly consistent in the sense that for all $L \in \mathbb{R}_+$,

$$\sup_{t \in [0,L]} |\hat{\lambda}_{SM}(t,T) - \lambda_{SM}(t)| \longrightarrow 0 \ \ (a.s.), \ \text{as } T \longrightarrow \infty$$

Proof. In Ouhbi and Limnios (1997b), it was proved that the estimator of the reliability, $\hat{R}(t)$, is uniformly strongly consistent in the sense that,

$$\sup_{t\in[0,L]} |\hat{R}(t,T) - R(t)| \longrightarrow 0 \qquad (a.s.),\ as\ T \longrightarrow \infty.$$

Thus to prove the uniform strong consistency of the failure rate estimator, it is sufficient to prove the same property for the numerator in (12.8). On the other hand, in Ouhbi and Limnios (1996b), it was seen that the estimator of the Markov renewal matrix $\hat{\psi}$ is uniformly strongly consistent on [0,L], for all $L \in \mathbb{R}_+$ in the sense that

$$\max_{i,j} \sup_{t\in[0,L]} |\hat{\psi}_{ij}(t,T) - \psi_{ij}(t)| \longrightarrow 0 \qquad (a.s.),\ \text{as}\ T \longrightarrow \infty.$$

To derive the almost sure convergence of the estimator of $h_j(t)$, it was seen in Ouhbi and Limnios (1997b) that, it can be written as,

$$\hat{h}_j(t,T) = \frac{\sum_{l=1}^{N_j} 1_{\{X_{jl}\in I_k\}}}{N_j\Delta} + O(N_j^{-1}) = \frac{\sum_{l=1}^{N_j} \xi_l}{N_j\cdot\Delta} + O(N_j^{-1}),$$

where $\xi_l = 1_{\{X_{jl}\in I_k\}}$ and $N_j = N_j(T)$ is the number of times of the occurrence of the event $\{J_k = j\}$ for $0 < k < N_T + 1$ and the random variable X_{jl} denotes the lth sojourn time in state j in the time interval [0,T], these variables are independent and identically distributed.

It is clear that,

$$E[\frac{\sum_{l=1}^{N_j} 1_{\{X_{jl}\in I_k\}}}{N_j\cdot\Delta}] - h_j(t) \longrightarrow 0.$$

Since, in our case $N_j \longrightarrow \infty$ $(a.s.)$, $\Delta \longrightarrow 0$ and $N_j\cdot\Delta^2 \longrightarrow \infty$ $(a.s.)$, as $T \longrightarrow \infty$, to give the proof of the almost sure convergence of $\hat{h}_j(t,T)$ to $h_j(t)$, by Theorem 2.1, p. 10, in Gut (1988), is to give the proof of the almost sure convergence of $\hat{h}_j(t,n) = \frac{\sum_{l=1}^{n}\xi_l}{n\Delta}$ to $h_j(t)$ as n tends to infinity, with $\Delta = n^{-\alpha}$.

Let us consider, the empirical distribution function defined by

$$\hat{H}_j(t,n) = \frac{1}{n}\sum_{l=1}^{n} 1_{\{X_{jl}\le t\}}.$$

We have

$$\begin{aligned} |\hat{h}_j(t,n) - E[\hat{h}_j(t,n)]| &= \left|\frac{\hat{H}_j(v_{k+1},n) - \hat{H}_j(v_k,n)}{\Delta} - \frac{H_j(v_{k+1}) - H_j(v_k)}{\Delta}\right| \\ &\le \frac{2}{\Delta}\sup_x |\hat{H}_j(x,n) - H_j(x)| \qquad (12.9) \end{aligned}$$

By Dvoretzky's inequality Dvoretzky et al. (1956), there exists positive constants C and $0 < \beta \leq 2$ such that, for every $\varepsilon > 0$,

$$\begin{aligned} P\{\sup_t |\hat{h}_j(t,n) - E[\hat{h}_j(t,n)]| > \varepsilon\} &\leq P\{\sup_t |\hat{H}_j(t,n) - H_j(t)| > \varepsilon\frac{\Delta}{2}\} \\ &\leq C\exp(\frac{-\beta}{4}\varepsilon^2 n^{1-2\alpha}) \end{aligned}$$

Since, for $0 < \alpha < \frac{1}{2}$, we have,

$$\sum_{n=0}^{\infty} \exp(-2\varepsilon^2 n^{1-2\alpha}) < \infty.$$

By the Borel-Cantelli's lemma, we conclude that:

$$\limsup_{n\longrightarrow\infty} \sup_t |\hat{h}_j(t,n) - E[\hat{h}_j(t,n)]| = 0 \ \ (a.s.)$$

We conclude, by Theorem 2.1, p. 10, in Gut (1988), that

$$\sup_t |\hat{h}_j(t,T) - E[\hat{h}_j(t,T)]| \longrightarrow 0 \ \ (a.s.), \ \text{ as } T \longrightarrow \infty.$$

On the other hand, since $\hat{h}_j(t,T)$ is asymptotically unbiased, we have almost sure point convergence. The almost sure uniform convergence on compact $[0, L]$ is a consequence of the density's continuity.

Thus the estimator of $[\hat{H}^U]'$ is uniformly strongly consistent on $[0, L]$, for all $L \in \mathbb{R}_+$. Now, by a generalization of a Helly-Bray theorem, cf. Baxter and Li (1994), it is easy to see that $\hat{\psi}^U * (I - \text{diag}(\hat{Q}'(t,T)\cdot \mathbf{1})^U$ is uniformly strongly consistent.

Finally, since the number of states is finite, the desired result is obtained.

Theorem 2
If for all $(i,j) \in E \times E$, $f_{ij}(\cdot)$ is C^2 then $(T^{1-\alpha})^{1/2}(\hat{\lambda}_{SM}(t,T) - \lambda_{SM}(t))$ converges in law, as T tends to infinity, to a zero mean normal random variable with variance

$$\sigma^2(t) \ = \ \frac{\sigma_1^2(t)}{(R(t))^2}, \tag{12.10}$$

where

$$\begin{aligned} \sigma_1^2(t) \ = \ & \sum_{j\in U} \mu_{jj}\{[(\sum_{i\in U} p(i)\cdot \psi_{ij}^{U'})^2 * H_j](t) \\ & -(\sum_{i\in U} p(i)\cdot [\psi_{ij}^{U'} * H_j](t))^2\} \end{aligned} \tag{12.11}$$

and μ_{ij} denotes the mean first passage time from state i to state j.

Proof. From Ouhbi and Limnios (1997b), we have the convergence in probability of $\hat{R}(t)$ to $R(t)$. So, it is sufficient to prove that

$$(T^{1-\alpha})^{1/2}[\hat{p}^U \cdot \hat{\psi}^U * \mathrm{diag}[\hat{H}^U]'(t) \cdot \mathbf{1} - p^U \cdot \psi^U * \mathrm{diag}[H^U]'(t) \cdot \mathbf{1}]$$

converges in law to a zero mean normal random variable with variance σ_1^2.

Firstly, remark that

$$\begin{aligned}
&(T^{1-\alpha})^{1/2}[\hat{p}^U \cdot \hat{\psi}^U * \mathrm{diag}[\hat{H}^U]'(t) \cdot \mathbf{1} - p^U \cdot \psi^U * \mathrm{diag}[H^U]'(t) \cdot \mathbf{1}] \\
&= (T^{1-\alpha})^{1/2}[\hat{p}^U \cdot \Delta\psi^U * \mathrm{diag}[\hat{H}^U]'(t) \cdot \mathbf{1} + p^U \cdot \psi^U * \mathrm{diag}[\Delta H^U]'(t) \cdot \mathbf{1}] \\
&= (T^{1-\alpha})^{1/2}[\hat{p}^U \cdot \Delta\psi^U * \mathrm{diag}[\Delta H^U]'(t) \cdot \mathbf{1} + p^U \cdot \psi^U * \mathrm{diag}[\Delta H^U]'(t) \cdot \mathbf{1} \\
&-p \cdot \Delta\psi^U * \mathrm{diag} H' \cdot \mathbf{1}].
\end{aligned}$$

It is clear that, if for all $(i,j) \in E \times E$, $f_{ij}(\cdot)$ is twice continuously differentiable then $(T^{1-\alpha})^{1/2}[p \cdot \Delta\psi^U * \mathrm{diag}H' \cdot \mathbf{1}](t)$ converges to zero, in probability as $T \longrightarrow \infty$. On the other hand, we have

$$\begin{aligned}
\sqrt{T}\Delta\psi_{ij}(t,T) &= \sqrt{T}(\hat{\psi}_{ij}(t,T) - \psi_{ij}(t)) \\
&= \sqrt{T}[\hat{\psi}_{ij} - (\hat{\psi} * \psi)_{ij} + (\hat{\psi} * \psi)_{ij} - \psi_{ij}](t) \\
&= \sqrt{T}\{[\hat{\psi} * (I - \psi)]_{ij} + [(\hat{\psi} - I) * \psi]_{ij}\}(t) \\
&= \sqrt{T}\{-(\hat{\psi} * Q * \psi)_{ij} + (\hat{\psi} * \hat{Q} * \psi)_{ij}\}(t) \\
&= \sqrt{T}[\hat{\psi} * \Delta Q * \psi]_{ij}(t) \\
&= \sqrt{T}[\hat{\psi} * \Delta Q * \psi * \Delta Q * \psi]_{ij}(t) \\
&\quad + \sqrt{T}[\psi * \Delta Q * \psi]_{ij}(t).
\end{aligned} \tag{12.12}$$

From (12.11), we see that to prove that $(T^{1-\alpha})^{1/2}\{\hat{p}^U \cdot \Delta\psi^U * \mathrm{diag}[\Delta H^U]'\}(t) \cdot \mathbf{1}$ converges in probability to zero as $T \longrightarrow \infty$, it is sufficient to prove that for all $(i,j,k,l) \in E \times E \times E \times E$, $(T^{1-\alpha})^{1/2}[\Delta F_{ij} * \Delta F'_{kl}](t)$ converges in probability to zero as $T \longrightarrow \infty$. Consider now the empirical distribution function defined by:

$$\hat{F}_{ij}(x,T) = \frac{1}{N_{ij}} \sum_{l=0}^{N_{ij}-1} 1_{\{J_l=i, J_{l+1}=j, X_{l+1} \leq x\}} = \frac{1}{N_{ij}} \sum_{l=1}^{N_{ij}} 1_{\{X_{ijl} \leq x\}}$$

where $N_{ij} = N_{ij}(T)$ is the number of transitions from state i to state j which occur in the time interval [0, T] and X_{ijl} denote the lth sojourn time in state i before going to state j. These variables are independent and identically distributed.

$$
\begin{aligned}
&(T^{1-\alpha})^{1/2}[|\Delta F_{ij} * \Delta F'_{kl}(t)|] = \\
= \quad & (T^{1-\alpha})^{1/2}[\frac{1}{N_{ij}\cdot\Delta\cdot N_{kl}}\sum_{n=1}^{N_{ij}}\sum_{m=1}^{N_{kl}}\int_0^t [1_{\{t-u<X_{klm}\leq t-u+\Delta\}}1_{\{X_{ijn}\in(u,u+du]\}} \\
- \quad & \Delta\cdot F'_{kl}(t-u)1_{\{X_{ijn}\in(u,u+du]\}} - 1_{\{t-u<X_{klm}\leq t-u+\Delta\}}dF_{ij}(u) \\
+ \quad & \Delta\cdot F'_{kl}(t-u)dF_{ij}(u)]. \qquad (12.13)
\end{aligned}
$$

Denote Z_{nm} the random variable given by,

$$
\begin{aligned}
Z_{nm} \quad = \quad & \int_0^t 1_{\{t-u<X_{klm}\leq t-u+\Delta\}}1_{\{X_{ijn}\in(u,u+du]\}} \\
& -\Delta\cdot Q'_{kl}(t-u)1_{\{X_{ijn}\in(u,u+du]\}} \\
& -1_{\{t-u<X_{klm}\leq t-u+\Delta\}}dF_{ij}(u) \\
& +\Delta\cdot F'_{kl}(t-u)dF_{ij}(u). \qquad (12.14)
\end{aligned}
$$

It can be easily seen that $E[Z_{nm}/X_{klm}] = 0$. By the Markov inequality, we have

$$
\begin{aligned}
P(\sqrt{T^{1-\alpha}}|\Delta F_{ij} * \Delta f_{kl}| \geq \epsilon) \quad & \leq \quad \frac{T^{1-\alpha}}{\epsilon^2}E[(\Delta F_{ij} * \Delta f_{kl})^2] \\
& \leq \quad \frac{4\sup_{u\leq t}|f_{kl}(u)|T^{1-\alpha}}{\epsilon^2}E[\frac{N_{ij}N_{kl}}{(N_{ij}N_{kl})^2}] \\
& = \quad \frac{4\sup_{u\leq t}|f_{kl}(u)|}{\epsilon^2}E[\frac{T^{1-\alpha}}{N_{ij}N_{kl}}],
\end{aligned}
$$

which tends to zero as $T \longrightarrow \infty$.

Finally, $(T^{1-\alpha})^{1/2}(\hat{\lambda}_{SM}(t,T) - \lambda_{SM}(t))$ has the same limit in law as

$$
(T^{1-\alpha})^{1/2}\frac{p^U\cdot\psi^U * \mathrm{diag}(\Delta[H^U]'(t))\cdot\mathbf{1}}{R(t)}.
$$

Let G be a vector defined by $G_j(t) = \sum_{i\in U} p_i \cdot \psi_{ij}(t)$.

On the other hand

$$\sqrt{T^{1-\alpha}}p^U \cdot \psi^U * \mathrm{diag}([\hat{H}^U]'(t)) \cdot \mathbf{1} - p^U \cdot \psi^U * \mathrm{diag}([H^U]'(t)) \cdot \mathbf{1}$$
$$= \sum_{j\in U}\sum_{i\in U} p_i \cdot \psi_{ij} * [\hat{H}_j' - H_j'](t)$$
$$= \sqrt{T^{1-\alpha}}\sum_{j\in U} G_j * [\frac{\hat{H}_j(\cdot+\Delta) - \hat{H}(\cdot)}{\Delta} - H_j'(\cdot)](t)$$
$$= \sqrt{T^{1-\alpha}}\sum_{j\in U}\int_0^t [\frac{\hat{H}_j(t-u+\Delta) - \hat{H}_j(t-u)}{\Delta} - H_j'(t-u)]dG_j(u)$$
$$= \sqrt{T^{1-\alpha}}\sum_{j\in U}\frac{1}{N_j\cdot\Delta}\sum_{l=1}^{N_j}[G_j(t-X_{jl}+\Delta) - G_j(t-X_j) - \Delta\cdot G_j * H_j'(t)]$$

So, it is easy to see that the asymptotic mean of $\sqrt{T^{1-\alpha}}(\hat{\lambda}_{SM}(t) - \lambda_{SM}(t))$ is zero. Therefore, one may apply the theorem of Anscombe on sums of random numbers of random variables (see Theorem 3.1 p. 15 in Gut (1988)) and then consider $N_{ij} = n$ not random because $\frac{T}{N_j}$ converges (a.s.) to μ_{jj}. So, it is sufficient to prove the asymptotic normality of:

$$\begin{aligned}\sqrt{n^{1-\alpha}}W(t) &= \sqrt{n^{1-\alpha}}\sum_{j\in U}\frac{1}{n\cdot\Delta}\sum_{l=1}^{n}[G_j(t-X_{jl}+\Delta) \\ &\quad - G_j(t-X_j) - \Delta\cdot G_j * H_j'(t)], \qquad (12.15)\end{aligned}$$

Consider the random variable

$$Z_l = \frac{1}{\Delta}\sum_{j\in U}[G_j(t-X_{jl}+\Delta) - G_j(t-X_j) - \Delta\cdot G_j * H_j'(t)].$$

It is clear that $(Z_l)_{l\leq n}$ are not independent but one can fix a state, say i_0 and consider, as in Pyke and Schaufele (1964), the random variables $Y_l = W(S_{i_0,l+1}) - W(S_{i_0,l})$ where $S_{i_0,l}$ is the ith time entry of the semi-Markov process to state i_0 and then $(Y_l)_{l\leq n}$ are independent. On the other hand it is easy to see that

$$\begin{aligned}E[(Z_1)^3] &\approx \sum_{j\in U}[(G_j)^3 * H'(t) + 3(G_j)^2 * H_j'(t) \times G_j * H_j(t) \\ &\quad + 3[G_j * H_j'](t) \times ([G_j * H_j'](t))^2 - [G_j * H_j'(t)]^3\end{aligned}$$

and

$$[var(Z_1)]^{\frac{3}{2}} \approx \sum_{j\in U}[(G_j)^2 * H'(t) - [G_j * H_j'(t)]^2$$

Consequently, the Lyapounov's condition of asymptotic normality, see Billingsley (1986), p.362, is valid for $\delta = 1$. Hence $\hat{\lambda}_{SM}(t)$ is asymptotically normal.

5. NUMERICAL APPLICATION

In this paragraph, we will apply our results to a three-state semi-Markov process. The transitions between states are given in fig. 12.1, below. The sojourn time in state 1 before going to state 2 is exponential with parameter $\lambda_1 = 0.1$. When state 2 is entered, the next state will be 1 or 3. If the next state to be visited is 3 then the sojourn time is Weibull, $W(\alpha_2, \beta_2)$, with $\alpha_2 = 0.1$ and $\beta_2 = 2$, or else the next state to be entered is 1 and the sojourn time in this state is Weibull, $W(\alpha_1, \beta_1)$ with $\alpha_1 = 0.3$ and $\beta_1 = 2$. When state 3 is entered, the next state to be visited is state 1 and the sojourn time in state 3 is exponential with parameter $\lambda_2 = 0.2$. We also suppose that states 1 and 2 are up states and state 3 is a down one, so $U = \{1, 2\}$ and $D = \{3\}$. The failure rate estimation for the system in fig. 12.1 is given in fig. 12.2.

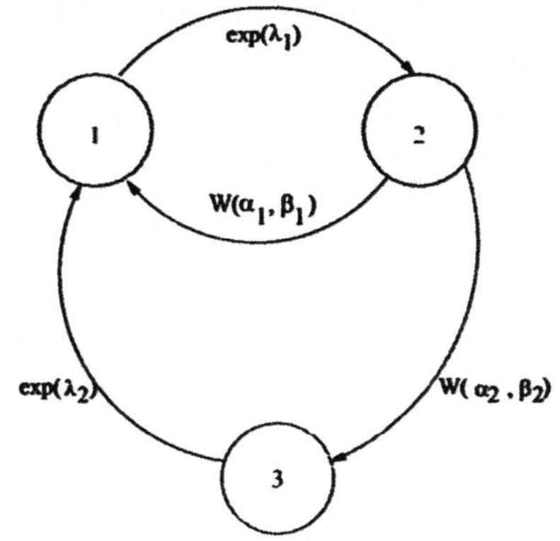

Figure 12.1 A three state semi-Markov system

References

[1] Baxter, L.A. and Li, L. (1994). Non-parametric confidence intervals for the renewal function and the point availability, *Scand. J. Statist.*, **21**, 277–287.

[2] Billingsley, P. (1986). *Probability and Measures*, J. Wiley & Sons, New York.

[3] Dvoretzky, A., Kiefer, J. and Wolfowitz, J. (1956). Asymptotic minimax character of the sample distribution function and the classical

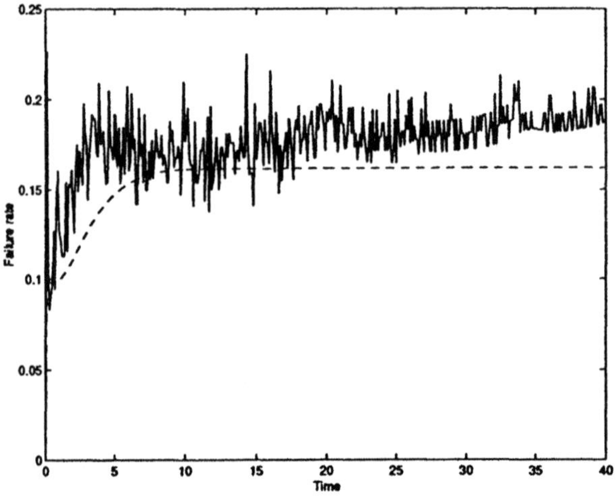

Figure 12.2 Estimation of the failure rate of the semi-Markov system

multinomial estimator, *Ann. Math. Statist.*, **27**, 642–669.

[4] Gut, A. (1988). *Stopped Random Walks. Limit Theorems and Applications*, Springer-Verlag.

[5] Limnios, N. (1997). Dependability analysis of semi-Markov systems, *Reliab. Eng. and Syst. Saf.*, **55**, 203–207.

[6] Limnios, N. and Oprişan, G. (1997). Semi-Markov process to regard of their applications, *World Energy System Journal*, **1**, no 1, 64–75

[7] Ouhbi, B. and Limnios, N. (1996). Estimation du taux de hasard associé au noyau du processus semi-Markovien et application en fiabilité, *XXVIIIèmes Journées de Statistique*, 27-30 May 96, Québec, Canada.

[8] Ouhbi, B. and Limnios, N. (1997a). Comportement asymptotique de la matrice de renouvellement markovien, *C. R. Acad. de Sci. Paris*, **325**, Série I, 921–924.

[9] Ouhbi, B. and Limnios, N. (1997b). Intervalles de confiance non-paramétriques pour la disponiblité des systèmes semi-markoviens, *Les XXIXèmes Journées de Statistique*, 26-30 Mai 1997, Carcassonne, France.

[10] Ouhbi, B. and Limnios, N. (1998). Estimation of kernel, availability and reliability of semi-Markov systems, In *Statistical and Probabilistic Models in Reliability*, D.C. Ionescu and N. Limnios (Eds), Birkhauser, Boston.

[11] Pyke, R. and Schaufele, R. (1964). Limit theorems for Markov renewal process, *Ann. Math. Statist.*, **35**, 1746–1764.

Chapter 13

ESTIMATION FOR SEMI-MARKOV MANPOWER MODELS IN A STOCHASTIC ENVIRONMENT

Sally McClean
University of Ulster, UK

Erin Montgomery
University of Ulster, UK

Abstract For purposes of manpower planning we often need to estimate and predict distributions of durations in various grades of a company hierarchy. We have developed a methodology for fitting a stochastic environment to manpower data for both the non-homogeneous Markov system in a stochastic environment (S-NHMS) and the non-homogeneous semi-Markov system in a stochastic environment (S-NHSMS). These models provide a means of describing changes in the environment between contiguous periods. We may therefore simultaneously model both the process underlying a particular homogeneous time period and the process governing movements between such periods, thus providing a description of time heterogeneity. We may then make predictions of future movements of staff through the system.

Keywords: Semi-markov models, stochastic environment, manpower planning.

1. INTRODUCTION

We are concerned with estimating distributions of durations in various grades of a company hierarchy for purposes of manpower planning. We will consider both continuous time Markov and semi-Markov models for the company hierarchy although it is typically the case that such durations have a peaked distribution representing cumulative inertia [11], indicating that a semi-Markov model is most appropriate. The distribution of such durations is also time inhomogeneous, since it varies with social, economic and demographic trends.

The database from which the durations were extracted consists of the career histories of some 50,000 nurses over an 11 year period [12, 8]. The data are the property of the Department of Health and Social Services of Northern Ireland. During the observation period any nurse withdrawing from or returning to active service was observed. We therefore have information on complete durations of employment, right censored durations where the individual entered service during the period and was still in service at the end of the period, left truncated durations where we observe staff who enter service prior to the start of the period and durations which are both left truncated and right censored who commenced service prior to the start of the period and were still in service at the end of the period [6, 7].

As is typical with manpower data, patterns of leaving and returning to service are highly time heterogeneous, in response to a changing social, economic and demographic environment. For example, during the 11 year period wastage rates of staff nurses fell from 26.9% per annum to 15.1%. This ever changing patterns means that we cannot base the estimation on data from the whole observation period. Rather we artificially truncate the data by viewing it through a series of windows during which patterns are fairly stable. Such a technique enables us to include information on durations which are as long as possible, thus providing an effective estimate of the tail of the distribution, within each window [9, 10]. This approach, which essentially uses census data for estimation, is a method familiar to demographers as discussed, for example, in [3]. Tuma and Hannan [15] have emphasised the need to use as much data as possible, otherwise seriously biased estimates will result. This technique, of basing the estimation on a time homogeneous window, means that, in addition to the types of statistical incompleteness mentioned earlier, the data are also subject to truncation.

The notion of a stochastic environment was introduced by Tsantas [13, 14] and Vassiliou [13], who presented the non-homogeneous Markov system in a stochastic environment (S-NHMS). This was later extended to the non-homogeneous semi-Markov model in a stochastic environment (S-NHSMS). These models provide a means of describing movements between contiguous periods in which time-homogeneous Markov or semi-Markov models prevail, along with a classification of such periods into states of what is known as the stochastic environment. We may therefore simultaneously model both the process underlying a particular homogeneous time period and the process governing movements between such periods, thus providing a description of time heterogeneity. We can thus make predictions of future movements of staff through the hierarchy.

More generally, such an approach is appropriate for event history data, e.g. Allison [1], Tuma & Hannan [15], or econometric duration data, e.g. Heckman & Singer [4]. Such repeatable durations are frequently encountered in the social sciences. However, the survival analysis literature has tended to focus on biomedical data where the event of interest is commonly death which is of course not repeatable, and the estimation is based on cohort data rather than on census data. Time heterogeneity is also more commonly a feature of social data than biomedical data, although medical data which derive from social processes, e.g. AIDS data, may also display time heterogeneity.

2. THE STOCHASTIC ENVIRONMENT

Tsantas [13, 14] and Vassiliou [13] defined the non-homogeneous Markov system in a stochastic environment (S-NHMS). This was later extended to the non-homogeneous semi-Markov model [16] in a stochastic environment (S-NHSMS). In each case we assume that we have a sequence of contiguous time windows where the behaviour of the company hierarchy in each window is described by a Markov or semi-Markov model, known as a scenario. We assume that the set of such scenarios is given by $S = \{S_1, \ldots, S_\nu\}$ where scenarios are chosen in accordance with a discrete time Markov model with corresponding transition matrix, the *compromise* matrix $\mathbf{C} = \{c_{ij}\}$. Then :

$$c_{ij} = Prob\,[S(t) = S_j \mid S(t-1) = S_i]$$

where $S(t)$ is the scenario in time period t.

We define the continuous time Markov model in terms of hazards of moving from grade G_f to grade G_j where there are k grades in all. We then define $\lambda_{fj}^{(u)}$ as the hazard (rate) of moving from grade G_f to grade G_j where scenario S_u is in place.

Analogously, for the semi-Markov model, we have the hazard of moving from grade G_f to grade G_j defined by $\lambda_{fji}^{(u)}$ for duration i, when scenario S_u is in place.

3. ESTIMATION

3.1 ESTIMATION FOR THE COMPROMISE MATRIX

The compromise matrix is the matrix of conditional probabilities describing transitions between the possible scenarios. The likelihood function required to determine the $\hat{c}_{ij}$'s is just that used to estimate probabilities for a Markov chain [2]. The likelihood function is therefore

$$L = \prod_{i=1}^{\nu} \prod_{j=1}^{\nu} c_{ij}^{n_{ij}} (1 - c_{ij})^{n_i - n_{ij}}$$

where ν denotes the total number of scenarios, n_{ij} is the number of transitions from scenario S_i to S_j and n_i is the total number of transitions from S_i, where $n_i = \sum_j n_{ij}$.

3.2 ESTIMATION FOR THE S-NHMS

We have previously described the estimation of parameters of continuous time Markov models when fitted to successive time windows where we then compared each pair of windows in turn in order to determine whether we may assume time homogeneity and merge them [10]. For the case of a stochastic environment we begin by fitting the model (Markov or semi-Markov) to the first time window; this is the first scenario. The model is then fitted to the second time window and tested to see if we can combine it with the first; if not this window becomes the second scenario. We then fit the model to the third window; if the first and second have been combined we then compare them with the third; if not we compare the third period with the second and if possible combine them, otherwise we compare the third period with the first window and if possible combine them. If none of the first, second and third windows can be combined, they are all regarded as separate scenarios. We continue in this manner until all windows have been assigned to scenarios. The compromise matrix is then estimated using the method described in the last section. Here the comparisons between windows may be carried out using likelihood ratio tests.

In each case, we must therefore derive likelihood functions and maximum likelihood estimators, assuming that we know the sequence of scenarios. We here denote the scenarios in the sequence which we want to use for estimation by $S(w)$ which is the state of the scenario matrix within the w-th time window. The estimation process then assigns a series of values to $S(w)$ which are chosen from the set of possible states of the compromise matrix. The likelihood function is given by :

$$L = \prod_{u=1}^{\nu} \prod_{w:S(w)=S_u} \prod_{f=1}^{m} \prod_{j=1}^{m+1} \left\{ \left(\lambda_{fj}^{(u)}\right)^{d_{fj}^{(w)}} e^{-\lambda_f^{(u)} T_f^{(w)}} \right\}$$

where $d_{fj}^{(u)}$ is the number of staff who move from grade G_f to grade G_j under scenario S_u and $T_f^{(u)}$ is the total time spent in grade G_f under scenario S_u. Differentiating and putting partial derivatives equal to zero yields the maximum likelihood estimators :

$$\hat{\lambda}_{fj}^{(u)} = \frac{\sum_{\{w|S(w)=S_u\}} d_{fj}^{(w)}}{\sum_{\{w|S(w)=S_u\}} T_j^{(w)}},$$

$(u = 1, \dots, \nu; f = 1, \dots, m; j = 1, \dots, m+1; f \neq j)$. Having evaluated the maximum likelihood estimators we may then calculator the maximum likelihood functions and using likelihood ratio tests derive a suitable sequence of scenarios. Once this sequence is constructed we use the formula in the previous section to derive the maximum likelihood estimator of the compromise matrix.

For the Markov model, using transition matrices based on the three grades Sister, Staff Nurse and Auxiliary as described by Gribbin and McClean [6] and McClean and Gribbin [7], the sequence of scenarios, where each corresponding time window was six months long, was found to be :

$$S_1 \rightarrow S_2 \rightarrow S_3 \rightarrow S_2 \rightarrow S_4 \rightarrow S_5 \rightarrow S_6 \rightarrow S_7 \rightarrow S_6 \rightarrow S_7 \rightarrow S_8 \rightarrow S_9 \rightarrow S_{10} \rightarrow S_9 \rightarrow S_8 \rightarrow S_9 \rightarrow S_{11} \rightarrow S_9.$$

The corresponding compromise matrix was then estimated to be :

$$\hat{C} = \begin{pmatrix} 0 & 1 & 0 & 0 & 0 & 0 & 0 & 0 & 0 & 0 & 0 \\ 0 & 0 & \frac{1}{2} & \frac{1}{2} & 0 & 0 & 0 & 0 & 0 & 0 & 0 \\ 0 & 1 & 0 & 0 & 0 & 0 & 0 & 0 & 0 & 0 & 0 \\ 0 & 0 & 0 & 0 & 1 & 0 & 0 & 0 & 0 & 0 & 0 \\ 0 & 0 & 0 & 0 & 0 & 1 & 0 & 0 & 0 & 0 & 0 \\ 0 & 0 & 0 & 0 & 0 & 0 & 1 & 0 & 0 & 0 & 0 \\ 0 & 0 & 0 & 0 & 0 & \frac{1}{2} & 0 & \frac{1}{2} & 0 & 0 & 0 \\ 0 & 0 & 0 & 0 & 0 & 0 & 0 & 0 & 1 & 0 & 0 \\ 0 & 0 & 0 & 0 & 0 & 0 & 0 & \frac{1}{3} & 0 & \frac{1}{3} & \frac{1}{3} \\ 0 & 0 & 0 & 0 & 0 & 0 & 0 & 0 & 1 & 0 & 0 \\ 0 & 0 & 0 & 0 & 0 & 0 & 0 & 0 & 1 & 0 & 0 \end{pmatrix}$$

As we can see from these estimates, transition patterns were changing fairly rapidly during the period. In the short term there is some evidence to suggest spells of homogeneity and periodicity but these were not maintained for long. We now move on to estimation of the compromise matrix for the S-NHSMS which should be more realistic than the S-NHMS considered in this section.

3.3 ESTIMATION FOR THE S-NHSMS

Estimation of the compromise matrix for the S-NHSMS proceeds much as for the S-NHMS where, again, we require the construction of an appropriate likelihood function from which the estimated hazard functions and corresponding homogeneity tests may be derived. However, since we no longer have the Markov property, we must now ensure that contiguous time windows are treated as a single large time window, not as separate windows as was the case for the S-NMMS. In this case, therefore, the index r (below) refers to the derived, possibly combined, window rather than the original one which we indexed by w. The likelihood function is then given by :

$$L = \prod_{u=1}^{\nu} \prod_{\{r|S(r)=S_u\}} \prod_{f=1}^{m} \prod_{j=1}^{m+1} \prod_{i=1}^{k_{fj}} \left\{ \left(\lambda_{fji}^{(u)}\right)^{\left[d_{fji}^{(r)} + \sum_{l=1}^{i} a_{fjil}^{(r)}\right]} \times \left(1 - \lambda_{fji}^{(u)}\right)^{\left[n_{fji}^{(r)} - d_{fji}^{(r)} - \sum_{l=1}^{i} a_{fjil}^{(r)}\right]} \right\}$$

where :

$n_{fji}^{(r)}$ are the number of individuals who are available to move from grade G_f to grade G_j at duration t_i during scenario S_r (the risk set);

$d_{fji}^{(r)}$ are the number of individuals who move from grade G_f to grade G_j at duration t_i during scenario S_r (the failures);

$a_{fjil}^{(r)}$ are the number of individuals who move from grade G_f to grade G_j at duration t_i, who are left truncated at duration t_l, during scenario S_r (the truncated failures);

k_{fj} is the number of time points at which a transition occurs from S_f to S_j.

Differentiating and putting partial derivatives equal to zero yields the maximum likelihood estimators :

$$\hat{\lambda}_{fji}^{(u)} = \frac{\sum_{\{r|S(r)=S_u\}} \left(d_{fji}^{(r)} + \sum_{l=1}^{i} a_{fjil}^{(r)}\right)}{\sum_{\{r|S(r)=S_u\}} n_{fji}^{(r)}}$$

$(u = 1, \ldots, \nu; f = 1, \ldots, m; j = 1, \ldots, m+1; f \neq j; i = 1, \ldots, k_{fj})$. These estimators are a generalisation of those described by Kalbfleish and Prentice [5] for simple competing risks models, McClean and Gribbin [7] for competing risks models subject to left truncation and right censoring and McClean, Montgomery and Ugwuowo [10] for contiguous time windows.

For the S-SHSMS, using transition matrices based on the three grades Sister, Staff Nurse and Auxiliary as before, the sequence of scenarios, where each corresponding time window was six months long, was found to be :

$$S_1 \to S_2 \to S_3 \to S_2 \to S_4 \to S_5 \to S_6 \to S_7 \to S_7 \to S_7 \to S_8 \to S_9 \to S_{10} \to S_9 \to S_9 \to S_9 \to S_{11} \to S_9 .$$

It is encouraging to note that this pattern is remarkably similar to that found for the S-NHMS; in fact slightly more time heterogeneity was found in this case. The corresponding compromise matrix was estimated to be :

$$\hat{C} = \begin{pmatrix} 0 & 1 & 0 & 0 & 0 & 0 & 0 & 0 & 0 & 0 & 0 \\ 0 & 0 & \frac{1}{2} & \frac{1}{2} & 0 & 0 & 0 & 0 & 0 & 0 & 0 \\ 0 & 1 & 0 & 0 & 0 & 0 & 0 & 0 & 0 & 0 & 0 \\ 0 & 0 & 0 & 0 & 1 & 0 & 0 & 0 & 0 & 0 & 0 \\ 0 & 0 & 0 & 0 & 0 & 1 & 0 & 0 & 0 & 0 & 0 \\ 0 & 0 & 0 & 0 & 0 & 0 & 1 & 0 & 0 & 0 & 0 \\ 0 & 0 & 0 & 0 & 0 & 0 & \frac{2}{3} & \frac{1}{3} & 0 & 0 & 0 \\ 0 & 0 & 0 & 0 & 0 & 0 & 0 & 0 & 1 & 0 & 0 \\ 0 & 0 & 0 & 0 & 0 & 0 & 0 & \frac{1}{2} & 0 & \frac{1}{4} & \frac{1}{4} \\ 0 & 0 & 0 & 0 & 0 & 0 & 0 & 0 & 1 & 0 & 0 \\ 0 & 0 & 0 & 0 & 0 & 0 & 0 & 0 & 1 & 0 & 0 \end{pmatrix}$$

We conclude from these results that in the short term there is evidence of time homogeneity but in the longer term the process is definitely time heterogeneous, as described by the S-NHSMS. There is little evidence of periodicity.

4. USING THE MODEL

The main advantage of using a model of the stochastic environment, which describes migrations between time windows as well as behaviour within specific time windows, is that we may obtain better predictions of the future behaviour of the system. Our previous approach could only make predictions within the current time window whereas our current approach allows us to predict environments for future windows.

We assume that we are interested in predicting grades of staff at a point u years ahead where the current duration is d years and each time window is w years wide where $sw \leq u < (s+1)w$ and the current state of the compromise matrix is C_r.

Then for the S-NHMS, we obtain :

$$\mathbf{P}(u) = \sum_{h_1} \cdots \sum_{h_s} c_{rh_1} \cdots c_{h_{s-1}h_s} \left(\prod_{\nu=1}^{s} \mathbf{P}_{h_\nu}(w) \right) \mathbf{P}_{h_{s+1}}(u - sw)$$

where $\mathbf{P}_h(t) = e^{-\Lambda_h t}$, $\Lambda_h = \lambda_{fj}^{(h)}$, $\mathbf{P}(u)$ is the matrix of probabilities of being in each state a time u years ahead. Here, we adopt the notation : $\lambda_{ff}^{(h)} = -\sum_{j \neq f} \lambda_{fj}^{(h)}$. For the S-NHSMS, we obtain the similar expression :

$$\mathbf{P}(u,d) = \sum_{h_1} \cdots \sum_{h_s} c_{rh_1} \cdots c_{h_{s-1}h_s} \left(\prod_{\nu=1}^{s} \mathbf{P}_{h_\nu}(w, d + (\nu - 1)w) \right) \times \mathbf{P}_{h_{s+1}}(u - sw, d + sw),$$

where $\mathbf{P}(u,d)$ is the matrix of probabilities of being in each state a time u years ahead when the duration is now d. These relationship between these transitions matrices and the hazards $\lambda_{fji}^{(h)}$ can be described in terms of the cumulative incidence functions, as describes by Bartholomew, Forbes and McClean [3].

Having estimated transition probabilities for each individual, from the current time point to a future time point, we may then predict future numbers in each grade.

5. CONCLUSION

We have developed a methodology for fitting a stochastic environment to manpower data for both the non-homogeneous Markov system in a stochastic environment (S-NHMS) and the non-homogeneous semi-Markov model in a stochastic environment (S-NHSMS). These models provide a means of describing movements between contiguous periods, in which time-homogeneous Markov or semi-Markov models prevail, in terms of the compromise matrix which is a transition matrix describing movements between successive stochastic environments. We may therefore simultaneously model both the process underlying a particular homogeneous time period and the process governing movements between such periods, thus providing a description of time heterogeneity. We can then make predictions of future movements of staff through the system.

Acknowledgments

Thanks are extended to the Department of Education for Northern Ireland for the provision of a grant during the course of the research which led to this paper.

References

[1] Allison, P. D. (1984). *Event History Analysis: Regression for Longitudinal Event Data*, Beverly Hills, CA: Sage.

[2] Bartholomew, D. J. (1977). The analysis of data arising from stochastic processes. In C O'Muircheartaigh and C Payne (eds.). *The Analysis of Survey Data*, **2**, Model Fitting, Wiley, 145–173.

[3] Bartholomew, D. J. and Forbes, A. F. and McClean, S. I. (1991). *Statistical Techniques for Manpower Planning*, Second edition, Wiley.

[4] Heckman, J. J. and Singer, B. (1984). Economic Duration Analysis. *J. Econometrics*, **24**, 63–132.

[5] Kalbfleisch, J. D. and Prentice, R. L. (1980). *The Statistical Analysis of Failure Time Data*, New York, Wiley.

[6] McClean, S. I. and Gribbin, J. O. (1987). Estimation for Incomplete Manpower Data. *Applied Stochastic Models and Data Analysis*, **3**, 13–25.

[7] McClean, S.I. and Gribbin, J. O. (1991). A non-parametric competing risks model for manpower planning. *Applied Stochastic Models and Data Analysis*, **7**, 327–341.

[8] McClean, S. I. and Reid, N. and Devine, C. and Gribbin, O. and Thompson, K. (1992). Using a Manpower Database to Analyse the Nurse Limbo Stock. *J. Advanced Nursing*, **17**, 992–1001.

[9] McClean, S. I. and Devine, C. (1995). A nonparametric maximum likelihood estimator for incomplete renewal data. *Biometrika*, **82:4**, 791–803.

[10] McClean, S. I. and Montgomery, E. J. and Ugwuowo, F. (1998). Non–homogeneous continuous–time Markov and semi–Markov models. *Applied Stochastic Models and Data Analysis*, **13**, 191–198.

[11] McGinnis, R. (1968). A Stochastic Model for Social Mobility. *Amer. Sociol. Rev.*, **33**, 712–721.

[12] Reid, N. (1986). Nurse Manpower: The Problems Ahead. *International Journal of Nursing Studies*, **23**, 187–197.

[13] Tsantas, N. and Vassiliou, P. C-G. (1993). The Non–homogeneous Markov System in a Stochastic Environment. *Journal of Applied Probability* **30**, 285–301.

[14] Tsantas, N. (1995). Stochastic Analysis of a non–homogeneous Markov System. *European Journal of Operational Research*, **85**, 670-685.

[15] Tuma, N. B. and Hannan, M. T. (1984). *Social Dynamics: Models and Methods.* New York, Academic Press.

[16] Vassiliou, P. C-G. (1998). The Evolution of the Theory of Non–homogeneous Markov Systems. *Applied Stochastic Models and Data Analysis*, **13**, 159–176.

Chapter 14

SEMI-MARKOV MODELS FOR LIFETIME DATA ANALYSIS

Rafael Pérez-Ocón
Universidad de Granada, Spain

Juan Eloy Ruiz-Castro
Universidad de Granada, Spain

M. Luz Gámiz-Pérez
Universidad de Granada, Spain

Abstract A semi-Markov model is applied to a data set of items that change stages over time. Two data sets are considered, in medical studies and in reliability: survival to breast cancer and the lifetime of optical fibres. The lifetimes of the subjects or items are considered, distribution functions are fitted to the staying times, and it is shown that the Weibull and log-normal distributions play an important role in these times. Covariates are incorporated to the model via the probability density functions of the staying times in the medical data set. The likelihood function is built and the parameters estimated. The transition probability functions and the survival functions for different risk groups are estimated from these values. A comparison with the empirical survival functions is made. Original computational programmes have been performed with *MATLAB* to estimate the transition probability functions, the survival functions, and the graphs of these functions.

Keywords: Semi-Markov process, Likelihood function, *MATLAB* programme, Survival function

1. INTRODUCTION

In lifetimes we are concerned with cohorts of items classified according to different states in the evolution of the item's lifetime. These states are determined by the specific characteristics of the problem considered. The items pass from one state to another, staying a random amount of time in each state. There are several important magnitudes that need to be known in these processes, such as the first time that a certain state is reached, the sojourn times in states, the entry times in a state, etc., all of which are of obvious interest. A model that has proven to be useful in these classes of problems is the homogeneous Markov process. Considerable progress has been made in the development and application of such models, particularly in biostatistics, in applications to longitudinal studies of diseases (see [4], and [8]; and references therein).

In homogeneous Markov processes it is assumed that the staying times in states are exponentially distributed, although this is not realistic in general. It is reasonable to assume that the probability of leaving a state depends on the time already spent there. One model taking into account this assumption is the semi-Markov model. [3] explained in detail the semi-Markov model in medical applications and applied it to the analysis of bone-marrow transplant. Time-dependent covariates were introduced by these authors, and the transition intensity functions were considered as in the Cox model, [1], with baseline hazard functions depending on the backward recurrence times. Semi-Markov models have also been considered by [11] and [14], among others, in manpower planning.

In this paper, certain research ideas are applied to analyze a data set. Two data sets are considered in order to apply a semi-Markov model. General distributions are introduced in the semi-Markov model as random variables that model the transition times in order to determine the best fit to the data set. In the study of the medical data set covariates are incorporated, so the non-homogeneity of the population is introduced in the changes among states. We have confirmed that the Weibull distribution plays an important role in the data set. In the study of reliability data, the items can be repaired twice at most, and replaced at the third failure. The mathematical expression of the survival function up to the fatal failure is built and plotted.

One problem when stochastic processes are involved in modelling studies is that the resulting expressions for the transition probability functions and the survival function have unmanageable mathematical expressions. Therefore, an important aspect of dynamic survival studies is emphasized in this paper is computation, which will allow the application of these models in longitudinal studies in a tractable way. The likelihood function associated in the general case is built, and the total parameters are estimated using the *MATLAB* programme when data are involved. Transition probability functions are calculated and the survival functions for different risk groups obtained.A comparison with the empirical functions is made. A goodness-of-fit test for censored data is applied to measure the difference among the empirical survival curves and the one obtained from the model. Life-tables for survival to different risk groups are given.

2. THE SEMI-MARKOV MODEL

2.1 TRANSITION PROBABILITY FUNCTIONS

Let $Z = \{Z(t); t \geq 0\}$ be a semi-Markov stochastic process with state space $E = \{1, 2, ..., m\}$. Details about this class of processes can be found in [2]. At different times, the system occupies well-defined states. If X_n denotes the state occupied in stage n, then $X = \{X_n; n \geq 0\}$ is a Markov chain, which is the Markov chain associated to the semi-Markov process Z. We denote by p_{ij} the transition probabilities in this chain, that is,

$$p_{ij} = P\{X_{n+1} = j | X_n = i\}.$$

The staying time in state i is denoted by T_i, $i = 1, \cdots, m$. The staying time in state i before going to state j has a distribution function $F_{ij}(t)$ and probability density function $f_{ij}(t)$. We write $S_{ij}(t) = 1 - F_{ij}(t)$ and

$$S_i(t) = 1 - F_i(t) = \sum_{j \neq i} p_{ij} S_{ij}(t). \tag{14.1}$$

The transition intensity functions are defined by

$$\lambda_{ij} = \lim_{h \to 0} \frac{P\left[T_{X_{n-1}} \in (t, t+h), X_n = j | T_{X_{n-1}} \geq t, X_{n-1} = i\right]}{h},$$

for $i \neq j, i, j \in E$, with $\lambda_{ii}(t) = -\lambda_i(t) = -\sum_{j \neq i} \lambda_{ij}(t)$.

These functions can be expressed by

$$\lambda_{ij}(t) = \frac{p_{ij} f_{ij}(t)}{S_i(t)}. \tag{14.2}$$

The transition probability functions of the process are

$$p_{ij}(t) = P[Z(t) = j | Z(0) = i], i, j = 1, \cdots, m, \ t > 0,$$

which can be expressed as

$$p_{ij}(t) = \sum_{r=1}^{m} \int_0^t p_{ir} f_{ir}(x) p_{rj}(t-x)dx + \delta_{ij} \sum_{n=1}^{m} p_{in} S_{in}(t), \qquad (14.3)$$

δ_{ij} being the Kronecker delta.The Laplace transform must be used to compute these probability functions.

A quantity of interest in these classes of processes is the semi-Markov kernel, defined by the family

$$Q = \{Q(i,j;t), i,j = 1,2,\cdots,m; t \geq 0\},$$

with

$$Q(i,j;t) = P\{X_{n+1} = j, T_i \leq t | X_n = i\}.$$

It is well known that $Q(i,j;t)/p_{ij}$ is the distribution function $G(i,j;t)$ of the conditioned random variable of the staying time in state i given that the process jumps from state i to state j. The methodology we follow in applications is related with the functions $f_{ij}(t)$.

2.2 MAXIMUM-LIKELIHOOD FUNCTION

Let us take the above semi-Markov model. Over a period of time, n items are observed, denoted by $h = 1,2,3,\cdots,n$. All the items begin in state 1. Item h has changed state m_h times in the instants $0 = t_{h,1} < t_{h,2} < \cdots < t_{h,m_h}$, the last time can be a censoring time. In these times the item has successively occupied the states $x_1^h, \cdots, x_{m_h}^h$.

The contribution to the likelihood function of an item that undergoes the transition $i \rightarrow j$ in a time t is the factor

$$\lambda_{ij}(t) S_i(t) = p_{ij} f_{ij}(t).$$

If an item is in a state i and is censored in a time t, the contribution to the likelihood function is the factor $S_i(t)$. Keeping this in mind, we build the likelihood function for all times and transition times observed,

$$L = \prod_{h=1}^{n} \left[\prod_{r=1}^{m_h-1} p_{x_{r-1}^h, x_r^h} f_{x_{r-1}^h, x_r^h}(t_{h,r} - t_{h,r-1}) \right]^{I(m_h)-1}$$

$$\left[p_{x_{m_h-1}^h, x_{m_h}^h} f_{x_{m_h-1}^h, x_{m_h}^h}(t_{h,m_h} - t_{h,m_h-1}) \right]^{\delta_h} \left[S_{x_{m_h-1}^h}(t_{h,m_h} - t_{h,m_h-1}) \right]^{1-\delta_h} \qquad (14.4)$$

δ_h being $= 1$ if item h fails and $= 0$ if it is censored, and $I(x)$ being the function $= 0$ if $x \leq 0$ and $= 1$ is $x > 0$.

3. SOME DATA ON SURVIVAL TIMES

We have applied this model to a cohort of patients of breast cancer. The subjects were 300 patients with breast cancer that had been given surgical treatment. Three states have been considered: no relapse (state 1), relapse(state 2), and death (state 3). All patients begin in state 1. The only observed transitions are $1 \to 2 \to 3$ and $1 \to 3$. State 3 is an absorbent state. The follow-up of this cohort spanned from 1973 up to 1996. The application of the semi-Markov model to this data set must be done incorporating covariates, as is usual in medical studies. In this cohort the covariates are the treatments applied to the patients: radiotherapy, hormonal therapy, and chemotherapy.

Let $\{Z(t;z); t \geq 0\}$ be the semi-Markov stochastic process defined in 2., where a covariate vector z has been introduced. The embedded Markov chain $\{X_n; n \geq 0\}$ does not depend on the covariate vector z, so we have the same transition probabilities in this process, $p_{ij} = P\{X_{n+1} = j | X_n = i\}$. We consider the notation given in 2., but now vector z is introduced, and the corresponding expressions for (14.1), (14.2) and (14.3) are now written as $S_{ij}(t;z) = 1 - F_{ij}(t;z)$, $\lambda_{ij}(t;z)$ and $p_{ij}(t;z)$, where $f_{ij}(t;z)$ substitutes $f_{ij}(t)$ and satisfies the expressions given in 2.. Expression (14.3) takes the same form.

The staying times in different states are censored, so we have applied non-parametric techniques to fit distribution functions to these random variables. The product-limit estimate and the empirical hazard-rate function have been built. The Weibull distribution has a good fit to the censored data, so we have considered a semi-Markov model with the density probability functions of transition time $i \to j$ given by

$$f_{ij}(t;z) = \frac{\beta_{ij}}{\alpha_{ij}} \left(\frac{t}{\alpha_{ij}}\right)^{\beta_{ij}-1} exp\{z\gamma_{ij}\} \exp\left\{ - \left(\frac{t}{\alpha_{ij}}\right)^{\beta_{ij}} \exp\{z\gamma_{ij}\} \right\}, t \geq 0,$$

(i,j) being one of the couples $(1,2), (1,3), (2,3)$, and γ_{ij} the parameter vector associated to the covariable vector $\gamma_{ij} = (\gamma_{ij}^1, \gamma_{ij}^2, \gamma_{ij}^3)$. If radiotherapy is applied, its effect on the transition $1 \to 2$ is given by γ_{12}^1, and the other components are the effects of the other treatments. The super-index indicates the treatment: $1, 2, 3$ for radiotherapy, hormonal therapy, and chemotherapy, respectively. The components of the vector γ_{ij} are 0 if the corresponding treatment is applied and 1 if it is not. The transition probabilities in the embedded Markov chain are $p_{23} = 1$, and $p_{13} = 1 - p_{12}$, so we must estimate 16 parameters: $p_{12}, \alpha_{12}, \beta_{12}, \gamma_{12}^1, \gamma_{12}^2, \gamma_{12}^3, \alpha_{13}, \beta_{13}, \gamma_{13}^1, \gamma_{13}^2, \gamma_{13}^3, \alpha_{23}, \beta_{23}, \gamma_{23}^1, \gamma_{23}^2$ and γ_{23}^3. We build the new likelihood function (14.3) with the covariate vector and estimate the parameters using the *MATLAB* programme. The values and standard errors of the estimation are given in 3.,and have been calculated with the Mathematica programme. The value of the logarithm of the likelihood function for this data set is $\log L = -1437.6629$.

Table 14.1 Estimated parameters and standard errors in the semi-Markov model with covariates

Estimated Parameters	Standard Errors
$\hat{p}_{12} = 0.5578$	0.0316
$\hat{\alpha}_{12} = 583.5900$	221.0932
$\hat{\beta}_{12} = 0.9075$	0.0913
$\hat{\gamma}_{12}^1 = 2.6817$	0.3260
$\hat{\gamma}_{12}^2 = -0.1420$	0.3223
$\hat{\gamma}_{12}^3 = 0.6785$	0.3017
$\hat{\alpha}_{13} = 76.1143$	14.5056
$\hat{\beta}_{13} = 1.4158$	0.1250
$\hat{\gamma}_{13}^1 = -1.0410$	0.2900
$\hat{\gamma}_{13}^2 = 0.2084$	0.2598
$\hat{\gamma}_{13}^3 = 0.5396$	0.2396
$\hat{\alpha}_{23} = 35.1793$	4.5544
$\hat{\beta}_{23} = 1.3334$	0.1143
$\hat{\gamma}_{23}^1 = -0.3938$	0.2052
$\hat{\gamma}_{23}^2 = 0.4273$	0.2108
$\hat{\gamma}_{23}^3 = 0.1550$	0.2100

The survival curves according to the model are plotted, which is done by performing a computational programme with *MATLAB*, available from the authors. The goodness-of-fit between the empirical survival curves and the corresponding one obtained from the model is performed applying a test for censored data [5], and the fit cannot be rejected in all treatments except in radiotherapy and chemotherapy together. An example is provided in 14.1 for the survival function of the group of patients treated with radiotherapy. The p-value for the survival fit to this treatment is 0.3687, so the goodness of fit cannot be rejected.

4. SOME DATA ON FAILURE TIMES

We illustrate the applicability of a semi-Markov model considering a data set of lifetimes for a group of optical fibres used in telecommunica-

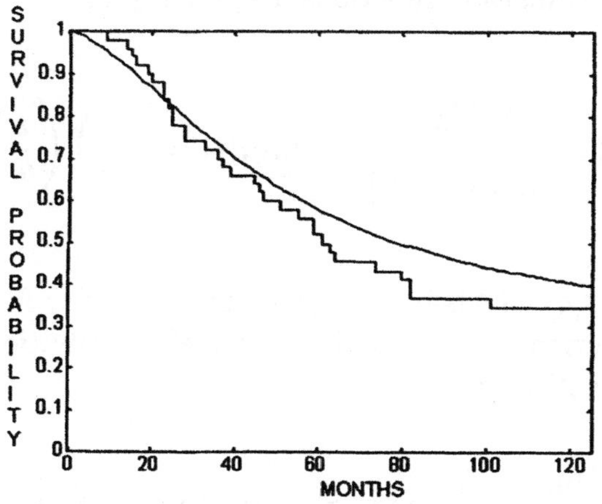

Figure 14.1 Empirical (step function) and theoretical (continuous line) survival curves for breast cancer patients treated with radiotherapy

tions. They are made of silica or a glass material and are brittle. Laboratory tests have demonstrated that the first failure time is distributed as a Weibull distribution [7]. We have been observed fibre optics run in pipe with amplifiers along the line. The total number of cables between two amplifiers is 302, in a dynamic environment. The follow-up of this line spanned from 1992 up to 1997 (six years). Sometimes cables are broken, which implies the failure of the fibre optics. Each broken cable can be repaired twice at most, and so the third failure is considered to be fatal and the fibres are replaced by new material. Repair is considered to be as good as new. Instantaneous repair times are assumed, and three states are considered in the cables taken as a whole: up to the first repair (state 1), up to the second repair (state 2), up to third repair (state 3), and up to the fatal failure (absorbent state 4). The transition among the states of the system are $1 \to 2 \to 3 \to 4$. When state 4 is reached, the broken cable is replaced by a new one, and can be considered an instantaneous state. Repair and substitution times are negligible. Other transitions times have not been observed.

We have analyzed the staying times in states before a transition for the data set. Different distributions have been fitted to the different data sets and the best fits to the staying times are the log-normal distribution for staying in state 1, and the Weibull distribution for staying times in states 2 and 3. The fit of these functions have been made following graphical methods, since so many data are censored. For these distribution times, the product-limit estimate, the hazard rate function and the cumulative function have been calculated, all available from the authors.

The intensity transition functions are, for $t \geq 0$,

$$\lambda_{12}(t) = \frac{\varphi\left(\frac{lnt-\mu}{\sigma}\right)}{\sigma t Q\left(\frac{lnt-\mu}{\sigma}\right)} \quad and \quad \lambda_{ij}(t) = \left(\frac{\gamma_i}{\theta_i}\right)\left(\frac{t}{\theta_i}\right)^{\gamma_i - 1},$$

for (i,j)=(2,3),(3,4),$\varphi(\cdot)$ being the probability density function and $Q(\cdot)$ the survival function of the standard normal distribution. The staying time survival functions are

$$S_1(t) = Q\left(\frac{lnt-\mu}{\sigma}\right) \quad \text{and} \quad S_i(t) = \exp\left\{-\left(\frac{t}{\theta_i}\right)^{\gamma_i}\right\}, i = 2, 3.$$

We have developed the calculations of the functions with four states. Based on these functions, transition probability functions are calculated. A total of six parameters are estimated using the *MATLAB* programme, applying expression (14.3) to this data set. We obtain $lnL = -1695.5726$. Similarly to in the medical data set, the survival curves and the fit between the empirical survival functions and the curves corresponding to the model are compared applying a test for censored data. The estimated parameters are given in table 14.2.

Table 14.2 Estimated parameters and standard errors in lifetime data of fibre optics

Estimated Parameters	Standard Errors
$\hat{\mu} = 8.4436$	0.1587
$\hat{\sigma} = 1.6250$	0.1396
$\hat{\theta}_2 = 968.5468$	155.3097
$\hat{\gamma}_2 = 0.8087$	0.0895
$\hat{\theta}_3 = 805.3440$	170.6004
$\hat{\gamma}_3 = 0.7316$	0.0954

For practical interest, we have calculated the lifetime distribution of the broken sections until absorption. This is the time elapsed before a new section replaces a broken section. The mathematical expression of this survival function is

$$\overline{F}\ (t) = Q\left(\frac{lnt-\mu}{\sigma}\right) + \frac{1}{\sigma}\int_0^t \frac{\varphi\left(\frac{ln\tau-\mu}{\sigma}\right)\exp\left(-\left(\frac{t-\tau}{\theta_2}\right)^{\gamma_2}\right)}{\tau} d\tau + \frac{\gamma_2}{\theta_2^{\gamma_2}\sigma}$$

$$\int_0^t \frac{1}{\tau}\varphi\left(\frac{ln\tau-\mu}{\sigma}\right)\left[\int_0^{t-\tau} u^{\gamma_2-1}\exp\left\{-\left[\left(\frac{u}{\theta_2}\right)^{\gamma_2}+\left(\frac{t-\tau-u}{\theta_3}\right)^{\gamma_3}\right]\right\}du\right]d\tau. \tag{14.5}$$

This function and the corresponding empirical one are given in figure 14.2. As can be seen, survival has very high values. After six years, the probability of survival of the cables between two amplifiers is more than 85%. During the first three years there is no difference between the theoretical and empirical curves. The test of [5] have been applied to this fit and it is shown that it cannot be rejected. Different survival functions for times in states 1, 2, and 3 have been obtained.

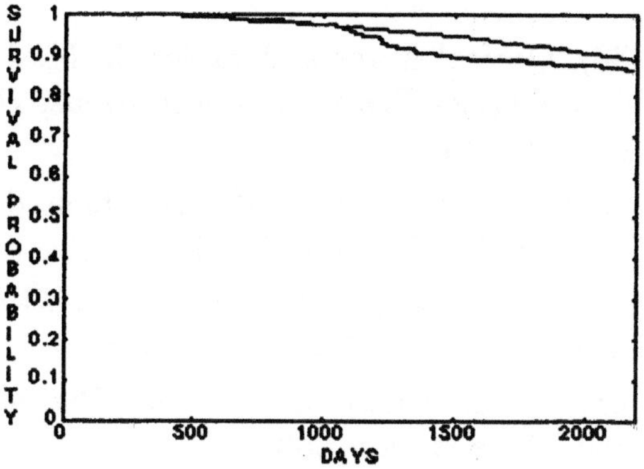

Figure 14.2 Empirical (step function) and theoretical (continuous line) survival curves for fibre optics cables

Acknowledgments

Thanks are due to Dr. Pedraza of the Departamento de Radiología y Medicina Física de la Universidad de Granada, for supplying the medical data. Thanks are also given to Dr. F. Pérez-Ocón of the Departamento de Óptica de la Universidad de Granada, for supplying the fibre optics data.

References

[1] Cox, D.R. (1972). Regression models and life tables, (with Discussion), *Journal of the Royal Statistical Society*, B, **34**, 187-220.

[2] Çinlar, E. (1975). Markov renewal processes: a survey, *Management Science*, **21**(7) 727-752.

[3] Dabrowska, D.M., Sun, G., and Horowitz, M.M. (1994). Cox Regression in a Markov Renewal Model: An application to the Analysis of Bone Marrow Transplan Data, *Journal of the American Statistical Association*, **89**,867-877.

[4] Gentleman, R.C., Lawless, J.F., Lindsey, J.C., and Yan, P. (1994). Multi-state Markov models for analyzing incomplete disease history data with illustrations for HIV disease, *Statistics in Medicine*,**13**, 805-821.

[5] Hollander, M. and Proschan, F. (1976). Testing to Determine the Underlying Distribution Using Randomly Censored Data. *Biometrics* **35**, 393-401.

[6] Janssen, J. (1983). *Semi-Markov Processes and Applications in Risk and Queueing Theories.* Cooperativa Universitaria Editrice Napoletana, Napoli.

[7] Koa, C.K. (1980). Optical fibres and cables, in M. J. Howes and D. V. Morgan (Eds.), *Optical Fibre Communications, Devices, Circuits and Systems*, 189-249 Wiley.

[8] Marshall, G. and Jones, R.H. (1995).Multi-state models and diabetic retinopathy, *Statistics in Medicine*, **14**,1975-1983.

[9] MATLAB Reference Guide. (1992). *Matlab.* High-Performance Numeric Computation and Visualization Software.

[10] MATLAB Users Guide. (1992). *Matlab.* High-Performance Numeric Computation and Visualization Software.

[11] McClean, S., and Montgomery, E. (1995). Estimation for continuyous time non-homogeneous Markov and semi-Markov manpower models, *Proc. of the Seventh Intern. Symp. of A.S.M.D.A.*, Dublin, 384-395.

[12] Nelder, J.A. and Mead, R. (1987). A simplex Method for Function Minimization, *Computer Journal*, **7**, 308-313.

[13] Pérez-Ocón, R., Ruiz-Castro, J.E., and Gámiz-Pérez, M.L. (1998). A Multivariate Model to Measure the Effect of Treatments in Survival to Breast Cancer, *Biometrical Journal*, **40**, 6, 703-715.

[14] Vassiliou, P-C.G. and Papadopoulou, A.A. (1992). Nonhomogeneous Semi-Markov Systems and Maintainability of the State Sizes, *Journal Applied Probability*, **29**, 519-534.

III
NON-HOMOGENEOUS MODELS

Chapter 15

CONTINUOUS TIME NON HOMOGENEOUS SEMI-MARKOV SYSTEMS

Aleka A. Papadopoulou
Aristotle University of Thessaloniki, Greece

Panagiotis C.G. Vassiliou
Aristotle University of Thessaloniki, Greece

Abstract In the present we introduce and define the non-homogeneous semi-Markov system in continuous time. We study the problem of finding the expected population structure in closed analytic form, in relation with the basic sequences of the system. Moreover, the problem of the asymptotic behavior of the system is studied under certain conditions. We first study the limiting behavior of the imbedded non-homogeneous semi-Markov chain and then we provide the asymptotic population structure in closed analytic form.

Keywords: Stochastic population models, semi-Markov processes, manpower systems, population structure.

1. INTRODUCTION

The concept of a non homogeneous semi-Markov system (NHSMS) in discrete time was introduced and defined in Vassiliou and Papadopoulou ([22]) and the asymptotic behavior of the same model was studied in Papadopoulou and Vassiliou ([23]). Continuing this effort, in the present we define the non homogeneous semi Markov system considering time as a continuous parameter. This effort is a sequel of the work created by Vassiliou ([16], [17], [18], [19], [20], [21]) in the theory of non-homogeneous Markov systems during the 80's. The motive for this work was to provide a more general framework for a number of Markov chain models in manpower systems. As a result a great variety of applied probability models can be accommodated in this general framework. Continuous time semi-Markov models have found important applications in manpower systems, especially in the work by McClean ([11], [12], [13], [14]), Mehlman ([15]), Bartholomew ([1], [2]), Bartholomew, Forbes and McClean ([3]), Janssen ([10]) and Howard ([8]).

2. THE MODEL

Consider a population which is distributed into a set of states $S = \{1, \ldots, k\}$ according to some characteristics. In the present, time t is considered to be a continuous parameter. The states are assumed to be exclusive and exhaustive, so that each member of the system may be in one and only one state at a specific time. The state of the system at any given time is described by the vector $\mathbf{N}(t) = [N_1(t), N_2(t), \ldots, N_k(t)]$ where $N_i(t)$ is the expected number of members of the system in the i-th state at time t. Let also $T(t)$ be a known continuous function of t denoting the expected number of members of the system at time t and $N_{k+1}(t)$ is the expected number of leavers of the system at time t. Suppose that the individual transitions between the states take place according to a non homogeneous semi-Markov chain which is the imbedded chain in the system. In this respect we define:

$$f_{ij}(t, t+\delta t) = \text{prob}\{\text{a member of the system selects state } j \text{ for its next transition during } (t, t+\delta t) \text{ entered state } i \text{ at time } t\} = r_{ij}(t)\delta t + O(\delta t)$$

where $r_{ij}(t)$ is the intensity for a member that entered state i at t to select state j for its next transition. Now let us define the probabilities :

$$\begin{aligned} p_{i,k+1}(t, t+\delta t) &= \text{prob}\{\text{a member of the system decides to leave the system during}(t, t+\delta t) \text{ entered state } i \text{ at time } t\} \\ &= r_{i,k+1}(t)\delta t + O(\delta t) \end{aligned}$$

where $k+1$ denotes the absorbing state corresponding to the leavers from each grade,

$$p_{oj}(t) = \text{prob}\{\text{a new member enters the system in state } j\text{, as a replacement of a member who entered his last state at time } t\}$$

We now define the idea of a "membership" (Vassiliou & Papadopoulou [22]), which will prove to be helpful in many ways in what follows. The root of the idea is in Bartholomew ([1]), p.73, from which we quote: "...

they thus provide a formal justification for our earlier remark that an open system in which gains and losses were equal could be treated as closed. Each person who leaves can be paired with a new entrant and the two changes treated as one. Thus a transition from grade i to grade j can either take place within the system or by loss from grade i and replacement to grade j with total probability $p_{ij}+p_{i,k+1}p_{oj}\dots$". Also on p.78 : "... Each loss is directly responsible for a recruitment, if the loss occurs from grade i the probability that the consequential recruitment is in grade j is $p_{i,k+1}p_{oj}$. The replacement of a leaver can thus be treated as if it were an internal transfer. The total probability of a move out of grade i which results in an addition to grade j is thus $p_{ij}+p_{i,k+1}p_{oj}\dots$". We consider that initially there are $T(0)$ memberships in the system and a member entering the system holds a particular membership which moves within the states with the member. When a member leaves, the membership is taken by a new recruit and moves within the system with the replacement, and so on. When the system is expanding, then new memberships are created in the system which behave like the initial ones. Let now $p_{ij}(t,t+\delta t)$ be the probability that a membership of the system selects state j for its next transition during $(t,t+\delta t)$ given that the membership entered state i at time t. Then

$$\begin{aligned} p_{ij}(t,t+\delta t) &= f_{ij}(t,t+\delta t)+p_{i,k+1}(t,t+\delta t)p_{oj}(t) \\ &= r_{ij}(t)\delta t + r_{i,k+1}(t)\delta t p_{oj}(t) + O(\delta t) \end{aligned}$$

Let $q_{ij}(t)=\lim_{\delta t\to 0}\frac{p_{ij}(t,t+\delta t)}{\delta t}$ and $\mathbf{Q}(t)=\{q_{ij}(t)\}$. Obviously $\mathbf{Q}(t)$ is an intensity matrix. Thus if we suppose that $\mathbf{Q}(t)$ has continuous elements and $\sup_{i\in S}\mid q_{ij}(t)\mid<\infty$ we have that $\mathbf{Q}(t)$ defines uniquely a Markov chain $\mathbf{P}(s,t)$ which determines the selection of the next state when the time comes for a transition. So, the (i,j) element of $\mathbf{P}(s,t)$ equals to

$$p_{ij}(s,t) = \text{prob}\{\text{a membership of the system selects state } j \text{ for its next transition during } (s,t) \ / \text{ entered state } i \text{ at time } s\}$$

Now, let us define as

$$\varphi_{ij}(s,t) = \text{prob}\{\text{a membership of the NHSMS which entered state } i \text{ at time } s, \text{ is in state } j \text{ at time } s+t\}$$

In the system, whenever a membership selects the next state j, holds for a time τ_{ij} in the previous state i. The holding times are positive continuous random variables which are governed by a probability density function $h_{ij}(.)$ i.e.

$$\text{prob}\{a \leq \tau_{ij} \leq b\} = \int_a^b h_{ij}(x)dx$$

Let us also define the waiting time density function for state i as :

$$\begin{aligned} {}^{\leq}w_i(s,t) &= \text{prob}\{\text{a membership which entered state } i \text{ at time } s \text{ holds} \\ &\quad \text{for time } \tau_i \leq t \text{ in } i \text{ before making its next transition}\} \\ &= \sum_{k \in S} \int_0^t p_{ik}(s, s+x) h_{ik}(x) dx \end{aligned}$$

With careful reasoning we could prove that

$$\varphi_{ij}(s,t) = \delta_{ij} \, {}^{>}w_i(s,t) + \sum_{k \in S} \int_0^t p_{ik}(s, s+x) h_{ik}(x) \varphi_{kj}(s+x, t-x) dx$$

and if we define $c_{ij}(s,x) = p_{ij}(s, s+x)h_{ij}(x)$

$$\varphi_{ij}(s,t) = \delta_{ij} \, {}^{>}w_i(s,t) + \sum_{k \in S} \int_0^t c_{ik}(s,x) \varphi_{kj}(s+x, t-x) dx \qquad (15.1)$$

Equation (15.1) in matrices is as follows

$$\mathbf{\Phi}(s,t) = {}^{>}\mathbf{W}(s,t) + \int_0^t \mathbf{C}(s,x) \mathbf{\Phi}(s+x, t-x) dx \qquad (15.2)$$

where $\mathbf{C}(s,x) = \mathbf{P}(s, s+x) \Diamond \mathbf{H}(x)$ is the Hadamard product of the two matrices. Obviously $\mathbf{\Phi}(s,0) = \mathbf{I}$.
The matrix $\mathbf{\Phi}(s,t)$ defines the interval transition probabilities for the memberships of the non homogeneous semi-Markov chain which is imbedded in our system. This semi-Markov chain is fully described by the probabilities $p_{ij}(s,t)$ and the probability density functions of the holding times $h_{ij}(x)$ as it is shown by equation (15.1). So, we have a non homogeneous semi-Markov system in which the individual transitions take place according to a non homogeneous semi-Markov chain. Using probabilistic arguments we could find probabilities $\varphi_{ij}(s,t)$ in closed

analytic form as follows

$$\begin{aligned}\Phi(s,t) &= {}^{>}\mathbf{W}(s,t) + \int_0^t \mathbf{C}(s,x_1)\{{}^{>}\mathbf{W}(s+x_1,t-x_1) + \mathbf{C}(s+x_1,t-x_1)\}dx_1 \\ &+ \sum_{k\geq 2}\int_0^t \int_0^{t-x_1} \int_0^{t-x_1-x_2} .. \int_0^{t-x_1..-x_{k-1}} \mathbf{C}(s,x_1)..\mathbf{C}(s+..+x_{k-1},x_k) \\ &\{{}^{>}\mathbf{W}(s+..+x_k,t-x_1-..-x_k) + \mathbf{C}(s+..+x_k,t-x_1-..-x_k)\} \\ &dx_k dx_{k-1}\dots dx_2 dx_1 \end{aligned} \tag{15.3}$$

Now, let us assume that our system is expanding and that $r_{0i}(x)$ is the conditional probability of a new membership, entering the system at time x, to enter state i. Then, the expected number of memberships which are in state j at time t equals to the expected number of memberships of the initial population which is in the system at time t in state j, added to the expected number of memberships which are created by expansion of the system at time $x \in (0,t)$ and are in state j at time t. From the above we can have that

$$N_j(t) = \sum_{i=1}^{k} N_i(0)\varphi_{ij}(0,t) + \sum_{i=1}^{k}\int_0^t \Delta T(x) r_{0i}(x)\varphi_{ij}(x,t-x)dx$$

where $\Delta T(x)$ is the instantaneous expected increase of the size of the system, for the interval $(x, x+\delta x)$. In matrix notation the above equation is of the form

$$\mathbf{N}(t) = \mathbf{N}(0)\Phi(0,t) + \int_0^t \Delta T(x)\mathbf{r}_0(x)\Phi(x,t-x)dx \tag{15.4}$$

where $\mathbf{N}(0)$ is the initial population structure and $\mathbf{r}_0(x)$ the recruitment probability vector of the new memberships. Relations (15.3) and (15.4) provide the expected population structure as a function of the basic sequences of the system, which apart from prediction purposes, is also useful for answering problems related to the asymptotic behavior of the NHSMS.

3. LIMITING BEHAVIOR

The problem of the asymptotic behavior is one of the most important problems in manpower systems (Bartholomew [1], [2]), non homogeneous Markov and semi-Markov systems (Vassiliou [16], [17], [18], [19], [20], [21], Conlisk [5], Feichtinger [6], Feichtinger & Mehlmann [7], Howard [8], Benevento [4], Papadopoulou & Vassiliou [23]). In what follows, we deal with the classical problem of finding the conditions under which in a continuous time NHSMS the sequence of the expected population structure converges and the closed analytic form of the limit $\mathbf{N}(\infty)$. In order to achieve this we have to study first the ergodic behavior of the imbedded non homogeneous semi-Markov chain which is fully described by the sequence $\Phi(s,t)$. We assume that the sequence $\mathbf{P}(s,s+x)$ converges as $s \to \infty$ to the corresponding homogeneous chain $\mathbf{P}(x)$ which is also ergodic with the limit equal to a stable stochastic matrix $\mathbf{P}^*$ and that the sequence of the p.d.f. of the holding times $\mathbf{H}(x)$ converges to zero matrix exponentially fast. According to the previous assumptions and following similar reasoning as in Papadopoulou & Vassiliou [23] we could prove, using relation (15.3), the existence of the single and double limits of $\Phi(s,t)$. Let now that $\Phi(t) = \lim_{s\to\infty} \Phi(s,t)$. From relation (15.2) we can get

$$\Phi(t) = {}^{>}\mathbf{W}(t) + \int_0^t \mathbf{C}(x)\Phi(t-x)dx \tag{15.5}$$

where ${}^{>}\mathbf{W}(t)$ equals to the limit $\lim_{s\to\infty} {}^{>}\mathbf{W}(s,t)$. If we take Laplace transforms in relation (15.5) we have

$$\Phi^L(z) = [\mathbf{I} - \mathbf{C}^L(z)]^{-1}\, {}^{>}\mathbf{W}^L(z) \tag{15.6}$$

Then from Howard [8], Papadopoulou and Vassiliou [23], the properties of Laplace transforms and relation (15.6) we can get that

$$\Phi^* = \lim_{t\to\infty, s\to\infty} \Phi(s,t) = \frac{\mathbf{P}^*\mathbf{M}}{\mathbf{1}[\mathbf{P}^*\Diamond\mathbf{M}]\mathbf{1}'} \tag{15.7}$$

where $\mathbf{M} = \lim_{z\to 0}\ {}^{>}\mathbf{W}^L(z)$ is a diagonal matrix with elements the limiting mean waiting times in the states.

Finally, from relations (15.4) and (15.7) we are able to give the limit $\mathbf{N}(\infty)$ of the expected population structure of the system in closed analytic form. If we take limits in (15.4) we have

$$\mathbf{N}(\infty) = \mathbf{N}(0)\Phi^* + \lim_{t\to\infty} \int_0^t \Delta T(x)\mathbf{r}_0(x)\Phi(x,t-x)dx \tag{15.8}$$

The second term of the second part in (15.8) equals to (Isaacson and Madsen [9]) :

$$\lim_{t\to\infty} \int_0^t \Delta T(x)\mathbf{r}_0(x)\Phi^* dx = \lim_{t\to\infty} \int_0^t \Delta T(x)dx\varphi$$

where φ is the row of the stable matrix Φ^* and $\lim_{t\to\infty}\int_0^t \Delta T(x)dx$ is the total expected increase of the population. Let us assume that the sequence $T(t)$ is convergent and $\lim_{t\to\infty} T(t) = T$, then $\lim_{t\to\infty}\int_0^t \Delta T(x)dx = T - T(0)$. Also, it is obvious that $\mathbf{N}(0)\Phi^*$ equals $T(0)\varphi$. From the above results we have from (8) that

$$\mathbf{N}(\infty) = T\varphi \tag{15.9}$$

4. ILLUSTRATION: A SPECIAL CASE

In this section, some of the previous results are illustrated numerically with hypothetical data representative of the kind of data we usually get from manpower systems.
Let us assume that we have an organization with three grades and the following probability matrices which correspond to a special case.

$$\mathbf{P} = \begin{pmatrix} 9/15 & 5/15 & 1/15 \\ 1/10 & 8/10 & 1/10 \\ 1/20 & 4/20 & 15/20 \end{pmatrix}$$

$$\mathbf{H}(t) = \begin{pmatrix} 2e^{-2t} & 3e^{-3t} & 4e^{-4t} \\ 2e^{-2t} & e^{-t} & 3e^{-3t} \\ 3e^{-3t} & 2e^{-2t} & e^{-t} \end{pmatrix}$$

Simulating equation (15.2) of the interval transition probabilities, using *Mathematica V3.0*, for $t = 50$ and *step* $= 0.3$ we get the limit of $\Phi(s,t)$:

$$\begin{pmatrix} 0.094458 & 0.61095 & 0.294592 \\ 0.094458 & 0.61095 & 0.294592 \\ 0.094458 & 0.61095 & 0.294592 \end{pmatrix} \tag{15.10}$$

Figures 15.1, 15.2, 15.3 and 15.4 give the plots of the numerical results for the interval transition probabilities corresponding to the data of our example. Calculating now equation

$$\Phi^* = \lim_{t\to\infty, s\to\infty} \Phi(s,t) = \frac{\mathbf{P}^*\mathbf{M}}{\mathbf{1}[\mathbf{P}^*\Diamond\mathbf{M}]\mathbf{1}'}$$

where

$$\mathbf{P}^* = \begin{pmatrix} 0.1731 & 0.5577 & 0.26992 \\ 0.1731 & 0.5577 & 0.26992 \\ 0.1731 & 0.5577 & 0.26992 \end{pmatrix}$$

and

$$\mathbf{M} = \begin{pmatrix} 0.4278 & 0 & 0 \\ 0 & 0.8583 & 0 \\ 0 & 0 & 0.8667 \end{pmatrix}$$

we get as

$$\Phi^* = \begin{pmatrix} 0.0942 & 0.6090 & 0.2968 \\ 0.0942 & 0.6090 & 0.2968 \\ 0.0942 & 0.6090 & 0.2968 \end{pmatrix}$$

The above matrix is in agreement with the limit in (15.10) which is the numerical result produced by the simulation.

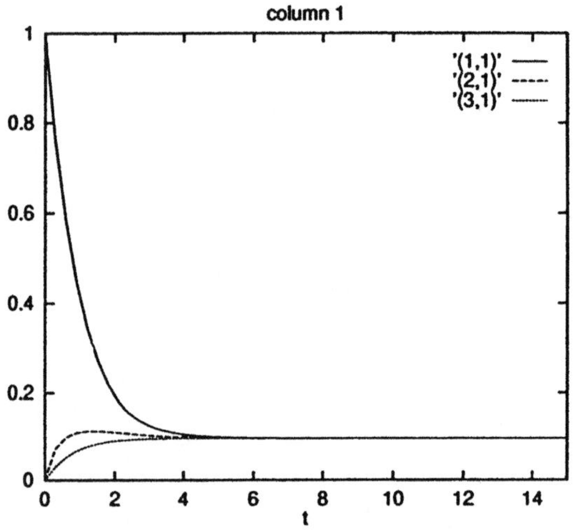

Figure 15.1 Plot of the interval transition probabilities φ_{11}, φ_{21}, φ_{31}.

References

[1] Bartholomew, D.J. (1982). *Stochastic Models for Social Processes*, Jonh Wiley, Chichester.

[2] Bartholomew, D.J. (1986). Social Applications of Semi-Markov Processes, In *Semi-Markov Models: Theory and Applications*, Plenum Press, New York.

[3] Bartholomew, D.J., A.F. Forbes & S.I. McClean (1991). *Statistical Techniques for Manpower Planning*, Jonh Wiley, Chichester.

[4] Benevento, R.V. (1986). Asymptotic analysis of some non homogeneous semi-Markov processes, In *Semi-Markov models: Theory and Applications*, Plenum Press, New York.

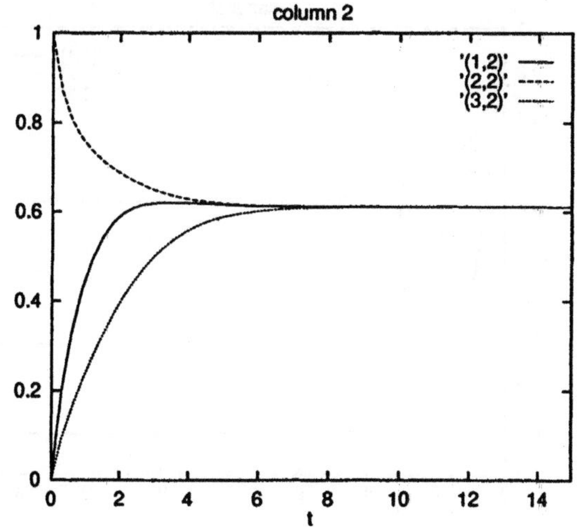

Figure 15.2 Plot of the interval transition probabilities φ_{12}, φ_{22}, φ_{32}.

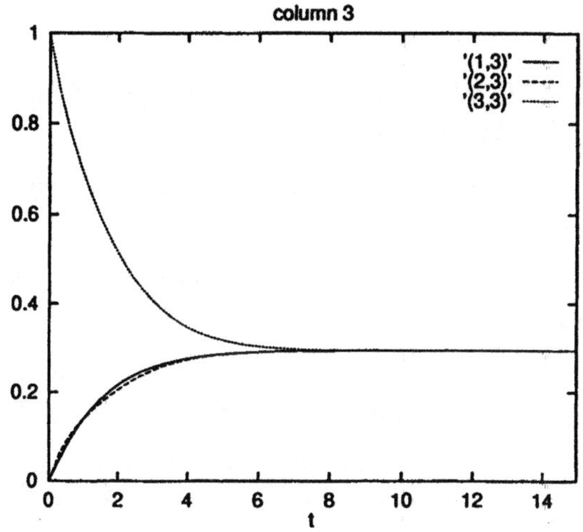

Figure 15.3 Plot of the interval transition probabilities φ_{13}, φ_{23}, φ_{33}.

[5] Conlisk, J. (1976). Interactive Markov chains, *J. Math. Soc.*, 4, 157 -185.

[6] Feichtinger, G. (1976). On the generalization of stable age distributions to Gani type person-flow models, *Adv. Appl. Prob.*, **8**, 433-445.

[7] Feichtinger, G. & Mehlmann, A. (1976). The recruitment trajectory corresponding to particular stock sequences in Markovian person-flow models, *Math. Oper. Res.*, 1, 175-184.

[8] Howard, R.A. (1971). *Dynamic Probabilistic Systems*, Jonh Wiley,

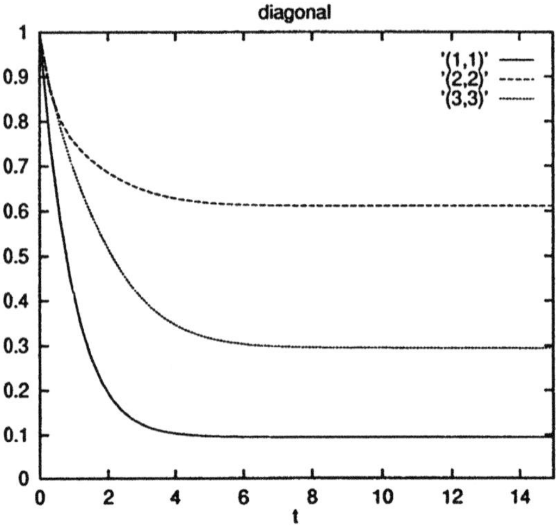

Figure 15.4 Plot of the interval transition probabilities φ_{11}, φ_{22}, φ_{33}.

Chichester.

[9] Isaacson, D. & R.W. Madsen (1976) *Markov Chains Theory & Applications*, Jonh Wiley, New York.

[10] Janssen, J. (1986). *Semi-Markov models: Theory and Applications*, Plenum Press, New York.

[11] McClean, S.I.(1976). The two stage model for personnel behaviour, *J.R.S S.*, **A139**, 205-217.

[12] McClean, S.I.(1978). Continuous time stochastic models for a multigrade population, *Journal of Applied Probability*, **15**, 26-32.

[13] McClean, S.I.(1980). A semi-Markovian model for a multigrade population, *Journal of Applied Probability*, **17**, 846-852.

[14] McClean, S.I. (1986). Semi-Markov models for Manpower planning, In *Semi-Markov Models: Theory and Applications*, Plenum Press, New York.

[15] Mehlmann, A. (1979). Semi-Markovian Manpower models in continuous time, *Journal of Applied Probability*, **16**, 416-422.

[16] Vassiliou, P.-C.G. (1981a). On the limiting behavior of a non-homogeneous Markov chain model in manpower systems, *Biometrica*, **68**, 557-561.

[17] Vassiliou, P.-C.G. (1981b). Stability in a non-homogeneous Markov chain model in manpower systems, *Journal of Applied Probability*, **18**, 924-930.

[18] Vassiliou, P.-C.G. (1982a). Asymptotic behavior of Markov systems, *Journal of Applied Probability*, **19**, 851-857.

[19] Vassiliou, P.-C.G (1982b). Predicting the service in the grade distribution in Manpower systems, *Eur. J. Operat. Res.*, **10**, 373-378.

[20] Vassiliou, P.-C.G (1984). Cyclic behavior and asymptotic stability of non-homogeneous Markov systems, *Journal of Applied Probability*, **21**, 315-325.

[21] Vassiliou, P.-C.G (1986). Asymptotic variability of non-homogeneous Markov systems under cyclic behavior, *Eur. J. of Oper. Res.*, **27**, 215-228.

[22] Vassiliou, P.-C.G and A.A. Papadopoulou (1992). Non homogeneous semi-Markov systems and maintainability of the state sizes, *Journal of Applied Probability*, **29**, 519-534.

[23] Papadopoulou, A.A. and P.-C.G. Vassiliou (1994). Asymptotic behavior of non homogeneous semi-Markov systems, *Linear Algebra and Its Applications*, **210**, 153-198.

Chapter 16

THE PERTURBED NON-HOMOGENEOUS SEMI-MARKOV SYSTEM

Panagiotis C.G. Vassiliou
Aristotle University of Thessaloniki, Greece

Helena Tsakiridou
Aristotle University of Thessaloniki, Greece

Abstract In the present we introduce and define for the first time the concept of a perturbed non-homogeneous semi-Markov system (P-NHSMS). We study the problem of the expected population structure as a function of the basic parameters of the system.

Keywords: Perturbed Non-Homogeneous Semi-Markov System (P-NHSMS).

1. INTRODUCTION

The roots of the theory of semi-Markov models for various population problems can be found in the important theoretical results for semi-Markov models in manpower systems. Such results can be found in the work by Mc Clean (1980, 1986, 1993), Mc Clean, E. Montgomery and F.Ugwuowo (1997), Mehlman (1979), Bartholomew (1982), Bartholomew, Forbes and Mc Clean (1991). Also, an interesting account of theoretical results and important applications of semi-Markov models can be found in Bartholomew (1982, 1986). The concept of a non-homogeneous semi-Markov system (NHSMS) was introduced and defined for the first time in Vassiliou and Papadopoulou (1992). This provided also a general framework for a number of semi-Markov chain models in manpower systems and a great variety of applied probability models (Janssen (1986)). The asymptotic behavior of the NHSMS was found in analytic form by Papadopoulou and Vassiliou (1994) and that provided also the asymptotic behavior of non-homogeneous semi-Markov chains which are a special case of NHSMS. In Meyer (1975) and Campbell and Meyer (1979) appeared for the first time the idea of a perturbation of a stochastic matrix. The concept of a perturbed non-homogeneous Markov system in discrete and continuous time was introduced in Vassiliou and Symeonaki (1997, 1999). The non-homogeneous semi-Markov system in a stochastic environment was studied in Vassiliou (1996). In the present we will introduced the concept of the perturbed non-homogeneous semi-Markov system and we will establish the expected population structure.

2. THE PERTURBED NON HOMOGENEOUS SEMI-MARKOV SYSTEM

Let a population the members of which could be sections of human societies, biological populations, various types of machine, etc. The population is stratified into k classes or states and let $S = \{1, 2, \ldots, k\}$ be the set of states that are assumed to be exclusive and exhaustive. Let $T(t)$ be the expected number of members of the system at time t. We consider that initially there are $T(0)$ memberships in the system and a member entering the system holds a particular membership which moves within the states with the member. When a member leaves, the membership is taken by a new recruit and moves within the system with the replacement and so on. When the system is expanding $[\Delta T(t) = T(t) - T(t-1) > 0]$, then new memberships are created in the system which behave like the initial ones. The state of the system at any given time t is represented by the vector $\mathbf{N}(t) = [N_1(t), N_2(t), \ldots, N_k(t)]$ where $N_i(t)$ is the expected number of members of the system in the i-th state at time t. Let $\mathbb{M}_{n,m}$ be the set of all $n \times m$ matrices with elements from $\mathbb{R}$. Also let $S\mathbb{M}_n$ the set of $n \times n$ stochastic matrices.

Definition 1 *We call a stochastic matrix* $\mathbf{Q}$ *fully regular (Gautmacher (1959)) if* $\mathbf{Q}$ *has 1 as an eigenvalue of algebraic multiplicity 1 and the only one of modulus 1.*

From Meyer (1975) we get the following definition.

Definition 2 *Let* $\mathbf{Q} \in S\mathbb{M}_n$*, be irreducible and fully regular. The matrix* $\mathcal{E} \in \mathbb{M}_n$ *is a perturbation for* $\mathbf{Q}$ *if the matrix* $\tilde{\mathbf{Q}} = \mathbf{Q} - \mathcal{E}$ *is also an irreducible and fully regular stochastic matrix.*

We also provide the following definition.

Definition 3 *Let* $\mathbf{p} \in S\mathbb{M}_{1,n}$*, then the vector* $\varepsilon \in \mathbb{M}_{1,n}$ *is a perturbation for* $\mathbf{p}$ *if the vector* $\tilde{\mathbf{p}} = \mathbf{p} - \varepsilon$ *is a stochastic vector.*

Now let $f_{ij}(t)$ =prob{a member of the system who entered in state i at time t to move in state j at its next transition}. Let $\mathbf{F}(t) = \{f_{ij}(t)\}_{t=o}^{\infty}$ and assume that $\mathbf{F}(t)$ is of the form (see figure 16.1).

$$\mathbf{F}(t) = \mathbf{F} - \mathcal{E}_f(t) \text{ for } t = 1, 2, \ldots$$
$$\text{with } \mathbf{F}(t)\mathbf{1}' \leq \mathbf{1}', \mathbf{F}\mathbf{1}' \leq \mathbf{1}' \text{ and } \mathbf{F}(t) \geq 0, \mathbf{F} \geq 0.$$

The matrix $\mathcal{E}_f(t)$ is chosen randomly from a finite set :

$$\mathcal{E}_f = \{\mathcal{E}_f(1), \mathcal{E}_f(2), \ldots, \mathcal{E}_f(\nu)\} \text{ with}$$
$$\text{prob}\{\mathcal{E}_f(t) = \mathcal{E}_f(i)\} = c_i(t) > 0 \text{ for } i = 1, 2, \ldots, \nu.$$

Let $p_{ik+1}(t)$ =prob{a member who entered state i at time t to leave

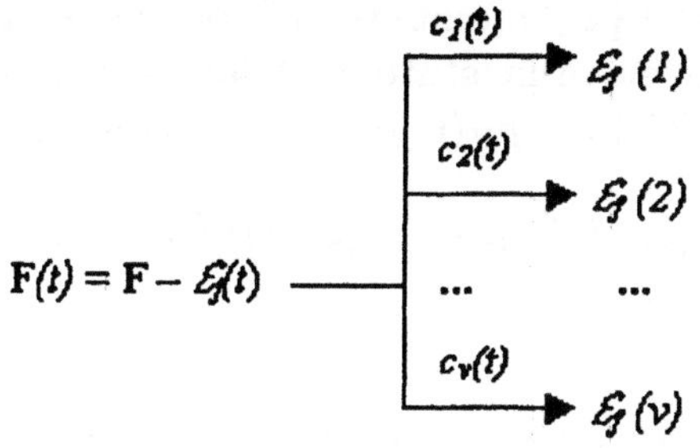

Figure 16.1 Internal transitions at time *t*.

the system at its next transition}, then we assume that for every $t = 0, 1, 2, \ldots,$ we have :

$$\begin{aligned}(\mathbf{I} - \mathbf{F} + \mathcal{E}_f(t))\mathbf{1}' &= \mathbf{p}'_{k+1}(t), \\ \text{where } \mathbf{p}'_{k+1}(t) &= [p_{1k+1}(t), p_{2k+1}(t), \ldots, p_{kk+1}(t)]'. \quad (16.1)\end{aligned}$$

Let $p_{oi}(t)$ =prob{a new member who enters the system in state i as a replacement of a member who entered its last state at time t}, and $\mathbf{p}_o(t) = [p_{o1}(t), p_{o2}(t), \ldots, p_{ok}(t)]$. Assume that for every t the vector $\mathbf{p}_o(t)$ is of the form (see figure 16.2).

$$\mathbf{p}_o(t) = \mathbf{p}_o - \varepsilon_o(t) \text{ with } \mathbf{p}_o(t)\mathbf{1}' = \mathbf{1}' \text{ and } \mathbf{p}_o\mathbf{1}' = \mathbf{1}' \text{ for every } t. \quad (16.2)$$

The vector $\varepsilon_o(t)$ is assumed to be a *perturbation vector* for p_o and is randomly chosen from a finite set $\mathcal{E}_o = \{\varepsilon_o(1), \varepsilon_o(2), \ldots, \varepsilon_o(m)\}$, with probability prob$\{\varepsilon_o(t) = \varepsilon_o(i)\} = c_{oi}(t)$. From (16.1), we get $\mathbf{p}_{k+1}(t) = \mathbf{p}_{k+1} - \varepsilon_{k+1}(t)$ where :

$$\mathbf{p}'_{k+1} = (\mathbf{I} - \mathbf{F})\mathbf{1}' \text{ and } \varepsilon'_{k+1}(t) = -\mathcal{E}_f(t)\mathbf{1}',$$

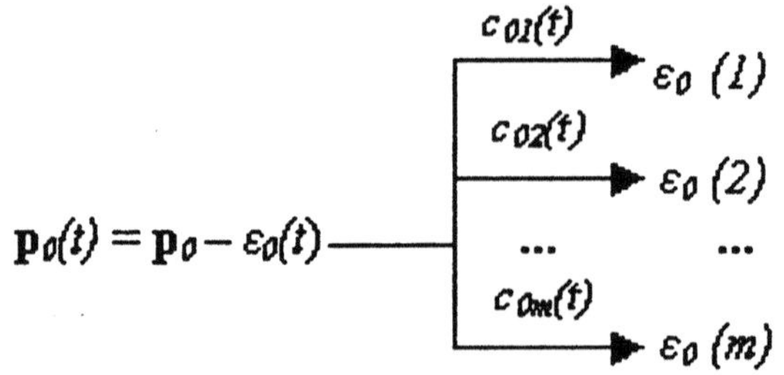

Figure 16.2 Replacement recruits at time t.

$$\text{prob}\{\varepsilon_{k+1}(t) = \varepsilon_{k+1}(i)\} = c_i(t).$$

Now let $p_{ij}(t)$ =prob{a "membership" of the system which entered state i at time t to move in state j at its next transition} $= f_{ij}(t) + p_{ik+1}(t)p_{oj}(t)$. Let $\mathbf{P}(t) = \{p_{ij}(t)\}_{i,j\in S}$ then (see also figure 16.3) :

$$\mathbf{P}(t) = (\mathbf{F} - \mathcal{E}_f(t)) + (\mathbf{p}_{k+1} - \varepsilon_{k+1}(t))'(\mathbf{p}_o - \varepsilon_o(t)) = \mathbf{P} - \mathcal{E}_p(t),$$

with

$$\mathcal{E}_p(t) = \mathcal{E}_f(t) + \mathbf{p}'_{k+1}\varepsilon_o(t) + \varepsilon'_{k+1}(t)\mathbf{p}_o - \varepsilon_{k+1}(t)\varepsilon_o(t) \text{ for } t = 0, 1, 2, \ldots$$

$$\text{prob}(\mathcal{E}_p(t) = \mathcal{E}_p(i)\} = c_{pi}(t) > 0 \text{ for } i = 1, 2, \ldots m\nu$$

We have $\mathcal{E}_p(t)\mathbf{1}' = \varepsilon'_{k+1}(t) + \mathcal{E}_f(t)\mathbf{1}' = \mathbf{0}$. Finally we assume that $\mathcal{E}_p(t)$

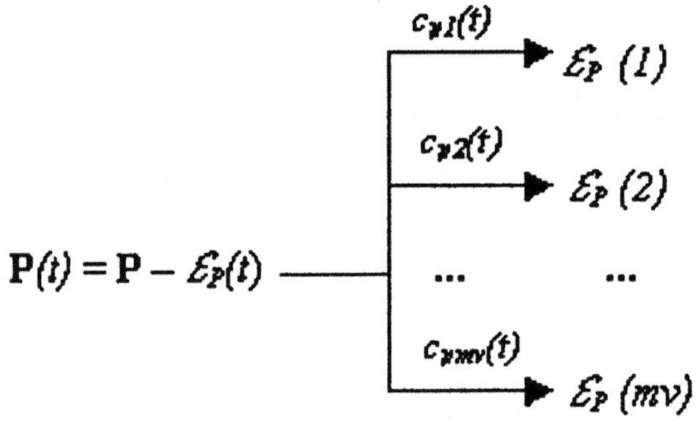

Figure 16.3 S.E. at time t.

is a *perturbation* for the matrix $\mathbf{P}(t)$.

Now we imagine that whenever a membership enters state i at time t then a stochastic mechanism is taken place and a matrix $\mathbf{P} - \mathcal{E}_p(t)$ is selected according to the above described way. Then the membership determines the next state j to which it will move according to state i's transition probabilities provided in $\mathbf{P} - \mathcal{E}_p(t)$. However, after j has been selected but before making the transition to j, the membership "holds" for a time τ_{ij} in state i. The holding times τ_{ij} are positive integer-valued random variables with probability mass function $h_{ij}(m,t)$ defined as : $h_{ij}(m,t)$ =prob$\{\tau_{ij} = m$ |the membership entered state i at time t and state j has been selected$\}$; where $m = 1, 2, \ldots,\ i, j \in S$ and $h_{ij}(0,t) = 0$ for every i, j, t. Now let $\mathbf{H}(m,t) = \{h_{ij}(m,t)\}_{i,j \in S}$ then we provide the following definition.

Definition 4 *The matrix $\mathcal{E}_H(m,t)$ is called a perturbation for the matrix $\mathbf{H}(m,t)$ if the matrix $\mathbf{H}(m,t) = \mathbf{H}(m) - \mathcal{E}_H(m,t)$ has the following properties :*

$$\mathbf{H}(m,t) \geq \mathbf{0} \text{ and } \sum_{m=0}^{\infty} \mathbf{H}(m,t) = \mathbf{U},$$
$$\text{with } \mathbf{H}(m) \geq \mathbf{0} \text{ and } \sum_{m=0}^{\infty} \mathbf{H}(m) = \mathbf{U}$$

where $\mathbf{U}$ is a matrix of 1's.

We assume that $\mathbf{H}(m,t)$ is of the form $\mathbf{H}(m,t) = \mathbf{H}(m) - \mathcal{E}_H(m,t)$ where $\mathcal{E}_H(m,t)$ is a perturbation matrix for $\mathbf{H}(m,t)$ and is randomly selected from the set of perturbation matrices

$$\mathcal{E}_H = \{\mathcal{E}_H(m,1), \mathcal{E}_H(m,2), \ldots, \mathcal{E}_H(m,l)\},$$

with probability :

$$prob\{\mathcal{E}_H(m,t) = \mathcal{E}_H(m,j)\} = c_{Hj}(t) > 0,\ j = 1, 2, \ldots, l.$$

For the new memberships created by expansion at time t let $r_{oi}(t)$ =prob$\{$ a new membership to enter state i given that it enters the system at time $t\}$ and $\mathbf{r}_o(t) = [r_{o1}(t), r_{o2}(t), \ldots, r_{ok}(t)]$. Assume that for every t the vector $\mathbf{r}_o(t)$ is of the form :

$$\mathbf{r}_o(t) = \mathbf{r}_o - \varepsilon_r(t) \text{ with } \mathbf{r}_o(t)\mathbf{1}' = \mathbf{1}' \text{ and } \mathbf{r}_o\mathbf{1}' = \mathbf{1}' \text{ for every } t.$$

The vector $\varepsilon_r(t)$ is assumed to be a perturbation vector for $\mathbf{r}_o$ and is randomly chosen from a finite set $\varepsilon_r = \{\varepsilon_r(1), \varepsilon_r(2), \ldots, \varepsilon_r(\mu)\}$, with probability prob$\{\varepsilon_r(t) = \varepsilon_r(i)\} = c_{ri}(t)$.

At this point it is useful to summarize somehow what is taken place when a membership enters a state at anytime, let us say t for example. Firstly a stochastic evolution takes place which we call *stochastic evolution 1 (S.E.1)* and the system is selecting one of the $\mathbf{P} - \mathcal{E}_p(i)$ for $i = 1, 2, \ldots, m\nu$ (see figure 16.4). At the same instantaneous moment at time t another stochastic evolution takes place which we call *stochastic evolution 2 (S.E.2)* and the system is selecting one of the $\mathbf{H}(m) - \mathcal{E}_H(m, j)$for $j = 1, 2, \ldots, l$. Thirdly at the same moment another stochastic evolution takes place which we call *stochastic evolution 3 (S.E.3)* and the system is selecting one of the $\mathbf{r}_o - \varepsilon_r(d)$ for $d = 1, 2, \ldots, \mu$. We assume that the three stochastic evolutions are independent which in fact seems quite realistic and is a well established inherent assumption. Now let $q_{ij}(s,t)$ =prob{a "membership" of the

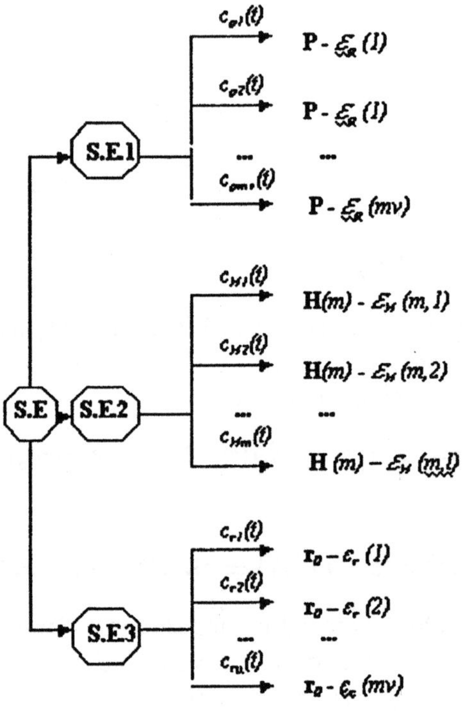

Figure 16.4 Stochastic Evolutions (S.E.) at time t.

system which entered state i at time s to be in state j after t steps}. Also let $\mathbf{Q}(s,t) = \{ q_{ij}(s,t)\}_{i,j \in S}$ the matrix of what we call interval transition probabilities. A population whose evolution is adequately described by a model as above is called a perturbed non-homogeneous semi-Markov system.

3. THE EXPECTED INTERVAL TRANSITION PROBABILITIES

We have seen in the previous section that when a membership enters a state at anytime, let us say t for example, then three stochastic mechanisms are taken place independently. The first selects a matrix $\mathbf{P} - \mathcal{E}_p(t)$ from the available perturbations and the second selects a matrix $\mathbf{H}(m) - \mathcal{E}_H(m,t)$ also from the respective perturbations. A third stochastic mechanism is taking place when a new membership is entering the system which selects a vector $\mathbf{r}_o - \varepsilon_r(t)$ from the available perturbations independently of the other two. Thus at each instant of time it is logical to introduce $E[\mathbf{P}(t)], E[\mathbf{H}(m,t)$ and $E[\mathbf{r}_o(t)]$. It could be proved that

$$E[\mathbf{P}(t)] = \mathbf{P} - E[\mathcal{E}_p(t)] \text{ where } : \tag{16.3}$$

$$\hat{E}_p(t) = E[\mathcal{E}_p(t)] = \sum_{h=1}^{\nu} \mathcal{E}_p(h) c_h(t) + \mathbf{p}'_{k+1} \left\{ \sum_{b=1}^{m} \varepsilon_o(b) c_{ob}(t) \right\}$$
$$+ \left\{ \sum_{h=1}^{\nu} \varepsilon'_{k+1}(h) c_h(t) \right\} \mathbf{p}_o - \sum_{h=1}^{\nu} \sum_{b=1}^{m} \varepsilon'_{k+1}(h) \varepsilon_o(b) c_{ob}(t) c_h(t). \tag{16.4}$$

$$E[\mathbf{H}(m,t)] = \mathbf{H}(m) - \sum_{\alpha=1}^{l} \mathcal{E}_H(m,a) c_{Ha}(t) = \mathbf{H}(m) - \hat{\mathbf{E}}_H(t), \tag{16.5}$$

$$E[\mathbf{r}_o(t)] = \mathbf{r}_o - \sum_{h=1}^{\mu} \varepsilon_r(h) c_{rh}(t) = \mathbf{r}_o - \hat{E}_r(t). \tag{16.6}$$

Now define $w_i(t,m)$ =prob{a "membership" of the system which entered state i at time t to stay m time units in state i before its next transition}. Let $\mathbf{W}(n,s)$ the $k \times k$ matrix which has zeros everywhere apart from the diagonal which has in position i the element :

$$\sum_{m=n+1}^{\infty} w_i(m,s) = 1 - \sum_{m=0}^{n} w_i(m,s).$$

Let $\mathbf{A} \diamond \mathbf{B}$ the Hadamard product of two matrices. From Vassiliou and Papadopoulou (1992) we get that

$$\mathbf{W}(n,s) = \sum_{m=n+1}^{\infty} \mathbf{I} \diamond \{\mathbf{P}(s) \diamond \mathbf{H}(m,s)\} \mathbf{U}$$

Using the fact that $E[E(X \mid Y)] = E[X]$ and that $\mathbf{P}(s)$ and $\mathbf{H}(m,s)$ will be selected from two independent stochastic evolutions we could prove that

$$E[\mathbf{W}(n,s)] = \sum_{m=n+1}^{\infty} \mathbf{I} \diamond \{E[\mathbf{P}(s) \diamond \mathbf{H}(m,s)]\mathbf{U}\}$$

$$= \mathbf{I} \diamond \{([\mathbf{P} - \hat{E}_p(s)] \diamond \sum_{m=n+1}^{\infty} [\mathbf{H}(m) - \hat{E}_H(s)])\mathbf{U}\}. \tag{16.7}$$

It is evident that the matrix $\mathbf{Q}(n,s)$ of the interval transition probabilities is a matrix with elements that are random variables thus it is important to consider the matrix $E[\mathbf{Q}(n,s)]$. Let us define the following events.

$\mathcal{A}_1 \equiv$ {a membership of the system,which entered state i at time s, to remain in state i at least n time units}.

$\mathcal{A}_2 \equiv$ {a membership of the system, which entered state i at time s, makes its next transition to state j and makes that transition after n time units}.

$\mathcal{A}_3 \equiv$ {a membership of the system, which entered state i at time s, to make one transition in the time interval $(s, s+n)$ at time $s+n_1-1$ and be in state j at time $s+n$} (see figure 16.5). $\mathcal{A}_4 \equiv$ {a membership

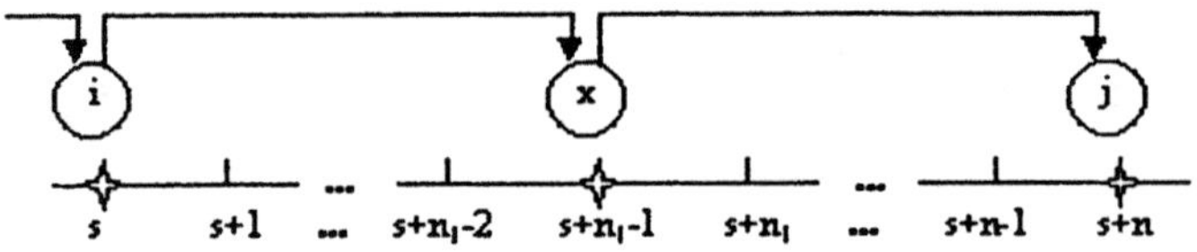

Figure 16.5 Event $\mathcal{A}_3$.

of the system, which entered state i at time s, makes ν transitions in the time interval $(s, s+n_1-1)$ and enters state j at time $s+n_1-1$ and be at time $s+n$ in state j} (see figure 16.6). The events $\mathcal{A}_1, \mathcal{A}_2, \mathcal{A}_3$

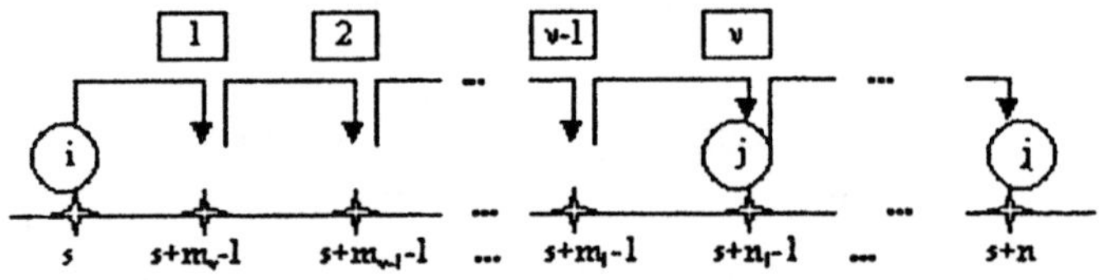

Figure 16.6 An example of event $\mathcal{A}_4$.

over all the possible values of n_1 and $\mathcal{A}_4$ over all the possible values of

$n_1$and n are all mutually exclusive and all the possible ones with which a membership enters state i at time s and is in state j at time $s+n$. Thus :

$$q_{ij}(n,s) = prob(\mathcal{A}_1) + prob(\mathcal{A}_2) + \sum_{n_1=2}^{n} prob(\mathcal{A}_3) + \sum_{n_1=2}^{n}\sum_{\nu=1}^{n_1-2} prob(\mathcal{A}_4),$$

and consequently since $q_{ij}(n,s)$ are random variables we have :

$$E[q_{ij}(n,s)] = E[prob(\mathcal{A}_1)] + E[prob(\mathcal{A}_2)] + \sum_{n_1=2}^{n} E[prob(\mathcal{A}_3)]$$

$$+ \sum_{n_1=2}^{n}\sum_{\nu=1}^{n_1-2} E[prob(\mathcal{A}_4)] \text{ for } i,j = 1,2,\ldots,k \text{ and } n,s = 1,.\quad (16.8)$$

Now we must have in mind that we want to find the matrix $E[\mathbf{Q}(n,s)]$ which obviously will emerge from the matrix form of the above equation. Let us start with the part $E[prob(\mathcal{A}_1)]$ then its contribution to $E[\mathbf{Q}(n,s)]$ has been already calculated as $E[\mathbf{W}(n,s)]$ and is in equation (16.7). Now :

$$prob(\mathcal{A}_2) = p_{ij}(s)h_{ij}(n,s). \qquad (16.9)$$

Its contribution to the matrix $\mathbf{Q}(n,s)$ will obviously be $\mathbf{P}(s) \diamond \mathbf{H}(n,s)$. Using the fact that $E[E[X \mid Y,Z]] = E[X]$ and that $\mathbf{P}(s)$ and $\mathbf{H}(n,s)$ will be selected from two independent stochastic evolutions we could prove that :

$$E[\mathbf{P}(s) \diamond \mathbf{H}(n,s)] = [\mathbf{P} - \hat{E}_p(s)] \diamond [\mathbf{H}(n) - \hat{E}_H(s)], \qquad (16.10)$$

which will be the contribution of $E[prob(\mathcal{A}_2)]$ to the matrix $E[\mathbf{Q}(n,s)]$. Now we have with very careful consideration of possible transitions prob($\mathcal{A}_3$) = prob{a membership of the system which enters state i at time s, makes a transition to state j at time $s+n_1-1$ and remains in j up to time $s+n$}+ prob{a membership of the system which enters state i at time s, makes a transition to any state x at time $s+n_1-1$ and enters state j at time $s+n$ with a new transition}=

$$\begin{aligned} &= p_{ij}(s)h_{ij}(n_1-1,s)w_j(n-n_1+1,s+n_1-1) \\ &+ \sum_x p_{ix}(s)h_{ix}(n_1-1,s)p_{xj}(s+n_1-1)h_{xj}(n-n_1+1,s+n_1-1). \end{aligned}$$

Its contribution to the matrix $\mathbf{Q}(n,s)$ will be

$$\begin{aligned} &\sum_{n_1=2}^{n}\{\mathbf{P}(s) \diamond \mathbf{H}(n_1-1,s)[\mathbf{W}(n-n_1+1,s+n_1-1) \\ &+ \ \mathbf{P}(s) \diamond \mathbf{H}(n_1-1,s)\mathbf{P}(s+n_1-1) \diamond \mathbf{H}(n-n_1+1,s+n_1-1)]\}. \end{aligned}$$

In order to find the contribution of the above part to the matrix $E[\mathbf{Q}(n,s)]$ we must find its expected value. Using the fact that $E[E(X \mid Y,Z)] = E[X]$ and the independence of the various stochastic evolutions we get that the expected value of the above formula is :

$$\sum_{n_1=2}^{n}[\mathbf{P}-\hat{E}_p(s)]\circ[\mathbf{H}(n_1-1)-\hat{E}_H(s)]\{\mathbf{I}\circ[[\mathbf{P}-\hat{E}_p(s+n_1-1)]$$
$$\circ\sum_{m=n+1}^{\infty}[\mathbf{H}(m-n_1+1)-\hat{E}_H(s+n_1-1)]\mathbf{U}]$$
$$+(\mathbf{I}+[\mathbf{P}-\hat{E}_p(s+n_1-1)][\mathbf{H}(m-n_1+1)-\hat{E}_H(s+n_1-1)])\}. \quad (16.11)$$

Now we consider the probability which is the (i,j) element of the matrix

$$[\mathbf{P}(s)\circ\mathbf{H}(m_{\nu-1}-1,s)][\mathbf{P}(s+m_\nu-1)\circ\mathbf{H}(m_{\nu-1}-m_\nu,s+m_\nu-1)]$$
$$\cdot\ldots\cdot[\mathbf{P}(s+m_1-1)\circ\mathbf{H}(n-m_1,s+m_1-1)],$$

then this is prob{a membership of the system which entered state i at time s makes ν transitions in the time interval $(s,s+n_1-1)$ at exactly the times $s+m_\nu-1, s+m_{\nu-1}-1,\ldots,s+m_1-1$ and enters j at time $s+n_1-1$}. We could also write the above product as

$$\prod_{r=-1}^{\nu-1}[\mathbf{P}(s+m_{\nu-r}-1)\circ\mathbf{H}(m_{\nu-r-1}-m_{\nu-r},s+m_{\nu-r}-1)]$$
$$\text{with } m_{\nu+1}=1 for \nu=1,2,\ldots,n_1-2 \text{ and } m_0=n_1.$$

Now if we sum the above product for all possible values of the holding times then we get :

$$\mathbf{S}_{n_1}(\nu,s,m_\nu)=\sum_{m_\nu=2}^{n_1-\nu}\left\{\sum_{m_{\nu-1}=1+m_\nu}^{n_1-\nu+1}\cdots\right.$$
$$\left.\left\{\sum_{m_1=1+m_2}^{n_1-1}\prod_{r=-1}^{\nu-1}[\mathbf{P}(s+m_{\nu-r}-1)\circ\mathbf{H}(m_{\nu-r-1}-m_{\nu-r},s+m_{\nu-r}-1)]\right\}\cdots\right\}$$

for every $n_1 \geq \nu+2$, while for $n_1 < \nu+2$ we have $\mathbf{S}_{n_1}(\nu,s,m_\nu)=\mathbf{0}$. Obviously the (i,j) element of $\mathbf{S}_{n_1}(\nu,s,m_\nu)$ is the probability prob{a membership of the system which entered state i at time s makes ν transitions in the time interval $(s,s+n_1-1)$ and enters state j at time $s+n_1-1$}.

It is straightforward to prove that the contribution of prob(A_4) to the matrix $\mathbf{Q}(n,s)$ is : $\mathbf{S}_{n_1}(\nu,s,m_\nu)[\mathbf{W}(n-n_1+1,s+n_1-1)+\mathbf{P}(s+n_1-1)\circ\mathbf{H}(n-n_1+1,s+n_1-1)]$. Using the fact that $E[E(X \mid Y,Z)] = E[X]$

and the independence of the various stochastic evolutions we get that the contribution of $E[prob(\mathcal{A}_4)]$ to the matrix $E[\mathbf{Q}(n,s)]$ is :

$$E[\mathbf{S}_{n_1}(\nu,s,m_\nu)]\{\mathbf{I}\diamond\{([\mathbf{P}-\hat{E}_p(s+n_1-1)]$$
$$\diamond\sum_{m=n+1}^{\infty}[\mathbf{H}(m-n_1+1)-\hat{E}_H(s+n_1-1)])\mathbf{U}\}$$

$$+[\mathbf{P}-\hat{E}_p(s+n_1-1)]\diamond[\mathbf{H}(n-\nu+1)-\hat{E}_H(s+n_1-1)]\} \qquad (16.12)$$

where

$$E[\mathbf{S}_{n_1}(\nu,s,m_\nu)]=\sum_{m_\nu=2}^{n_1-\nu}\{\sum_{m_{\nu-1}=1+m_\nu}^{n_1-\nu+1}\cdots\{\sum_{m_1=1+m_2}^{n_1-1}\prod_{r=-1}^{\nu-1}$$

$$[\mathbf{P}-\hat{E}_p(s+m_{\nu-r}-1)]\diamond[\mathbf{H}(m_{\nu-r-1},m_{\nu-r})-\hat{E}_H(s+m_{\nu-r}-1)]\}\ldots\} \quad (16.13)$$

Now from (16.7), (16.8), (16.10), (16.11), (16.12), we get that :

$$E[\mathbf{Q}(n,s)]=\mathbf{I}\diamond\{([\mathbf{P}-\hat{E}_p(s)]\diamond\sum_{m=n+1}^{\infty}[\mathbf{H}(m)-\hat{E}_H(s)])\mathbf{U}\}+$$
$$[\mathbf{P}-\hat{E}_p(s)]\diamond[\mathbf{H}(n)-\hat{E}_H(s)]+$$
$$\sum_{n_1=2}^{n}[\mathbf{P}-\hat{E}_p(s)]\diamond[\mathbf{H}(n_1-1)-\hat{E}_H(s)]\{\mathbf{I}\diamond[[\mathbf{P}-\hat{E}_p(s+n_1-1)]$$
$$\diamond\sum_{m=n+1}^{\infty}[\mathbf{H}(m-n_1+1)-\hat{E}_H(s+n_1-1)]\mathbf{U}]$$
$$+(\mathbf{I}+[\mathbf{P}-\hat{E}_p(s+n_1-1)][\mathbf{H}(m-n_1+1)-\hat{E}_H(s+n_1-1)]])\}+$$
$$\sum_{n_1=2}^{n}\sum_{\nu=1}^{n_1-2}E[\mathbf{S}_{n_1}(\nu,s,m_\nu)]\{\mathbf{I}\diamond\{([\mathbf{P}-\hat{E}_p(s+n_1-1)]$$

$$\diamond\sum_{m=n+1}^{\infty}[\mathbf{H}(m-n_1+1)-\hat{E}_H(s+n_1-1)])\mathbf{U}\}$$
$$+[\mathbf{P}-\hat{E}_p(s+n_1-1)]\diamond[\mathbf{H}(n-\nu+1)-\hat{E}_H(s+n_1-1)]\}.$$

4. THE EXPECTED POPULATION STRUCTURE

The first important step in population system theory and in stochastic modelling for social processes (Bartholomew(1982), Bartholomew,

Forbes and Mc Clean (1991)) is to provide the expected population structure as a function of the basic parameters of the system. For a P-NHSMS the basic parameters that determine uniquely the process are the sequences $\{\mathbf{P} - \mathcal{E}_p(t)\}_{t=0}^{\infty}$, $\{\mathbf{H}(m) - \mathcal{E}_H(m,t)\}_{m,t=0}^{\infty}$, $\{\mathbf{r}_o - \varepsilon_r(t)\}_{t=0}^{\infty}$ and $\{T(t)\}_{t=0}^{\infty}$.

In order to establish the relationship of $E[N_j(t)]$, $j = 1, 2, \ldots, k, t = 1, 2, \ldots$ with the basic parameters of the system in analytic form we consider the population divided into two classes. In the first belong the initial memberships that survived up to time t and are in state j. In the second we have all the new memberships that were created due to the expansion of the population after $t = 0$ and which are in state j at time t. With careful probabilistic analysis we could prove that

$$E[N_j(t)] = \sum_{i=1}^{k} N_i(0)E[q_{ij}(0,t)] + \sum_{m=1}^{k}\sum_{i=1}^{k} \Delta T(m)[\mathbf{r}_o - \varepsilon_r(t)]E[q_{ij}(t-m,m)]$$

for $j = 1, 2, \ldots, k$. The above equation in matrix notation could be written as :

$$E[\mathbf{N}(t)] = \mathbf{N}(0)E[\mathbf{Q}(0,t)] + \sum_{m=1}^{k} \Delta T(m)[\mathbf{r}_o - \varepsilon_r(t)]E[\mathbf{Q}(t-m,m)], \tag{16.14}$$

where the expected value of the matrix of the interval transition probabilities is given in (16.9). The relation (16.14) apart from being the only way to predict the population structure is important in solving a large number of problems in the theory of P-NHSMS.

References

[1] Bartholomew, D.J. (1982). *Stochastic models for social processes*, Wiley, Chichester, 3rd ed.

[2] Bartholomew, D.J., Forbes, A.F. and Mc Clean, S.I. (1991). *Statistical Techniques for Manpower Planning*, Wiley, Chichester.

[3] Campbell, S. L. and C.D. Meyer (1979). *Generalized inverses of Linear Transformations*, Pitman, London.

[4] Jansen, J. (1986). *Semi-Markov models : Theory and applications.* ed. J. Janssen, Plenum Press, New York.

[5] Mc Clean, S.I. (1980) A semi- Markov model for a multi-grade population with Poisson recruitment, *J. Appl. Prob.*, **17**, 846-852.

[6] Mc Clean, S. I. (1986). Semi-Markov models for manpower planning, *In Semi-Markov models : Theory and Applications*, ed. J. Janssen, Plenum Press, New York.

[7] Mc Clean, S.I. (1993). Semi-Markov models for human resource modelling. *IMA Journal of mathematics Applied in Business and Industry*, **4**, 307-315.

[8] Mc Clean, S.I., E. Montgomery and F.Ugwuowo (1997). Non - homogeneous Continuous Time Markov and Semi - Markov Manpower Models. *Appl. Stochastic Models and Data Analysis*, **13**, 191-198.

[9] Mehlman, A. (1979). Semi-Markovian manpower models in continious time, *J. Appl. Prob*, **6**, 416-422.

[10] Meyer, C.D. (1975). The role of the group generalized inverse in the theory of finite Markov chains. *Siam Review* ,**17**, 443-464.

[11] Papadopoulou, A.A and P.-C.G. Vassiliou (1994). Asymptotic behavior of non-homogeneous Semi- Markov Systems, *Linear Algebra and its applications*, **210**, 153-198.

[12] Vassiliou, P.-C. G. and A.A. Papadopoulou (1992), Non-homogeneous semi-Markov systems and maintainably of the state sizes, *J. Appl. Prob.*, **29**, 519-534.

[13] Vassiliou, P.-C. G. (1996). The non-homogeneous semi- Markov system in a stochastic environment, *In Applied Probability in Honor of J.M. Gani.* Ed. C.C. Heyde, Yu V. Prohorov, R. Pyke, S.T. Rachev. Springer

[14] Vassiliou, P.-C. G. and M.A. Symeonaki (1999) The perturbed non-homogeneous Markov system, *Linear Alg. and its Appl.* (to appear).

[15] Vassiliou, P.-C. G. and M.A. Symeonaki (1998). The perturbed non-homogeneous Markov system in continuous time, *Appl. Stochastic Models and Data Analysis*, **13**, 207-216.

IV

QUEUEING SYSTEMS THEORY

Chapter 17

SEMI-MARKOV QUEUES WITH HEAVY TAILS

Soren Asmussen
Lund University, Sweden

Abstract We consider queues with a sub-exponential heavy-tailed service time distribution B. Classical results stating that the tails of the steady-state waiting time, resp. the maximal waiting time within a regenerative cycle, are proportional to $\int_x^\infty \overline{B}(y)\,dy$, resp. $\overline{B}(y) = 1 - B(y)$, are extended to various semi-Markov queues.

Keywords: Cycle maximum, M/G/1 queue, Markov-modulation, rare event, regular variation, sub-exponential distribution.

1. INTRODUCTION

The present paper is concerned with the evaluation of rare events probabilities for queues when some underlying distribution is heavy-tailed. Typical examples are the probabilities of large delays in a queueing system with heavy-tailed service time distributions, or the probability of buffer overflow in a fluid ATM model with many i.i.d. ON–OFF sources with a heavy-tailed distribution of the ON period.

Heavy tails are currently receiving much attention and are often being linked up with long-range dependence and self-similarity. The motivation comes to a large extent from recent statistical analysis of data from areas such as LAN's (local area networks; e.g. Leland *et al.* [21], Willinger *et al.* [30]) and WAN's (wide area networks; e.g. Paxson & Floyd [26]); see also Resnick [28]. In this setting, it has been noted repeatedly (starting from Taqqu & Levy [29]) that a possible explanation of long-range dependence is heavy tails of quantities like packet lengths which in turn in simple queueing systems have their analogues in service times.

In the present paper, we do not go into the complications of long-range dependence but study the impact of heavy tails in the simplest setting, queues with heavy-tailed service time distributions. The classical result (von Bahr [12], Pakes [25], Embrechts & Veraverbeke [15]) is the tail approximation

$$\mathbb{P}(W > x) \sim \frac{1}{|\mu|}\int_x^\infty \overline{B}(y)\,dy, \quad x \to \infty, \tag{17.1}$$

for the tail of the steady-state GI/GI/1 waiting time W where A, B are the distributions of the inter-arrival and service times, $\overline{B}(x) = 1 - B(x)$ is the tail and and μ_A, μ_B are the respective means, $\mu = \mu_B - \mu_A$ (here and in the following, $\sim$ means that the ratio is one in the limit $x \to \infty$). A more recent result (Asmussen [4], Heath, Resnick & Samorodnitsky [18]) gives the similar asymptotics for the cycle maximum:

$$\mathbb{P}(M_C > x) \sim \mathbb{E}C \cdot \overline{B}(y), \quad x \to \infty, \tag{17.2}$$

where $C = \inf\{n > 0 : W_n = 0\}$ is the regenerative cycle (number of customers served in a busy cycle) for the waiting time process $\{W_n\}$ and $M_C = \max_{n<C} W_n$. In performance evaluation, $\mathbb{P}(M_C > x)$ is identified with the probability of buffer overflow within a cycle. The cycle maximum is also a key vehicle for studying extreme value behavior, i.e. asymptotic properties of quantities like the maximum $\max_{k=0,\dots,n} W_n$ up to time n and the time $\tau(x) = \inf\{n : W_n > x\}$ of first exceedance of a large level x, see Asmussen [5]. For example, (17.2) implies that $\overline{B}(x)\tau(x)$ has an asymptotical standard exponential distribution, cf. e.g. Keilson [20] or Glasserman & Kou [17].

The heuristics behind (17.1), (17.2) is, as explained in more detail in Section 2, that with heavy tails rare events occur as consequence of one big service time. This heuristics readily leads to the form of the tail asymptotics, see again Section 2, and also suggests that similar results should hold true far beyond the GI/GI/1 setting. However, a rigorous mathematical proof in complete generality is most often much more intricate. This is where semi-Markov assumptions play a role by providing a framework which is reasonably general but yet specific enough to allow the desired results to be established.

The present paper is a survey (obviously somewhat biased by the author's research interests) of the state-of-the-art of results like (17.1), (17.2) for various types of semi-Markov models. We start in Section 2 by giving some preliminaries and explaining the 'one big jump' heuristics' for simple queues, that is, in a random walk setting. Section 3 then surveys the extension to a Markov-modulated framework, using regeneration as a key tool, and in Section 4 we look at some queueing systems where the added complications of boundary modifications occur; in particular, the set–up includes Neuts' Markov chains of M/G/1 type ([23]) with heavy tails (for tail asymptotics in the light–tailed case, see Abate, Choudhury & Whitt [1]).

2. SUB-EXPONENTIAL DISTRIBUTIONS. RANDOM WALK RESULTS

Let G be a distribution on $(0, \infty)$ and let $Y_1, Y_2, \ldots$ be i.i.d. with common distribution G.

An established common framework for the heavy–tailed setting is that $G \in \mathcal{S}$, the sub-exponential class defined by the requirement

$$\lim_{x \to \infty} \frac{\overline{G^{*n}}(x)}{\overline{G}(x)} = \lim_{x \to \infty} \frac{\mathbb{P}(Y_1 + \cdots + Y_n > x)}{\mathbb{P}(Y_1 > x)} = n \tag{17.3}$$

for all n [actually, it suffices that (17.3) holds for $n = 2$]. Since for any distribution G, $\mathbb{P}(\max(Y_1, \ldots, Y_n) > x) \approx n\mathbb{P}(Y_1 > x)$, this formal definition contains what is the crucial intuition behind heavy tails: *the only way the sum can get large is by one of the summands getting large* (in contrast, in the light–tailed case all summands are large if the sum is so). The main examples (listed in order of tail heaviness) are

regularly varying : $\overline{G}(x) = L(x)/x^{\alpha}$ where L is slowly varying, $L(tx)/L(x) \to 1$ as $x \to \infty$ for all $t > 0$; for example this covers Pareto–, stable– and log-gamma distributions;

log-normal : the distribution of $e^{\mu + \sigma U}$ where U is standard normal;

Weibull with decreasing failure rate, $\overline{G}(x) = e^{-x^{\beta}}$ with $0 < \beta < 1$.

See Feller (1971) and Bingham, Goldie & Teugels [13] for the regularly varying case, and Embrechts, Klüppelberg & Mikosch [14] for the general sub-exponential class.

A property of sub-exponential distributions that is often used is

$$\overline{G}(x - y) \sim \overline{G}(x) \tag{17.4}$$

for y fixed and $x \to \infty$.

The derivation of the limit result (17.1) uses the representation $W \stackrel{\mathcal{D}}{=} M$ where $M = \sup_{0,1,2,\ldots} S_n$ with $\{S_n\}$ a random walk with increments X_k distributed as the difference $U - T$ between a service time U and an independent inter-arrival time T (thus the mean is $\mu = \mu_B - \mu_A$ which is negative under the stability condition $\rho = \mu_B/\mu_A < 1$ which we assume throughout). An easy heuristics using the 'one big service time' heuristics goes as follows. The random walk is typically at $n\mu$ at time n so that it will be above x for the first time w.p. $\overline{F}(x - n\mu)$ where F is the distribution of $X = U - T$. Since it is easy to deduce from (17.4) that $\overline{F}(x) \sim \overline{B}(x)$ when B is sub-exponential, we get

$$\begin{aligned} \mathbb{P}(W > x) &= \mathbb{P}(M > x) \sim \sum_{n=0}^{\infty} \overline{F}(x - n\mu) \sim \sum_{n=0}^{\infty} \overline{B}(x - n\mu) \\ &= \int_0^{\infty} I(x - n\mu < y)\, B(dy) \sim \frac{1}{|\mu|} \int_0^{\infty} (y - x)^+\, B(dy) \\ &= \frac{1}{|\mu|} \int_x^{\infty} \overline{B}(y)\, dy. \end{aligned} \tag{17.5}$$

The similar argument for (17.2) is even simpler: just note that $\mathbb{E}C$ is the expected number of customers for which there is the possibility to have a big service time, and that $\overline{B}(x) \sim \overline{B}(x-y)$ (cf. (17.4)) is the probability that the next waiting time will exceed x (here y is the waiting time of the customer with the big service time).

The rigorous proofs are, however, much more intricate. In particular, the argument in [15], which is the only one around, uses Wiener–Hopf arguments (the decomposition of M as a sum of ladder steps and a study of the ladder height distribution somewhat similar to (17.5)), and the proof of (17.2) in [4] involves in turn (17.1) and a non–standard level crossing idea. Nevertheless, the heuristics suggests that the results should be true in much greater generality. The rest of the paper gives two specific examples in a semi–Markov setting.

3. MARKOV–MODULATED M/G/1 QUEUES AND RELATED MODELS

We consider queues with dependent inter-arrival times, allowing also for possible dependence between the arrival process and the service times. We let W be some steady–state characteristics (say the stationary waiting time, the workload or the queue length) and assume the representation $W \stackrel{\mathcal{D}}{=} M$ where $M = \max S_t$ for some process $\{S_t\}$ in discrete or continuous time; it is well known that such a representation holds well beyond the simple random walk / GI/G/1 example in Section 2, see e.g. Asmussen [3] and references there.

Various criteria for (17.1) to be true subject to some dependence were recently given by Asmussen, Schmidli & Schmidt [11]. We give here one of them, Theorem 1 based upon a regenerative assumption, and apply it to some semi–Markov queues.

Assume that $\{S_t\}_{t\geq 0}$ has a regenerative structure in the sense that there exists a renewal process $\chi_0 = 0 \leq \chi_1 \leq \chi_2 < \ldots$ such that

$$\{S_{\chi_0+t} - S_{\chi_0}\}_{0\leq t<\chi_1-\chi_0}, \; \{S_{\chi_1+t} - S_{\chi_1}\}_{0\leq t<\chi_2-\chi_1}, \; \cdots$$

(viewed as random elements of the space of D–functions with finite lifelengths) are i.i.d. and the distribution of $\{S_{\chi_k+t} - S_{\chi_k}\}_{0\leq t<\chi_{k+1}-\chi_k}$ is the same for all $k = 1, 2, \ldots$. The zero–delayed case corresponds to $\chi_0 = \chi_1 = 0$ and we write then $\mathbb{P}_0$, $\mathbb{E}_0$ etc. We let F^* denote the $\mathbb{P}_0$–distribution of S_1^*, assume $\mu_{F^*} < 0$ and $\mathbb{E}_0\chi < \infty$ where $\chi = \chi_2 - \chi_1$ is the generic cycle. See Fig. 17.1 where the filled circles symbolize a regeneration in the path.

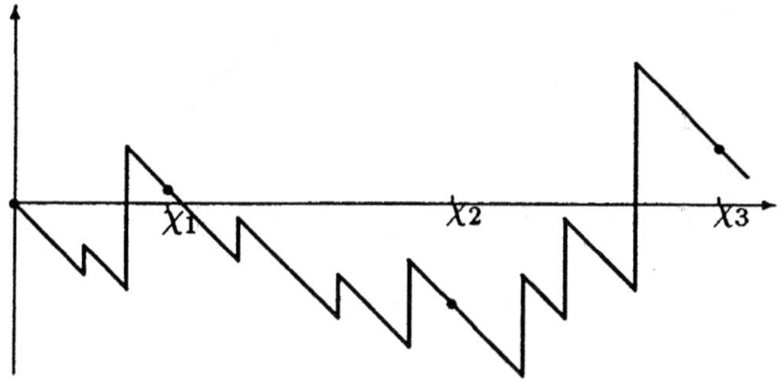

Figure 17.1

On the figure, $\{S_t\}$ has a sample path structure corresponding to the netput process of the workload process. However, this is only one example.

Define

$$S_n^* = S_{\chi_{n+1}}, \quad M_n^* = \max_{k=0,\ldots,n} S_k^*, \quad M^* = \max_{n=0,1,\ldots} S_n^*, \qquad M = \sup_{t\geq 0} S_t.$$

The idea is now to observe that in the zero–delayed case, $\{S_n^*\}_{n=0,1,\ldots}$ is a random walk. Thus the assumption

$$\overline{F}^*(x) \;=\; \mathbb{P}_0(S_1^* > x) \;\approx\; \overline{G}(x) \tag{17.6}$$

for some sub-exponential G makes (17.1) applicable so that

$$\mathbb{P}(M^* > u) \;\approx\; \frac{1}{|\mu_{F^*}|}\int_u^\infty \overline{F}^*(y)\,dy\,, \quad u \to \infty. \tag{17.7}$$

Imposing suitable conditions on the behavior of $\{S_t\}$ within a cycle will then ensure that M and M^* are sufficiently close to be tail equivalent. The one we focus on is

$$\mathbb{P}_0\left(M_1^{(\chi)} > x\right) \sim \mathbb{P}_0(S_1^* > x)\,. \tag{17.8}$$

where

$$M_n^{(\chi)} = \sup_{0 \le t < \chi_{n+1}-\chi_n} S_{\chi_n+t} - S_{\chi_n} = \sup_{0 \le t < \chi_{n+1}-\chi_n} S_{\chi_n+t} - S_n^*$$

Since clearly $M_1^{(\chi)} \ge S_1^*$, the assumption means that $M_1^{(\chi)}$ and S_1^* are not too far away. See Fig. 17.2.

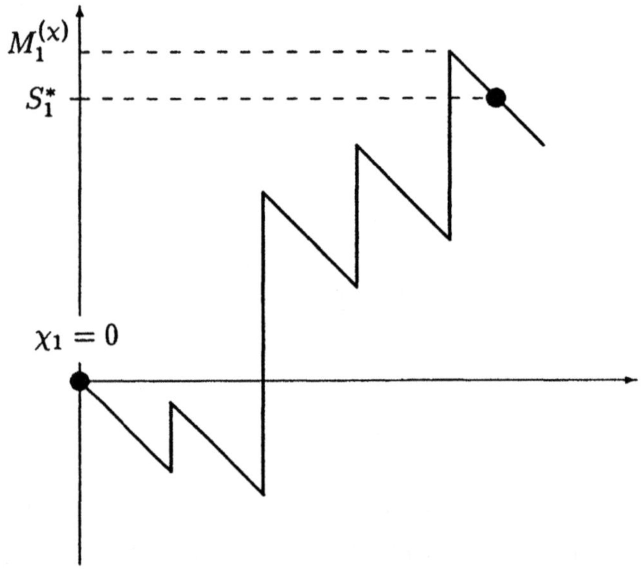

Figure 17.2

Theorem 1 *Assume that* (17.6) *and* (17.8) *hold. Then*

$$\mathbb{P}_0(M > u) \approx \frac{1}{|\mu_{F^*}|}\int_u^\infty \overline{F}^*(y)\,dy\,.$$

Proof. Since $M \ge M^*$, it suffices to by (17.7) to show $\liminf \mathbb{P}(M^* > u)/\mathbb{P}(M > u) \ge 1$. Define

$$\vartheta^*(u) = \inf\{n = 1, 2, \ldots : S_n^* > u\}\,,$$
$$\beta(u) = \inf\left\{n = 1, 2, \ldots : S_n^* + M_{n+1}^{(\chi)} > u\right\}$$

(note that $\{M > u\} = \{\beta(u) < \infty\}$). Let $a > 0$ be fixed. We shall use the estimate

$$\mathbb{P}_0(M > u,\, M^{(\chi)}_{\beta(u)} \le a) \;=\; o(\mathbb{P}_0(M > u)) \tag{17.9}$$

which follows since

$$\begin{aligned}\mathbb{P}_0(M > u,\, M^{(\chi)}_{\beta(u)} \le a) \;&\le\; \mathbb{P}_0\left(\bigcup_{n=1}^{\infty}\{M^*_n \in (u-a, u)\}\right)\\ \le\; \mathbb{P}(M^* \in (u-a,u))/\mathbb{P}(M^* = 0) \;&=\; o(\mathbb{P}_0(M^* > u))\end{aligned}$$

(using (17.4) for the last identity). Given $\epsilon > 0$, choose a such that $\mathbb{P}_0(S^*_1 > x) \ge (1-\epsilon)\mathbb{P}_0(M^{(\chi)}_1 > x)$, $x \ge a$. Then

$$\begin{aligned}\mathbb{P}_0(M^* > u) \;&\sim\; \mathbb{P}_0\left(M^* > u,\, S^*_{\vartheta(u)} - S^*_{\vartheta(u)-1} > a\right)\\ &=\; \sum_{n=1}^{\infty}\mathbb{P}_0\left(M^*_n \le u,\, S^*_{n+1} - S^*_n > a \vee (u - S^*_n)\right)\\ &\ge\; (1-\varepsilon)\sum_{n=1}^{\infty}\mathbb{P}_0\left(M^*_n \le u,\, M^{(\chi)}_{n+1} > a \vee (u - S^*_n)\right)\\ &\ge\; (1-\varepsilon)\sum_{n=1}^{\infty}\mathbb{P}_0\left(\max_{0<t\le\chi_{n+1}} S_t \le u,\, M^{(\chi)}_{n+1} > a \vee (u - S^*_n)\right)\\ &=\; (1-\epsilon)\mathbb{P}_0\left(M > u,\, M^{(\chi)}_{\beta(u)} > a\right) \;\approx\; (1-\epsilon)\mathbb{P}_0\left(M > u\right).\end{aligned}$$

Letting first $u \to \infty$ and next $\epsilon \downarrow 0$ yields $\liminf \mathbb{P}_0(M^* > u)/\mathbb{P}_0(M > u) \ge 1$. □

As a specific example, consider the workload process $\{W_t\}$ of a semi-Markov queue; in this setting, $\{S_t\}$ is the time-reversed of the netput process $\left\{\breve{S}_t\right\}$ generating $\{W_t\}$ as

$$W_t \;=\; \breve{S}_t - \inf_{0\le v\le t} \breve{S}_v\,.$$

Thus $\{S_t\}$ evolves in a semi-Markovian environment $\{(R_n, J_n)\}$, such that $\{J_n\}$ is an irreducible and aperiodic Markov chain with a finite number p of states, and that given $\{J_n\}$ the R_n are independent with distribution F_i for R_k when $J_k = i$. We allow a delay x, i.e. R_0 has conditional tail function $\overline{F}_i(y+x)/\overline{F}_i(x)$ rather than $\overline{F}_i(y)$ when $J_0 = i$. On the time interval $[R_0 + \cdots + R_{n-1}, R_0 + \cdots + R_n)$ upwards jumps of $\{S_t\}$ occur according to a Poisson process with intensity λ_i and have distribution B_i when $J_n = i$, and when $J_n = i$, $J_{n+1} = j$, a jump occurs at time $R_0 + \cdots + R_n$ and has distribution B_{ij} (we allow B_i or B_{ij} to have an atom at zero or even to be degenerate at 0 so that some jumps may be dummy). In between jumps, $\{S_t\}$ has a negative drift at a unit rate. We write m_i, μ_i, μ_{ij} for the means of F_i, B_i, resp. B_{ij}, and assume that

$$\rho = \sum_{i=1}^{p} \left\{ \pi_i \lambda_i \mu_i + \frac{\pi_i}{m_i} \sum_{j=1}^{p} q_{ij}\mu_{ij} \right\} < 1 .$$

Here q_{ij} are the transition probabilities of $\{J_n\}$ and π_i the (time) stationary probabilities for $\{J_t\}$, where $J_t = J_{n-1}$ if $t \in [R_0 + \cdots + R_{n-1}, R_0 + \cdots + R_n)$; note that the π_i always exist at least in the sense of ergodic averages which is sufficient here. It is well–known that $\pi_i = \pi_i^0 m_i / \sum_{j=1}^{p} \pi_i^0 m_i$ where π_i^0 are the stationary probabilities for the embedded Markov chain $\{J_n\}$. We further assume that

$$\frac{\overline{B}_i(x)}{\overline{H}(x)} \to c_i, \quad \frac{\overline{B}_{ij}(x)}{\overline{H}(x)} \to c_{ij} \tag{17.10}$$

for some sub-exponential distribution H and some constants $c_i, c_{ij} \in [0, \infty)$. We let $\mathbb{P}_{i,x}$ refer to the case where $J_0 = i$ and the age of R_0 is x.

Theorem 2 *Assume that* $\int_0^\infty e^{\delta x} F_i(dx) < \infty$ *for some* $\delta > 0$ *and all* i, *and that*

$$c = \sum_{i=1}^{p} \pi_i^0 \left\{ \lambda_i m_i c_i + \sum_{j=1}^{p} q_{ij} c_{ij} \right\} > 0 .$$

Then

$$\mathbb{P}_{i,x}(M > u) \sim \frac{c}{(1-\rho)\sum_{i=1}^{p} \pi_i^0 m_i} \int_u^\infty \overline{H}(x)\, dx .$$

In connection with the history of Theorem 2, we note that the model as special cases comprises Markov–modulated Poisson arrivals (F_i exponential, B_{ij} degenerate at 0) and discrete–time Markov random walks (F_i degenerate at 1, $\lambda_i = 0$). For the Markov–modulated case, it was shown in Asmussen et al. [7] by a rather long argument that $\mathbb{P}_{i,x}(M > u) \sim C_i \int_u^\infty \overline{H}(y)\, dy$. It is readily conjectured that C_i does not depend on i, and this was shown analytically in Asmussen & Højgaard [8] when the B_i do not depend on the environment i. In Jelenkovic & Lazar [19] the Markov random walk case is treated and it is shown in full generality that $C_i = C$; the proof follows Asmussen et al. [7] initially, $C_i = C$ is then shown by an algebraic argument. The added generality of the present Theorem 2 allows in particular to incorporate a general Markovian arrival process (Asmussen & Koole [9], Neuts [22]).

In the proof of Theorem 2 we need the following auxiliary result.

Lemma 1 *Let $N = (N_1, \dots, N_q)$ be a random vector in $\{0,1,2,\dots\}^q$, $\tau \geq 0$ a random variable and $\mathcal{F}$ a σ–algebra such that N and τ are $\mathcal{F}$–measurable. Let $\{B_i\}_{i=1,\dots,q}$ be a family of distributions on $[0,\infty)$ and define*

$$K_\tau = \sum_{i=1}^{q}\sum_{j=1}^{N_i} X_{ij} - \tau$$

where conditionally upon $\mathcal{F}$ the X_{ij} are independent with distribution B_i for X_{ij}. Assume $\mathbb{E}e^{\delta N_i} < \infty$ for some $\delta > 0$ and all i, and that for some distribution H on $[0,\infty)$ such that $H \in \mathcal{S}$ and some $c_1, \dots, c_q$ in $[0,\infty)$ it holds that $\overline{B}_i(x) \sim c_i\overline{H}(x)$, $i = 1, \dots, q$, and $\sum_{i=1}^q c_i > 0$. Then

$$\mathbb{P}(K_\tau > x) \sim c\overline{H}(x) \quad \text{where } c = \sum_{i=1}^{q} c_i \mathbb{E}N_i \,.$$

Proof. The key ingredients are 1) relation (17.4), and that for a subexponential H 2) $\overline{H^{*n}}(x) \leq C_\epsilon(1+\epsilon)\overline{H}(x)$, 3) $\overline{B_1 * B_2}(x) \sim (c_1+c_2)\overline{H}(x)$ whenever $\overline{B}_j(x) \sim c_j\overline{H}(x)$, $j = 1,2$ with $c_1 + c_2 > 0$. For details, see [11]. □

Proof of Theorem 2 We use regeneration in the form of Theorem 1. Fix some state i^*, start the generic cycle with $J_0 = i^*$ and define its length τ by $\tau = \inf_{n\geq 1}\{R_0 + \cdots + R_{n-1} : J_n = i^*\}$. Then

$$\mathbb{E}\tau = \frac{1}{\pi^0_{i^*}} \sum_{i=1}^{p} \pi^0_i m_i \,.$$

Let further t_i be the time spent in F_i–intervals before τ, let N_i be the number of arrivals in such intervals and N_{ij} the number of transitions from i to j. Then, noting that the expected number of F_i–intervals before τ is $\pi^0_i/\pi^0_{i^*}$, we have

$$\mathbb{E}t_i = \frac{\pi^0_i m_i}{\pi^0_{i^*}}, \quad \mathbb{E}N_i = \frac{\pi^0_i \lambda_i m_i}{\pi^0_{i^*}}, \quad \mathbb{E}N_{ij} = \frac{\pi^0_i q_{ij}}{\pi^0_{i^*}} \,.$$

With K_0 defined as the accumulated jumps in $[0,\tau)$ and $K_\tau = K_0 - \tau$, it follows easily from Lemma 1 that $\mathbb{P}(K_0 > x) \sim c\overline{H}(x)$, $\mathbb{P}(K_\tau > x) \sim c\overline{H}(x)$. Using a similar estimate for a possibly delayed first cycle, the proof is complete. □

4. STRUCTURED MARKOV CHAINS OF THE M/G/1 TYPE

We consider a bivariate Markov chain $\{I_n, L_n\}$ on the state space $\mathcal{S} = (E \times \{1, 2, \ldots\}) \cup (E_0 \times \{0\})$, where E and E_0 are finite (and disjoint for notational convenience), and with irreducible, aperiodic and positive recurrent transition matrix $\boldsymbol{P}$ given by

$$\boldsymbol{P} = \begin{bmatrix} \boldsymbol{M} & \boldsymbol{H}(1) & \boldsymbol{H}(2) & \boldsymbol{H}(3) & \cdots \\ \boldsymbol{K}(1) & \boldsymbol{B}(0) & \boldsymbol{B}(1) & \boldsymbol{B}(2) & \cdots \\ \boldsymbol{K}(2) & \boldsymbol{B}(-1) & \boldsymbol{B}(0) & \boldsymbol{B}(1) & \cdots \\ \boldsymbol{K}(3) & \boldsymbol{B}(-2) & \boldsymbol{B}(-1) & \boldsymbol{B}(0) & \cdots \\ \vdots & \vdots & \vdots & \vdots & \ddots \end{bmatrix}, \qquad (17.1)$$

where the $\boldsymbol{B}(k)$ have dimension $E \times E$, the $\boldsymbol{K}(k)$, $\boldsymbol{H}(k)$ and $\boldsymbol{M}$ have dimensions $E \times E_0$, $E_0 \times E$ and $E_0 \times E_0$ respectively. We refer to L_n as the *level* and to I_n as the *phase*.

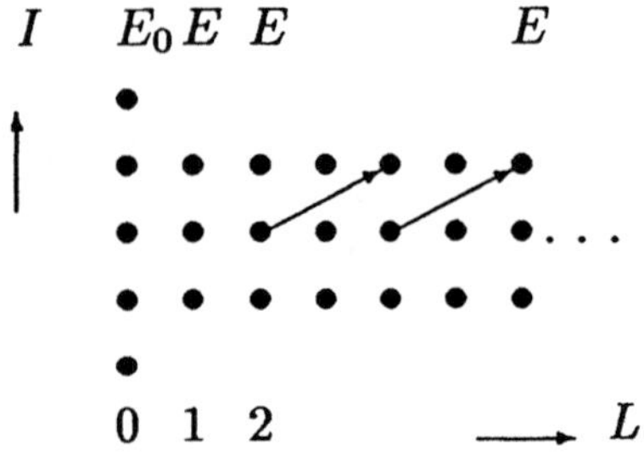

Figure 17.3

The intuitive content, cf. Fig. 17.3, is that away from the set $E_0 \times \{0\}$ of boundary states, $\{I_n, L_n\}$ moves as a Markov additive process; for example, the two marked transitions on Fig. 17.3 have the same probabilities. If the Markov additive process attempts a transition to a level ≤ 0, the level is reset to 0 and the phase to a state in E_0 according to the rule specified by the $\boldsymbol{K}(\ell)$. The process may then go to another boundary state according to $\boldsymbol{M}$, or leave the boundary according to some $\boldsymbol{H}(\ell)$. The set–up cover many interesting and non–standard queueing models. If $\boldsymbol{B}(-k) = \boldsymbol{K}(k) = \boldsymbol{0}$ for $k \geq 2$, we obtain the particularly important case of Neuts' Markov chain of M/G/1 type, see [23] (updates on the literature are in Neuts [24] and Ramaswami [27]).

We assume that $\boldsymbol{B} = \sum_{k=-\infty}^{\infty} \boldsymbol{B}(k)$ is irreducible and that each element of $\sum_{k=-\infty}^{\infty} |k| \boldsymbol{B}(k)$ is finite. Let π denote the stationary distribution of $\boldsymbol{B}$. We assume positive recurrence which is well–known to be equivalent to

$$\sum_{k=1}^{\infty} kH(k)e < \infty, \quad m = \pi \sum_{k=1}^{\infty} kB(k)e < 0 \tag{17.2}$$

where e is the vector with all components equal to 1.

Write $B_{ij}(k)$ for the element in the ith row and the jth column of $B(k)$ etc. We shall use

Condition SE *There exists a discrete sub-exponential distribution p such that*

$$\frac{\overline{B_{ij}}(n)}{\overline{p}(n)} \to c_{ij} \in [0, \infty)$$

for all $i, j \in E$, and

$$\frac{\overline{H_{ij}}(n)}{\overline{p}(n)} \to c_{ij} \in [0, \infty)$$

for all $i \in E_0$, $j \in E$ when $n \to \infty$ and such that $c_{ij} > 0$ for at least one pair i, j where $i \in E_0 \cup E$ and $j \in E$.

In many cases the $B_{ij}(k)$ have the form

$$\int_0^{\infty} e^{-\lambda x} \frac{(\lambda x)^k}{k!} A(dx)$$

for some distribution A, cf. [23]. In this case (and some more general ones), Condition SE and the form of the $p(k)$ follows from Asmussen, Klüppelberg & Sigman [6] when A is itself (continuous) sub-exponential.

Our main results are the following:

$$\mathbb{P}(M_{C^{reg}} > n) \approx c_1 \mathbb{E}(C^{reg}) \overline{p}(n) \tag{17.3}$$

$$\mathbb{P}(L > n) \approx c_2 \sum_{k=n}^{\infty} \overline{p}(k) \tag{17.4}$$

where I, L is the steady–state limit of I_n, L_n, $M_{C^{reg}}$ is the maximum of L_n over a regenerative cycle (see below) and

$$c_1 = \sum_{i \in E_0 \cup E} \sum_{j \in E} \mathbb{P}(I = i) c_{ij}, \quad c_2 = \frac{c_1}{|m|}$$

with m given by (17.2). We refer to [10] for the precise statement, which involves some slight regularity conditions beyond subexponentiality, and will instead at this place concentrate on explaining some intuitive ideas in the proofs.

Recall that in the setting of Section 2, the asymptotics of the cycle maximum were obtained by using the asymptotics of the steady–state characteristics. For the present model, neither is available so we go the other way round and consider first the cycle maximum. The key ideas are semi–regeneration and random walk comparisons.

Consider first the regenerative structure. Taking the regenerative cycle as a busy cycle

$$C = \inf\{n > 0 : L_n = 0 \mid L_0 = 0\}$$

faces the difficulty that the distribution depends on the phase at the start. Instead we fix a state $i_0 \in E_0$ and let

$$C^{reg} = \inf\{n > 0 : L_n = 0, I_n = i_0 \mid L_0 = 0, I_0 = i_0\}$$

Let $\tau(x) = \inf\{n > 0 : L_n > x\}$. Then for $x_0 < x$,

$$\begin{aligned} \mathbb{P}(M_{C^{reg}} > x) &= \mathbb{P}_{i_0,0}(M_{C^{reg}} > x,\, \tau(x_0) = \tau(x)) \\ &\quad + \mathbb{P}_{i_0,0}(M_{C^{reg}} > x,\, \tau(x_0) < \tau(x)). \end{aligned} \tag{17.5}$$

Here the first term is easily estimated:

Proposition 1 $\displaystyle \lim_{x_0\to\infty} \lim_{x\to\infty} \frac{\mathbb{P}_{i_0,0}(M_{C^{reg}} > x,\, \tau(x_0) = \tau(x))}{\overline{p}(n)} = c_1 \mathbb{E}(C^{reg}).$

Proof. As $x \to \infty$ with x_0 fixed,

$$\begin{aligned} &\mathbb{P}_{i_0,0}(M_{C^{reg}} > x,\, \tau(x_0) = \tau(x)) \\ &= \mathbb{E} \sum_{n=0}^{C^{reg}-1} \sum_{i \in E_0 \cup E} I(\tau(x_0) > n, I_n = i, L_{n+1} > x) \\ &\sim \mathbb{E} \sum_{n=0}^{C^{reg}-1} \sum_{i \in E_0 \cup E} \sum_{j \in E} I(\tau(x_0) > n, I_n = i) c_{ij} \overline{p}(x)\,, \end{aligned}$$

using (17.4) in the last step. Furthermore, as $x_0 \to \infty$

$$\mathbb{E} \sum_{n=0}^{C^{reg}-1} I(\tau(x_0) > n, I_n = i) \sim \mathbb{E} \sum_{n=0}^{C^{reg}-1} I(I_n = i) = \mathbb{E}(C^{reg}) \mathbb{P}(I = i).$$

□

Thus, it suffices for (17.3) to show that

$$\lim_{x_0\to\infty} \lim_{x\to\infty} \frac{\mathbb{P}_{i_0,0}(M_{C^{reg}} > x,\, \tau(x_0) < \tau(x))}{\overline{p}(n)} = 0 \tag{17.6}$$

Obviously it may occur that $L_n = 0, I_n \notin E_0$ for some $n < C^{reg}$ and thus a regenerative cycle may comprise several busy cycles. The process $\{L_n, I_n\}$ is semi–regenerative ([2] Ch. X). An easy argument based upon this property shows that (17.6) in turn will hold if

$$\lim_{x_0 \to \infty} \lim_{x \to \infty} \frac{\mathbb{P}_{i,0}(M_C > x,\ \tau(x_0) < \tau(x))}{\overline{p}(n)} = 0 \tag{17.7}$$

for all $i \in E_0$.

To obtain (17.7), we fix a state $i^* \in E$ and let L_n^* be $L_{\sigma^*(n)}$ where $\sigma^*(n)$ is the time of the nth visit of $\{I_n\}$ to i^*. Then $\{L_n^*\}_{n:\sigma^*(n)<C}$ is a segment of a random walk $\{S_n^*\}$, and from Condition SE, one easily gets:

1. The tail of the increment distribution F^* of $\{S_n^*\}$ is sub-exponential and asymptotically of the form $c_3\overline{p}(x)$;
2. The tail of $L_1^* = S_1^*$ is sub-exponential and asymptotically of the form $c_4\overline{p}(x)$

(here $c_3 + c_4 > 0$). In obvious notation, a modification and extension of basic lemmas in [4] then yields

$$\lim_{x_0 \to \infty} \lim_{x \to \infty} \frac{\mathbb{P}(M_{C^*}^* > x,\ \tau^*(x_0) < \tau^*(x))}{\overline{p}(x)} = 0 \tag{17.8}$$

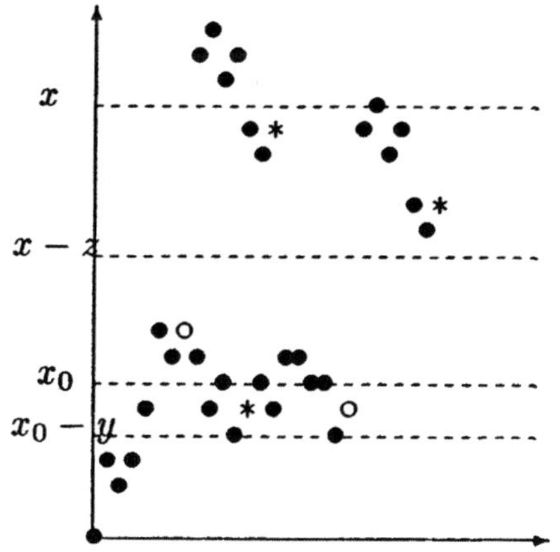

Figure 17.4

The proof of (17.7) now proceeds by a random walk comparison indicated in Fig. 17.4. The $*$'s indicate a value of L_n for which $I_n = i^*$, i.e., a value of $L_n^* = S_n^*$. The first $\circ$ is the value of $L_{\tau(x)-1}$, say occurring at phase j, and the upper sample path following $\circ$ corresponds to $\{L_n\}$ itself. The lower one until the second $\circ$ is then an auxiliary segment of $\{L_n\}$ ending in phase j in a level given by the second $\circ$ at most y below the level at the first $\circ$ and having visited phase i^* in between without having been at level 0; by choosing y large enough, such a sample path has strictly positive probability, say ϵ_1, for all large enough x_0. After the second $\circ$, the upper sample path following the first $\circ$ of $\{L_n\}$ is pasted in with a shift of the level. It follows that the probability of the lower sample path is at least ϵ_1 times the probability of the upper sample path. By choosing z large enough, there is further a probability at least $\epsilon_2 > 0$ that the level at the next visit to phase i^* after the up-crossing of level x will be at least $x - z$. From this we obtain

$$\mathbb{P}(M_{C^*}^* > x,\ \tau^*(x_0) < \tau^*(x)) \ \geq\ \epsilon_1\epsilon_2\mathbb{P}_{i,0}(M_C > x,\ \tau(x_0) < \tau(x))$$

and (17.7) follows by combining with (17.8).

Next consider (17.4). By regenerative process theory,

$$\begin{aligned} \mathbb{E}C^{reg} \cdot \mathbb{P}(L > x) &= \mathbb{E}_{i_0,0} \sum_{n=0}^{C^{reg}-1} I(L_n > x) \\ &= \mathbb{P}(M_{C^{reg}} > x) \cdot \mathbb{E}_{i_0,0} \left[\sum_{n=\tau(x)}^{C^{reg}-1} I(L_n > x) \middle| M_{C^{reg}} > x \right] \end{aligned} \qquad (17.9)$$

But it is contained in the proof of (17.7) that the distribution of $L_{\tau(x)}$ given $M_{C^{reg}} > x$ is asymptotically p conditioned to be above level x. In particular, by (17.4) $y = L_{\tau(x)}$ is substantially larger than x so that the expected time spent above level x before hitting level 0 is asymptotically $(y - x)/|m|$. Thus (17.9) becomes approximately

$$\mathbb{E}(C^{reg})c_1\overline{p}(x) \cdot \frac{1}{\overline{p}(x)} \sum_{y=x+1}^{\infty} \frac{y-x}{|m|} p(y) \ \sim\ \mathbb{E}(C^{reg})\frac{c_1}{|m|} \sum_{y=x+1}^{\infty} \overline{p}(x)\,.$$

References

[1] J. Abate, G.L. Choudhury & W. Whitt (1994) Asymptotics for steady-state tail probabilities in structured Markov queueing models. *Stochastic Models* **10**, 99–143.

[2] S. Asmussen (1987) *Applied Probability and Queues.* John Wiley & Sons, Chichester New York.

[3] S. Asmussen (1995) Stationary distributions via first passage times. *Advances in Queueing: Models, Methods & Problems* (J. Dshalalow ed.), 79–102. CRC Press, Boca Raton, Florida.

[4] S. Asmussen (1998a) Sub-exponential asymptotics for stochastic processes: extremal behavior, stationary distributions and first passage probabilities. *Ann. Appl. Probab.* **8**, 354–374.

[5] S. Asmussen (1998b) Extreme value theory for queues via cycle maxima. *Extremes* **1**(2) (in print).

[6] S. Asmussen, C. Klüppelberg & K. Sigman (1998) Sampling at subexponential times, with queueing applications. *Stoch. Proc. Appl.* (to appear).

[7] S. Asmussen, L. Fløe Henriksen & C. Klüppelberg (1994) Large claims approximations for risk processes in a Markovian environment. *Stoch. Proc. Appl.* **54**, 29-43.

[8] S. Asmussen & B. Højgaard (1996) Finite horizon ruin probabilities for Markov-modulated risk processes with heavy tails. *Th. Random Processes* **2**, 96–107.

[9] S. Asmussen & G. Koole (1993) Marked point processes as limits of Markovian arrival streams. *J. Appl. Probab.* **30**, 365-372.

[10] S. Asmussen & J.R. Møller (1998) Tail asymptotics for M/G/1 type queueing models with sub-exponential increments. Submitted to *QUESTA*.

[11] S. Asmussen, H. Schmidli & V. Schmidt (1998) Tail approximations for non–standard risk and queueing processes with sub-exponential tails. *Adv. Appl. Probab.* (to appear).

[12] B. von Bahr (1975) Asymptotic ruin probabilities when exponential moments do not exist. *Scand. Act. J.* **1975**, 6-10.

[13] N. H. Bingham, C. M. Goldie & J. L. Teugels (1987). *Regular Variation.* Encyclopedia of Mathematics and its Applications **27**. Cambridge University Press.

[14] P. Embrechts, C. Klüppelberg & T. Mikosch (1997) *Modelling Extremal Events for Finance and Insurance.* Springer, Heidelberg.

[15] P. Embrechts & N. Veraverbeke (1982). Estimates for the probability of ruin with special emphasis on the possibility of large claims. *Insurance: Mathematics and Economics* **1**, 55-72.

[16] W. Feller (1971). *An Introduction to Probability Theory and its Applications* **II** (2nd ed.). Wiley, New York.

[17] P. Glasserman & S.-G. Kou (1995) Limits of first passage times to rare sets in regenerative processes. *Ann. Appl. Probab.* **5**, 424–445.

[18] D. Heath, S. Resnick & G. Samorodnitsky (1997) Patterns of buffer overflow in a class of queues with long memory in the input stream. *Ann. Appl. Probab* **7**, 1021–1057.

[19] P.R. Jelenkovic & A.A. Lazar (1998) Sub-exponential asymptotics of a Markov–modulated random walk. *J. Appl. Probab.* **35**(2) (to appear).

[20] J. Keilson (1966) A limit theorem for passage times in ergodic regenerative processes. *Ann. Math. Statist.* **37**, 866–870.

[21] W. Leland, M. Taqqu, W. Willinger & D. Wilson (1994) On the self–similar nature of Ethernet traffic. *IEEE/ACM Trans. Netw.* **2**, 1–15.

[22] M.F. Neuts (1977) A versatile Markovian point process. *J. Appl. Probab.* **16**, 764–779.

[23] M.F. Neuts (1989) *Structured Stochastic Matrices of M/G/1 Type and Their Applications.* Marcel Dekker.

[24] M.F. Neuts (1995) Matrix-analytic methods in queueing theory. *Advances in Queueing: Models, Methods and Problems* (J.H. Dshalalow, ed.), 265–292. CRC Press, Boca Raton.

[25] A.G. Pakes (1975) On the tail of waiting time distributions. *J. Appl. Probab.* **12**, 555–564.

[26] V. Paxson & S. Floyd (1995) Wide area traffic: the failure of Poisson modelling. *IEEE/ACM Trans. Netw.* **3**, 226–244.

[27] V. Ramaswami (1997) Matrix analytic methods: A tutorial overview with some extensions and new results. *Matrix-Analytic Methods in Stochastic Models* (S.R. Chakravarthy & A.S. Alfa, eds.), 261–296. Marcel Dekker.

[28] S.I. Resnick (1997) Heavy tail modeling and teletraffic data. *Ann. Statist.* **25**, 1805–1869.

[29] M.S. Taqqu & J. Levy (1986) Using renewal processes to generate long–range dependence and high variability. *Dependence in Probability and Statistics* (E. Eberlein & M.S. Taqqu, eds.), 73–89. Birkhäuser, Boston.

[30] W. Willinger, M. Taqqu, W.E. Leland & D.V. Wilson (1995) Self–similarity in high–speed packet traffic: analysis and modeling of Ethernet traffic measurements. *Statistical Science* **10**, 67–85.

Chapter 18

MARKOV RENEWAL MODELLING OF POISSON TRAFFIC AT INTERSECTIONS HAVING SEPARATE TURN LANES

Rudy Gideon
University of Montana, USA

Ronald Pyke
University of Washington, USA

Abstract Consider a two-lane road which is intersected on one side by a single-lane secondary road. A single car waiting on the secondary road may merge into either the nearest or the farthest lane. It is assumed that the traffic in each lane is independent of the other lanes and that the inter-arrival times of cars at the intersection in their respective lanes is exponential. The main purpose of this paper is to study the queue size on the secondary road when the secondary road has a special right-turn lane (or left-turn lane in some countries) which allows some cars to merge into the nearest lane of the main road even when other cars waiting to enter the far lane are present. The problem is approached by first setting up a four-state Markov Renewal process to describe the traffic on the main road. Next the merging process on the secondary road is described as a Markov Renewal process with a random environment. The event that the queue is empty is studied, and conditions are stated under which this event is recurrent or transient. Finally, the quantities which occur in the conditions for recurrence of an empty queue are derived explicitly for a one-car right turn lane.

Keywords: Markov renewal processes, semi-Markov processes, traffic models, separate turn lane, road crossings, Poisson process.

1. INTRODUCTION

Adams [1] in 1937 and Garwood [3] in 1940 were among the first to discuss pedestrian and vehicle crossing problems in probabilistic terms. In 1951 Tanner [14] gave an account of these problems assuming Poisson traffic. Later writers who do not assume Poisson traffic often assume that the arrival times at the intersection are generated by a renewal process. Papers which dealt early on with such problems include those written by Jewell [6], Newell [7], [8], Oliver [9], Raff [13], Weiss and Maradudin [15], and Gaver [4]. Chapter five of Haight's book [5] on traffic flow remains a useful summary of many of these papers among others.

In this paper we study the generalization of Tanner's vehicle crossing problem in which the road being crossed is a two-lane road. Specifically we consider a two-lane road intersected by a side road from one direction only. Traffic on the side road must halt, whereas traffic on the two-lane road runs without interruption. The general problem concerns the merging of the traffic from the side road into the two-lane road. The paper by Troutbeck (1986) gives a good overview of the problem with relevant references along with its new derivations and extensive simulations.

For notation, let S_1 and S_2 represent, respectively, the closest and the farthest traffic lane on the primary two-lane road and let S_3 denote the secondary side road. Let $t_{i0}, t_{i1}, \ldots, t_{in}, \ldots$ represent the arrival times of cars on S_i at the intersection, $i = 1, 2, 3$. We assume that the arrival times of cars in their respective lanes are independent of the arrival times of vehicles in the other two lanes. Let $x_{i0} = t_{i0}$ and $x_{in} = t_{in} - t_{i,n-1}, i = 1, 2, 3$, be the inter-arrival times. Assume that x_{in}, $i = 1, 2, 3$ and $n = 0, 1, \ldots$, is a sequence of independent and identically distributed non-negative random variables.

We assume that at the intersection a vehicle arriving on S_3 may do one of two things; it may turn right into the near lane S_1, or it may go left into the far lane S_2. A car waiting on S_3 will turn right only if there is at least distance I between the intersection and the first car on S_1. A car waiting on S_3 will turn left only if there is simultaneously at least distance I and distance J between the intersection and the first car on S_1 and S_2, respectively.

It is convenient and without any loss of generality, to think of the traffic in S_2 as flowing in the same direction as that in S_1 (as shown in Figure 18.1). This enables one to superimpose the two lanes when studying the merging of cars into S_2. We call the event of a car merging into S_2 (S_1) an L (R) event, and we call a car of this kind an L (R) car.

The time it takes for a car to merge is assumed to be a random variable (rv), T, with distribution function (df), F, such that $F(0) = 0$. Cars arriving on S_3 if unable to merge will form a queue on S_3 for which service is given on a first come first served basis with one car served at a time. When a separate turn lane is introduced below, then two cars may be served at once.

We now make three major assumptions:

(1) Each car that arrives is labeled an R car or an L car, and the labeling is done by a Bernoulli process with probability p for the R label.

(2) $I = J > 0$.

(3) Traffic in each lane is a Poisson process, implying, in particular that the d.f.'s G_i are exponential with parameter λ_i, $i = 1, 2, 3$.

If we look at the arrival times of cars on S_1 and S_2 at the intersection, without regard to the lane to which the car belongs, it is known that the arrival times of the pooled traffic also form a Poisson process with parameter $\lambda_1 + \lambda_2 := \lambda$, say.

In Section 2 we introduce the 4-state Markov Renewal process (MRP) that models the state of traffic on the primary road comprising S_1 and S_2. An MRP is a stochastic process in which the movement of the process from one state to another is such that the choice of the next state and the length of time to get there depends only on the last state visited. It is assumption (3) applied to S_1 and S_2 that permits the treatment of the traffic as an MRP. Throughout the remainder of this paper we use the notation and basic results in Pyke (1961a, b).

In Section 3 we extend the Markov Renewal model to incorporate the traffic on the secondary road S_3, on which we first construct a separate right turn lane. This turning lane is assumed to be of finite length and the resulting finite number of states and the extended MRP are described. The main result in this section gives sufficient conditions for the ergodicity (positive recurrence) of the empty state on S_3 at the intersection. the derivation uses the concept of maximal traffic (related to the capacity of the intersection as studied in Troutbeck (1986)). Additional quantities are calculated in Section 4 for the MRP model of the intersection with the r-car right turn lane. Then, in Section 5, further results are given for the case of $r = 1$.

Before proceeding, let us state some results from Tanner [14] that are used below. Suppose that a car in S_3 arrives at the intersection with no cars waiting before it. Let $F_1(t)$ be the probability that this car's waiting time will be less than or equal to t given that the car wishes to go right. Let $F_{12}(t)$ be the corresponding probability for an L car. From Tanner [14], we have for $t > 0$,

$$F_1(t) = e^{-\lambda_1 I} \sum_{j=0}^{\tau} (-e^{-\lambda_1 I})^j \left[\frac{\lambda_1^{j+1}(t - jI)^{j+1}}{(j+1)!} + \frac{\lambda_1^j (t - jI)^j}{j!} \right] \quad (18.1)$$

where $\tau = [t/I]$, the integral part of t/I. Note that for t between 0 and I

$$F_1(t) = e^{\lambda_1 I}(\lambda_1 t + 1).$$

Clearly, the same formula holds for F_{12} with λ substituted for λ_1. Let F_2 denote the df of (18.1) with λ_2 in place of λ_1.

It is convenient, since I remains fixed throughout the paper, to let I be the unit of distance; that is, $I = 1$. Using this convention it follows also from Tanner[14] that the expected wait of an R car is $\lambda_1^{-1}(e^{\lambda_1} - \lambda_1 - 1)$, the variance is $\lambda_1^{-2}(e^{2\lambda_1} - 2\lambda_1 e^{\lambda_1} - 1)$ and the Laplace transform of F_1 is

$$\int_{0-}^{\infty} e^{-zx} dF_1(x) = (z + \lambda_1)e^{z-\lambda_1}(ze^z + \lambda_1 e^{-\lambda_1})^{-1}. \quad (18.2)$$

The expected wait of any car on S_3 which arrives at the intersection when the queue is empty therefore equals

$$p\lambda_1^{-1}(e^{\lambda_1} - \lambda_1 - 1) + (1 - p)\lambda^{-1}(e^{\lambda} - \lambda - 1).$$

2. THE 4-STATE MRP ON S_1 AND S_2

The four states for the primary road's Markov Renewal process are described as follows : The MRP is said to be *in* state

$\emptyset$ if there is at least one car in $(0, I]$ on S_1,

L if there are no cars in $(0, I]$ on both S_1 and S_2,

and that it *enters* state

R_L when the first car on road S_2 is at I and there are no cars in $[0, I]$ on road S_1,

$R_\emptyset$ when there is one car at 0 and no cars in $[0, 1]$ on road S_1 but there is a car in $(0, 1]$ on road S_2.

Although the above notation is more descriptive, we shall when convenient interchangeably identify the states $\emptyset, R_L, R_\emptyset$, and L as $1, 2, 3, 4$, respectively. Figure 18.1 shows our traffic model as it enters state $R_\emptyset$.

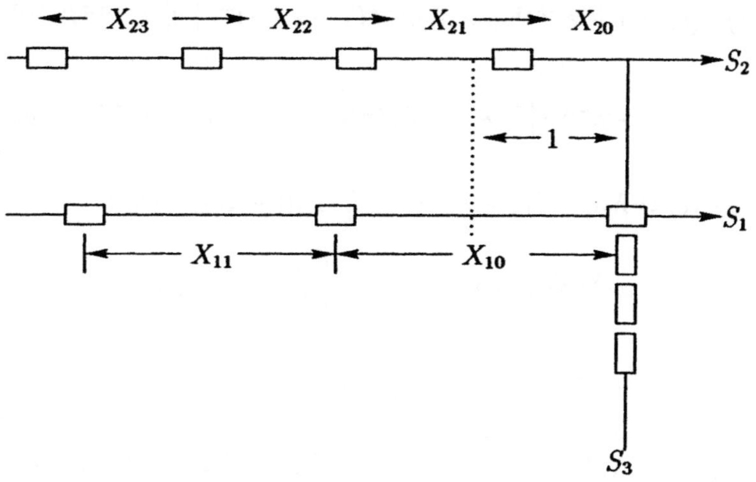

Figure 18.1 Secondary road S_3 intersects primary roads S_1 and S_2. The traffic is shown as it enters state R_φ.

For $n \geq 1$, let J_n represent the nth state visited, and X_n the holding time of the $(n-1)^{\text{th}}$ state. Thus J_n may equal 1, 2, 3, or 4 and $0 < X_n < +\infty$. For the 4-state MRP process $\{(J_n, X_n) : n = 0, 1, 2, \ldots\}$ we define $X_0 = 0$ and let $P(J_0 = i) = a_i$ denote the initial probability of state i, ($i = 1$, 2, 3, or 4). Also set

$$\begin{aligned} Q_{J_{n-1},j}(x) &= P(J_n = j, X_n \leq x | J_0, X_0, J_1, X_1, \ldots, J_{n-1}, X_{n-1}), \\ P_{ij} &= Q_{ij}(+\infty), H_i(t) = \sum_{j=1}^{4} Q_{ij}(t), \eta_i = \int_0^\infty x dH_i(x), \end{aligned}$$

and write μ_j for the mean recurrence time of state j (for which μ_{jj} is the usual notation). Observe that for the present model, $P_{11} = P_{12} = P_{22} = P_{23} = P_{32} = P_{33} = P_{43} = P_{44} = 0$. Let F_2 be the same as F_1, except that λ_2 replaces λ_1. In the following formulae all of the df's are understood to be zero on $(-\infty, 0)$.

Lemma 1 *The transition functions of the MRP are given by*

(a) $Q_{13}(t) = (1 - e^{-\lambda_2})F_1(t-1)$, *(b)* $Q_{14}(t) = e^{-\lambda_2} F_1(t-1)$,

(c) $Q_{21}(t) = 1 - e^{-\lambda_1 t} - e^{-\lambda_1} \int_o^{t-1} \lambda_1 e^{-\lambda_1 x} F_2(x) dx$,

(d) $Q_{24}(t) = e^{-\lambda_1 t} F_2(t-1) + e^{-\lambda_1} \int_0^{t-1} \lambda_1 e^{-\lambda_1 x} F_2(x) dx$,

(e) $Q_{31}(t) = 1 - e^{-\lambda_1 t} - \lambda_1\lambda_2\lambda^{-1}(1-e^{-\lambda_2})^{-1}\Big[\int_0^t e^{-\lambda_1 x}F_2(x)dx$

$$-e^{-\lambda}\int_0^{t-1} e^{-\lambda_1 x}F_2(x)dx - e^{-\lambda t}\int_{t-1}^{t} e^{\lambda_2 x}F_2(x)dx\Big],$$

(f) $Q_{34}(t) = \lambda_2 e^{-\lambda_1 t}(1-e^{-\lambda_2})^{-1}\int_0^1 F_2(t-u)e^{-\lambda_2 u}du$

$$+\lambda_1\lambda_2\lambda^{-1}(1-e^{-\lambda_2})^{-1}\Big[\int_0^t e^{-\lambda_1 x}F_2(x)dx$$

$$-e^{-\lambda}\int_0^{t-1} e^{-\lambda_1 x}F_2(x)dx - e^{-\lambda t}\int_{t-1}^{t} e^{\lambda_2 x}F_2(x)dx\Big],$$

(g) $Q_{41}(t) = \lambda_1\lambda^{-1}(1-e^{-t\lambda})$, (h) $Q_{42}(t) = \lambda_2\lambda^{-1}(1-e^{-t\lambda})$.

Proof. To indicate the proofs of these formulae, it suffices to state verbally the conditions for $t \geq 1$ that must exist on S_1 and S_2 in order for the various transitions to occur. Then upon conditioning this event on either the arrival of a car in S_1 or S_2, it is straightforward to obtain the indicated formulae. Equation (e) is worked out in detail below. For the case $0 < t < 1$, (a), (b), (c), (d), (g), (h) are immediate. Parts (e) and (f) for $0 < t < 1$ are obtained by a process similar to the case $t \geq 1$ for (e) and (f).

(a) A car in S_1 must pass the intersection before time t with at least unit time before the following car in S_1, but less than unit time before the next car arrives in S_2.

(b) Same as (a) except more than unit time must elapse before the next car arrives in S_2.

(c) If $t > 1$ and there is a car in S_1 during the time interval $(1,2]$ then the event occurs. If there are no cars in S_1 during $(1,2]$, then in order for the event to occur there must be a car in S_1 during $(2, t+1]$.

(d) Consider two cases: (1) no cars in $(1, t+1]$ on S_1 and (2) at least one car in $(1, t+1]$ on S_1. In case (2) the event cannot occur if there is a car in $(1,2]$ on S_1.

(e) Assume that the first car on S_2 is at $u \leq 1$. Then consider two cases: (1) the event occurs if there is a car in $(1, 1+u]$ on S_1, and (2) when there is no car in $(1, 1+u]$ on S_1, condition the probability on the position of the first car on S_1 in $(1+u, 1+t]$.

(f) Again consider two cases: (1) no cars on S_1 in $(1, t+1]$, and (2) at least one car on S_1 in $(1, t+1]$. Note that the event cannot occur if there is a car in $(1, 1+u]$ on S_1 where u is the position of the first car on S_2, $(u \leq 1)$.

(g) Fix a car on S_1 at v; then no cars can occur in $(1, v]$ on S_2 if the event is to occur.

(h) Interchange S_1 and S_2 in (g).

We illustrate the necessary derivations by working out the details for case (e) about transitions from state $R_\emptyset$ to $\emptyset$. From the above description of the conditions that must prevail at the intersection we see that for $t \geq 1$,

$$\begin{aligned} Q_{31}(t) &= \frac{\lambda_2}{1-e^{-\lambda_2}} \int_0^1 e^{-\lambda_2 u} \Big[1 - e^{-\lambda_1 u} \\ &\qquad \int_u^t \lambda_1 e^{-\lambda_1 x} \{1 - F_2(x-u)\} dx \Big] du \\ &= 1 - e^{-\lambda_1 t} - \int_0^1 \lambda_2 e^{-\lambda u} (1 - e^{-\lambda_2})^{-1} \int_0^{t-u} \lambda_1 e^{-\lambda_1 x} F_2(x) dx du, \end{aligned}$$

and this yields the desired form after an interchange of integration. Similarly for $0 < t < 1$,

$$\begin{aligned} Q_{31}(t) &= \frac{\lambda_2}{1-e^{-\lambda_2}} \Big\{ \int_t^1 e^{-\lambda_2 u} (1 - e^{-\lambda_1 t}) du + \int_0^t \Big[1 - e^{-\lambda_1 u} \\ &\qquad + \int_u^t \lambda_1 e^{-\lambda_1 x} \{1 - F_2(x-u)\} dx \Big] e^{-\lambda_2 u} du \Big\}. \end{aligned}$$

However,

$$\begin{aligned} &\int_0^t \int_u^t F_2(x-u) e^{-\lambda_1 x} e^{-\lambda_2 u} dx du \\ &= \lambda^{-1} \Big[\int_0^t e^{-\lambda_1 x} F_2(x) dx - e^{-\lambda t} \int_0^t e^{\lambda_2 x} F_2(x) dx \Big] \end{aligned}$$

and

$$\begin{aligned} &\int_t^1 \lambda_2 e^{-\lambda_2 u} du + (1 - e^{-\lambda_1 t})^{-1} \\ &\int_0^t \Big[1 - e^{-\lambda_1 u} + \int_u^t \lambda_1 e^{-\lambda_1 x} dx \Big] \lambda_2 e^{-\lambda_2 u} du = (1 - e^{-\lambda_2}). \end{aligned}$$

which leads to the desired expression. □

The following corollaries are straightforward.

Corollary 1 *The transition probabilities are given by*

(a) $P_{14} = e^{-\lambda_2}$, $P_{13} = 1 - P_{14}$

(b) $P_{24} = \lambda e^{-\lambda}(\lambda_1 + \lambda_2 e^{-\lambda})^{-1}$, $P_{21} = 1 - P_{24}$

(c) $P_{34} = \lambda_2 e^{-\lambda_2}(1 - e^{-\lambda})(1 - e^{-\lambda_2})^{-1}(\lambda_1 + \lambda_2 e^{-\lambda})^{-1}$, $P_{31} = 1 - P_{34}$

(d) $P_{41} = \lambda_1/\lambda = \lambda_1(\lambda_1 + \lambda_2)^{-1}$, $P_{42} = 1 - P_{41}$.

Corollary 2 *For the H_i's we have the following formulae:*

(a) $H_1(t) = F_1(t-1)$

(b) $H_2(t) = 1 - e^{-\lambda_1 t}[1 - F_2(t-1)]$

(c) $H_3(t) = 1 - e^{-\lambda_1 t} + e^{-\lambda_1 t}(1 - e^{-\lambda_2})^{-1} \int_0^1 F_2(t-u)\lambda_2 e^{-\lambda_2 u} du$

(d) $H_4(t) = 1 - e^{-\lambda t}$.

Corollary 3 *The mean holding times are given by*

$$\begin{aligned}
\eta_1 &= \lambda_1^{-1}(e^{\lambda_1} - 1), \qquad \eta_2 = (e^{\lambda} - 1)(\lambda_2 + \lambda_1 e^{\lambda})^{-1} \\
\eta_3 &= \lambda_1^{-1} + \lambda_2\lambda_1^{-1}e^{-\lambda_2}(e^{\lambda_1} - e^{-\lambda_2})(1 - e^{-\lambda_2})^{-1}(\lambda_2 e^{-\lambda_2} + \lambda_1 e^{\lambda_1})^{-1}, \\
\eta_4 &= \lambda^{-1}.
\end{aligned}$$

Let I denote the 4×4 identity matrix and $P = (P_{ij})$, the matrix of transition probabilities. Let C_i be the cofactor of the $(i,i)^{\text{th}}$ element of the matrix $I - P$, and $C = \sum_{i=1}^{4} C_i$. From Lemma 4.1 of [12], the mean recurrence time of state j is given by $\mu_j = \mu_j^* \sum_{i=1}^{4} \eta_i/\mu_i^*$ for $j = 1, 2, 3, 4$. Furthermore, by a result of Mihoc (pp. 114–116 of [2]), the mean recurrence times of the corresponding Markov Chain are given by $\mu_i^* = C_i C^{-1}$. For the present case the C_i's are calculated directly from $I - P$. They are

$$\begin{aligned}
C_1 &= 1 - P_{42}P_{24}, \quad C_2 = 1 - P_{13}P_{31} - P_{14}(P_{13}P_{34} + P_{14}) \\
C_3 &= 1 - P_{42}P_{24} - P_{21}P_{14}P_{42} - P_{14}P_{41}, \quad C_4 = 1 - P_{13}P_{31}.
\end{aligned}$$

An alternative derivation of μ_1 is possible if one observes that the length of time between successive $\emptyset$ events is the sum of three rv's; one rv that is always equal to one, an rv with df F_1, and an exponential rv of parameter λ_1. Thus

$$\mu_1 = 1 + \lambda_1^{-1}(e^{\lambda_1} - \lambda_1 - 1) + 1/\lambda_1 = e^{\lambda_1}/\lambda_1. \qquad (18.3)$$

One may calculate μ_4 in the same way, except that one must consider the roads S_1 and S_2 superimposed into one road, and therefore replace F_1 by F_{12} and λ_1 by λ. Thus

$$\mu_4 = e^{\lambda}/\lambda. \tag{18.4}$$

We consider now the problem of obtaining the asymptotic (or stationary) probabilities of this MRP. Let $S_n = \sum_{i=0}^{n} X_i$, and $N(t) = \sup\{n \geq 0; S_n \leq t\}$. Thus $N(t)$ is the number of transitions the process has made from the start until time t. Let $Z_t = J_{N(t)}$; that is, Z_t represents the state the system is in at time t. Let $Z_t^+ = J_{N(t)+1}$ and $V_t = S_{N(t)+1} - t$. Then Z_t^+ represents the next state visited after time t, whereas V_t represents the time after t until this visit occurs. Define

$$R_{jk}^{(i)}(x;t) = P[Z_t = j, Z_t^+ = k, V_t \leq x | Z_0 = i],$$

and let $\widetilde{a}_j = \eta_j \mu_j^{-1}$ and $\widetilde{Q}_{jk}(t) = \eta_j^{-1} \int_0^t \{P_{jk} - Q_{jk}(y)\}dy$. Since our MRP is recurrent, non-lattice, and $\eta_i < +\infty$ for each i, it is known (cf. Pyke [11]) that

$$\lim_{t \to \infty} R_{jk}^{(i)}(x;t) = \mu_j^{-1} \int_0^x [P_{jk} - Q_{jk}(y)]dy = \widetilde{a}_j \widetilde{Q}_{jk}(x).$$

Now define a new process $[(\widetilde{J}_n, \widetilde{X}_n); n = 0, 1, 2 \ldots]$ which is the same as $[(J_n, X_n); n = 0, 1, 2, \ldots]$ except that the new process has possibly different initial probabilities and first transition probabilities. (All probabilities for this new process will be indicated by the use of the superscript $\sim$.) Consider for these initial probabilities the limiting values of $R_{jk}^{(i)}(x;t)$ which are given above. Thus let

$$\widetilde{X}_0 = 0, \;\; P(\widetilde{J}_0 = i) = \eta_j \mu_j^{-1},$$

and

$$P(\widetilde{J}_1 = k, \widetilde{X}_1 \leq x | \widetilde{J}_0) = \widetilde{Q}_{\widetilde{J}_o, k}(x).$$

It was proved in [11] that the process $[(\widetilde{Z}_t, \widetilde{V}_t) : t \geq 0]$ is stationary in the sense that

$$P(\widetilde{Z}_t = j, \widetilde{Z}_t^+ = k, \widetilde{V}_t \leq x) = \sum_{i=1}^{4} a_i \widetilde{R}_{jk}^{(i)}(x;t) = \widetilde{a}_j \widetilde{Q}_{jk}(x)$$

which is constant in t. Also $P(\widetilde{Z}_t = j) = \widetilde{a}_j$.

To see the application of this to our problem, consider the questions: If an R car arrives at some time t what are the stationary probabilities of merging instantly, and leaving before a given length of time, say x, but not instantly?

The probability of leaving instantly is, of course, just the probability that the MRP is not in state $\emptyset$, which is

$$P(\widetilde{Z}_t \neq \emptyset) = 1 - P(\widetilde{Z}_t = \emptyset) = 1 - \eta_1 \mu_1^{-1} = e^{-\lambda_1}.$$

Under the assumptions of stationarity, the last equation results from using the calculated formulas for η_1 and μ_1 given in Corollary 2 and equation (18.3).

The other mentioned probability is the probability of arriving in state $\emptyset$ at time t and leaving state $\emptyset$ before $t+x$. Thus it equals

$$\begin{aligned}\sum_{k=3,4} P(\widetilde{Z}_t = 1, \widetilde{Z}_t^+ = k, \widetilde{V}_t \leq x) &= \widetilde{a}_1[\widetilde{Q}_{13}(x) + \widetilde{Q}_{14}(x)] \\ &= \lambda_1 e^{-\lambda_1}\left[x - \int_0^{x-1} F_1(y)dy\right]\end{aligned}$$

Finally, by summing these two results, one obtains the stationary probability that a car which arrives at a given time t merges before time $t+x$, namely,

$$e^{-\lambda_1}(1+\lambda_1 x) - \lambda_1 e^{-\lambda_1} \int_0^{x-1} F_1(y)dy, \qquad (18.5)$$

a formula due to Raff [13].

3. AN R-CAR RIGHT TURN LANE AND ERGODICITY

In this section the manner in which cars on S_3 merge into S_1 and S_2 is examined. As described before, cars arriving on S_3 are labeled as either L cars or R cars. The L cars merge during an L state only, whereas the R cars can merge during the L, $R_\emptyset$, or R_L states. No cars merge during state $\emptyset$.

Recall that T is a rv representing the amount of time required by a car to merge. We assume that a car merges instantaneously at the beginning of its merging interval, but that the next merging event cannot occur until time T has elapsed. Let these merging intervals be called gaps, and assume these gaps to be independent and identically distributed with df F. Thus cars merge at the beginnings of gaps, gaps occur only during merging states, and for each merging state there is a sequence of independent and identically distributed gaps in which cars may merge.

Divide the four states of the MRP on S_1 and S_2 into two phases. Let one phase be state L, and the other phase consist of the other three states. Call these phases the L and the $R-\emptyset$ phases, respectively. Thus the process on S_1 and S_2 proceeds alternately from one phase to the other, generating an alternating MRP.

Movement of traffic at an intersection is facilitated by the construction of a right turn lane. We say we have an r-car right turn lane ($r = 0, 1, 2, \ldots$) if

(1) S_3 has two lanes intersecting S_1 and S_2, called S_{3R} and S_{3L}, (cf. Figure 18.2)

(2) the maximum capacity of each of S_{3R} and S_{3L} is r cars,

(3) L cars merge from S_{3L} only, and R cars from S_{3R} only,

(4) S_3 is a single lane before it *branches* into S_{3R} and S_{3L}.

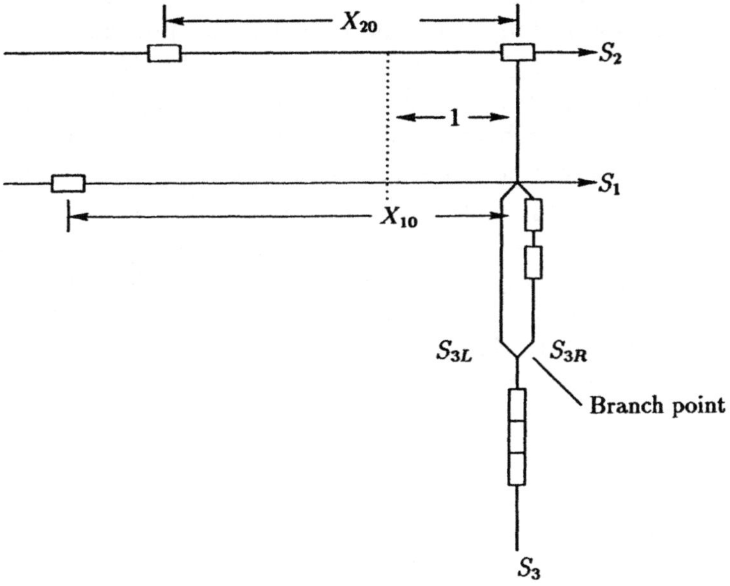

Figure 18.2 An intersection with a 3-car turn lane on S_3. The traffic is shown as it enters state L_{02}.

In order to prove the ergodic theorem of this section, we introduce a modified traffic model for which it is assumed that all of the cars that will ever arrive on S_3 have already arrived so that the queue length of S_3 is infinite. Whenever we assume this property about the queue on

S_3, we shall say that we are assuming *maximal traffic*. When we assume maximal traffic on S_3, we visualize the infinite line of traffic as a queue at the branch of S_3 into S_{3R} and S_{3L}. We emphasize the fact that both models are coupled by assuming that the nth car in the queue in maximal traffic has the same label (R or L) as the nth car to arrive in the queue on S_{3R} or S_{3L} in actual traffic. Cars on S_3 do not enter S_{3R} or S_{3L} until a merging gap occurs. When a gap occurs during an L phase, cars are successively and instantaneously labeled as R or L cars according to the Bernoulli process of (1) and advanced into S_{3R} or S_{3L} until either (i) both an R and an L car appear, or (ii) the entrance of S_{3L} is blocked by $r+1$ R cars, or (iii) the entrance to S_{3R} is blocked by $r+1$ L cars. In case (ii) a single R car merges, in case (iii) a single L car merges, whereas in case (i) two cars (an R car and an L car) merge during the gap.

Assume the last gap of an L phase has just occurred. The number of cars in S_{3L} and S_{3R} at this time is information that must be attached to the following $(R-\emptyset)$ phase. Let (i,j) be a tuple with the first coordinate representing the number of cars in S_{3L}, and the second the number in S_{3R}. Then by definition, from an L phase the process can enter an $(R-\emptyset)_{i,j}$ phase for $i,j \leq r$. For short, write R_{ij} for $(R-\emptyset)_{i,j}$.

During an $R-\emptyset$ phase, only R cars merge and any L cars must wait until the next L phase before merging. For each gap, cars are called from the branch only if no R cars are available in S_{3R} and there are less than $r+1$ L cars in S_{3L}. Thus at the end of an $R-\emptyset$ phase attach the information about the number of cars in S_{3R} and S_{3L} to the following L phase, labeling it as L_{ij}. Note that for $j>0$, the system can go from R_{ij} to $L_{ij'}$ for $0 \leq j' \leq j$, and that from R_{i0} the system can go to $L_{i',0}$ for any $i \leq i' \leq r+1$. Also, the only recurrent states are those with $ij=0$. We therefore introduce a new MRP to describe the state of the three lanes at the intersection. The states for this process are taken to be L_{0j}, R_{0j}, R_{j0}, $j=0,1,2,\ldots,r$, and L_{i0}, $i=1,2\ldots,r+1$. Figure 18.2 shows our traffic model as it enters state L_{02} with a 3-car right turn lane. In section 5, we specialize to the case of $r=1$, for which the possible states are L_{00}, L_{01}, L_{10}, L_{20} and R_{00}, R_{01}, and R_{10}.

We require the following definitions and notation. Define $\nu(t)$ to be the actual number of cars in the queue on S_3 at time t. Let $\mu_{00}^T(\mu_{00}^N)$ be respectively the expected length of time (the expected number of cars merging from S_3 into S_1 and S_2) between consecutive L_{00} events *when maximal traffic is assumed on* S_3. Let $D^*(t)$ be the number of cars that have merged into S_1 and S_2 from S_3 by time t under maximal traffic, and let $D(t)$ be the corresponding quantity for actual traffic on S_3. Define $A(t)$ to be the number of cars that arrived on S_3 in actual traffic before time t, so that A is a Poisson process with parameter λ_3. Let W_0 be the time until the start of the first L_{00} event, and let W_i be the time between the ith and $(i+1)^{\text{st}} L_{00}$ event, $i = 1, 2, \ldots$. The initial conditions are assumed to be the same whenever maximal and actual traffic processes are compared.

Lemma 2 $E(W_1) < +\infty$.

Proof. By Corollary 3 the expected holding time of any state in the underlying MRP (with states L, $R_\emptyset$, R_L, and $\emptyset$) is finite. Using Theorem 5.1 of Pyke [10] we have that their expected first passage times are finite. This implies that in the MRP with states L_{0j}, L_{j0}, R_{0j}, and R_{j0} the expected holding times are finite and so again the above reference implies that the first passage times of the states L_{0j}, L_{j0}, R_{0j} and R_{j0} are finite. In particular, $E(W_1) < +\infty$. □

The following lemma verifies the intuitively obvious remark that more cars will merge under maximal traffic than under normal traffic.

Lemma 3 $D(t) \leq D^*(t)$ *for all* t.

Proof. Assume the result is false. Let τ be the first time where $D(t) > D^*(t)$. Then regardless of initial conditions τ is positive and must be the beginning of a gap. Let $t_1, t_2, \ldots, t_n = \tau$ be the times not exceeding τ at which gaps start, with $t_1 < t_2 < \ldots < t_n$. First observe that $D(t_{n-1}) = D^*(t_{n-1})$. This follows from the fact that the difference $D(t) - D^*(t)$ can only increase by steps of one. To see this recall that the only way for $D(t)$ to gain on $D^*(t)$ during an $R - \emptyset$ phase is to have S_3 blocked under maximal traffic by L cars, and in this case $D(t)$ gains one car per gap. The only way for $D(t)$ to gain on $D^*(t)$ during an L phase is to have only one car merge during a gap under maximal traffic on S_3 but two in actual traffic; hence, in either case $D(t_{n-1}) = D^*(t_{n-1})$.

Define $D_L^*(t)$ and $D_R^*(t)$ to be the number of L cars and R cars, respectively, having merged from S_3 by time t under maximal traffic on S_3. Similarly, define $D_L(t)$ and $D_R(t)$ for actual traffic on S_3. We now show that $D_L(t_{n-1}) = D_L^*(t_{n-1})$, and hence $D_R(t_{n-1}) = D_R^*(t_{n-1})$. To prove this we prove that $D_L(t) \leq D_L^*(t)$ and $D_R(t) \leq D_R^*(t)$ for all $t \leq t_{n-1}$. Then since $D(t_{n-1}) = D_L(t_{n-1}) + D_R(t_{n-1}) \leq D_L^*(t_{n-1}) + D_R^*(t_{n-1}) = D^*(t_{n-1})$ we have that $D_L(t_{n-1}) = D_L^*(t_{n-1})$.

To prove this first inequality, note that if at some time t_k, $k \leq n-1$, $D_L(t_k)$ is strictly greater than $D_L^*(t_k)$ for the first time, then $D_L(t_{k-1})$ equals $D_L^*(t_{k-1})$ because at most one L car merges in one gap. Then for $D_L(t_k)$ to be strictly greater than $D_L^*(t_k)$ the $r+1$ cars on S_3 ready to merge at time t_k under maximal traffic must be R cars, whereas at time

t_k on S_3 in actual traffic among the $r+1$ cars ready to merge there must be at least one L car. However, at time t_{k-1}, $D_R(t_{k-1}) \leq D_R^*(t_{k-1})$ since $D(t_i) \leq D^*(t_i)$, $i = 1, 2, \ldots, n-1$. Recall that the sequence of labels R and L on the cars on S_3 is assumed to be the same for both maximal and actual traffic models. Hence, since the same number of L cars have merged by time t_{k-1}, but $D_R(t_{k-1}) \leq D_R^*(t_{k-1})$, there must be at least $r+1$ R cars next on S_3 in actual traffic. Thus $D_L(t) \leq D_L^*(t)$, $t \leq t_{n-1}$. By an analogous argument $D_R(t) \leq D_R^*(t)$, $t \leq t_{n-1}$.

Thus $D(t_n) > D^*(t_n)$ implies that $D_L(t_{n-1}) = D_L^*(t_{n-1})$ and $D_R(t_{n-1}) = D_R^*(t_{n-1})$, and these conditions obviously imply that $D(t_n) \leq D^*(t_n)$. This contradiction proves the lemma. □

The main ergodicity result of this paper is as follows.

Theorem 1 *(a) If $\lambda_3 \mu_{00}^T < \mu_{00}^N$, then*

$$Pr[\nu(t) > 0 \text{ for all } t \geq 0 | \nu(0) = k] = 0.$$

(b) If $\lambda_3 \mu_{00}^T > \mu_{00}^N$, then

$$Pr[\nu(t) > 0 \text{ for all } t \geq 0 | \nu(0) = k] > 0.$$

Proof. First, for $i = 0, 1, 2, \ldots$, let D_i^* represent the total number of merges under maximal traffic on S_3 during W_i, and $N_0(t)$ represent the number of times the process has entered state L_{00} during $(0, t]$.

The proof of (a) is outlined as follows. Under maximal traffic the queue length is assumed to be infinite. However, it still makes sense to discuss the event

$$C_m = [m + A(t) > D^*(t) \quad \text{for all } t \geq 0]$$

for any integer $m \geq 0$ where A is the arrival Poisson process with parameter λ_3 assumed to be independent of the process D^*. Clearly $t^{-1}A(t) \to \lambda_3$ with probability one as $t \to \infty$. Also, since

$$\sum_{i=0}^{N_0(t)-1} D_i^* \leq D^*(t) \leq \sum_{i=0}^{N_0(t)} D_i^*$$

it follows from the strong law of large numbers for MRP, (see Pyke and Schaufele (1964)), that $t^{-1}D^*(t) \to \mu_{00}^N / \mu_{00}^T$ with probability one as $t \to \infty$. Therefore under the hypothesis of (a), $Pr(C_m) = 0$ for all m.

Consider now the behavior of the queue under actual traffic. Since all integer values of $\nu(t) \geq 0$ are possible we may assume without loss of generality that $\nu(0) = m > r$. If in actual traffic at least $r+1$ cars are present in the queue on S_3 prior to each gap in a time interval $[0,T]$, then $D(t) = D^*(t)$ for each $t \in [0,T]$. Thus if $\nu(0)+A(t) \geq D(t)+r+1$ for all $t \in [0,T]$, then departures are the same as under maximal traffic. But the preceding paragraph shows that with probability one the inequality $\nu(0)+A(s) < D^*(s)+r+1$ and hence $\nu(0)+A(s) < D(s)+r+1$ must hold for some s. Thus the event of queue length less than $r+1$ (notice that $\nu(t) = \nu(0)+A(t)-D(t)$) is a recurrent event. Since the probability is positive of achieving a zero queue length given $\nu(0) < r+1$ the proof of (a) is complete.

To prove (b) observe first that the assumption $\lambda_3\mu_{00}^T > \mu_{00}^N$ implies that $Pr(C_m) > 0$ for each $m \geq 0$, the proof being similar to that given above under (a). However, since $D(t) \leq D^*(t)$ by Lemma 3 we have

$$Pr[\nu(t) > 0 \text{ for all } t \geq 0 | \nu(0) = k] \geq Pr(C_k) > 0$$

as required. □

4. COMPUTATIONS FOR AN R-CAR RIGHT TURN LANE

The main object of this section is to compute, under the maximal traffic model, some of the quantities for an r-car right turn lane, ($r = 0,1,2\cdots$) that are needed for the computation of μ_{00}^N and u_{00}^T. Having computed these, Theorem 1 can then be applied to check if the queue on S_3 is ergodic or not. The computations of other quantities are too prohibitive in the general case to include here, so they are carried out for $r = 1$ in Section 5 in order to illustrate the methods that are extendable to cases of $r > 1$.

Let M_r denote the position of the $r^{\text{th}}L$ car in the queue on S_3. Since the labeling of cars is done by a series of Bernoulli trails, [see assumption (3)], M_r is a negative binomial random variable with parameters p and r. Let N denote the number of R states during an $R-\emptyset$ phase. Let V_i represent the number of gaps during the ith R state of an $R-\emptyset$ phase, $i = 1,2,\ldots,N$; note that the first R state can be either an $R_\emptyset$ or an R_L state. Let Y denote the number of R cars merging during an R_{00} state, and let $m_r(p)$ denote the expectation of Y computed under parameters r and p. Let $h(k) = Pr(\sum_{i=1}^{N} V_i = k)$, $g(k) = P(V_i = k)$ for $i = 1,2,3,\ldots,N$, and $g_1(k) = Pr(V_1 = k)$. Let H, G, and G_1 be the corresponding df's and $\widetilde{h}$, $\widetilde{g}$, and $\widetilde{g}_1$ the generating functions of h, g and g_1, respectively. Finally, let ν_1, ν_2, and ν_3 represent the expected number of visits to states $\emptyset$, R_L, and $R_\emptyset$ respectively during an $R-\emptyset$ phase.

Lemma 4 *For this model, we have that*

$$\begin{aligned} E(N) &= P_{42}(P_{24}+P_{21}P_{14})+P_{41}P_{13}(P_{34}+P_{31}P_{14}) \\ &\quad +P_s(2-P_{13}P_{31})(1-P_{13}P_{31})^{-2}, \\ \nu_1 &= (1-P_{42}P_{24})(P_{14}+P_{13}P_{34})(1-P_{13}P_{31})^{-2}, \\ \nu_2 &= P_{42}, \quad \nu_3 = P_1(1-P_{13}P_{31})^{-2} \\ \textit{where}\quad P_1 &= (P_{34}+P_{31}P_{14})(1-P_{24}P_{42}), \quad \textit{and} \\ P_s &= (P_{42}P_{21}+P_{41}P_{13}P_{31})(P_{34}+P_{31}P_{14})P_{13}. \end{aligned}$$

Proof. To compute $E(N)$, observe that $Pr(N=0)=P_{41}P_{14}$, $Pr(N=1)=P_{42}(P_{24}+P_{21}P_{14})+P_{41}P_{13}(P_{34}+P_{31}P_{14})$, and $Pr(N=n+1)=P_s(P_{31}P_{13})^{n-1}$, $n=2,3,4,\ldots$. To compute μ_1 note that the probability of no visit to a $\emptyset$ state is $P_{42}P_{24}$, and the probability of exactly n visits to $\emptyset$ is $(1-P_{42}P_{24})(P_{14}+P_{13}P_{34})(P_{13}P_{31})^{n-1}$. To calculate ν_2 use the fact that the probability of no visit to an R_L state is $1-P_{42}$, of one visit is P_{42}, and of two or more is zero. To calculate ν_3 use the fact that the probability of exactly n visits to an $R_\emptyset$ is $P_1(P_{13}P_{31})^{n-1}$, $n=1,2,3,\ldots$ Then in all cases, the formulas are obtained by summing geometric series. □

Lemma 5 *For* $r\geq 1,\ 0\leq p\leq 1,\ q=1-p$,

$$E(Y)\equiv m_r(p)=\left(\frac{q^r}{(r-1)!}\right)\frac{d^{r-1}}{dp^{r-1}}\{p^r(1-p)^{-2}[1-\tilde{h}(p)]\}.$$

Proof. Since $Y=\min[\sum_{i=1}^{N}V_i,\ M_r-r]$, we have that

$$\begin{aligned} Pr(Y=n) &= Pr(M_r=n+r)Pr(\sum_{i=1}^{N}V_i>n) \\ &+ Pr(M_r\geq n+r)Pr(\sum_{i=1}^{N}V_i=n). \end{aligned}$$

Let $\varphi(s)=\sum_{n=0}^{\infty}s^nPr(Y=n)$. It follows that

$$\varphi(s) = \sum_{n=0}^{\infty}s^n\binom{r+n-1}{n}p^nq^r(1-H(n))+\sum_{n=0}^{\infty}s^nh(n)\sum_{k=n}^{\infty}\binom{r+k-1}{k}p^kq^r.$$

Using the fact that

$$\frac{1}{(r-1)!}\frac{d^{r-1}}{dp^{r-1}}(p^{r+k-1}) = \binom{r+k-1}{k}p^k,$$

one gets

$$\begin{aligned}\varphi(s) &= \frac{q^r}{(r-1)!}\left[\frac{d^{r-1}}{dp^{r-1}}(p^{r-1}(1-sp)^{-1}\right.\\ &+ \left. p^r(1-s)(1-p)^{-1}(1-sp)^{-1}\widetilde{h}(sp))\right].\end{aligned}$$

This completes the proof since $m_r(p) = (d/ds)\varphi(s)|_{s=1}$. □

The first three values of $m_r(p)$ may be shown to be

$$\begin{aligned} m_0(p) &= pq^{-1}(1-\widetilde{h}(p)),\ m_1(p) = 2pq^{-1}(1-\widetilde{h}(p)) - p^2\widetilde{h}'(p),\\ m_2(p) &= 3pq^{-1}(1-\widetilde{h}(p)) - p^2(3-p)\widetilde{h}'(p) - p^3 2^{-1}q\widetilde{h}''(p)\end{aligned}$$

where primes denote derivatives. We also have $\lim_{p\to 1} m_r(p) = \widetilde{h}'(1)$; that is, if all cars are R cars, then the expected number of cars merging during an R_{00} state is just the expected number of gaps during an R_{00} state, as it should be.

So far, $m_r(p)$ is expressed in terms of the function h; that is, in terms of the distribution of the total number of gaps during an R_{00} state. Using the fact that $V_2, V_3, \ldots$ are independent and identically distributed, $m_r(p)$ can be expressed in terms of the functions g_1 and g; that is, in terms of the number of gaps in an R_L and an $R_\emptyset$ state. The relationship between g, g_1 and h is shown in

Lemma 6 *For* $s \geq 0$,

$$\widetilde{h}(s) = Pr(N=0) + Pr(N=1)\widetilde{g}_1(s) + P_s\widetilde{g}_1(s)\widetilde{g}(s)(1-P_{13}P_{31}\widetilde{g}(s))^{-1}$$

where P_s *is as in Lemma 4.*

Proof. The df of N is given in the proof of Lemma 4. Clearly

$$\begin{aligned} h(k) &= Pr(\sum_{i=1}^{N} V_i = k) = \sum_{n=0}^{\infty} Pr(\sum_{i=1}^{n} V_i = k)Pr(N=n)\\ &= Pr(N=0)\delta_{k0} + P(N=1)g_1(k)\\ &+ \sum_{n=2}^{\infty} g_1 * g^{(n-1)}(k)P_s(P_{31}P_{13})^{n-2}\end{aligned}$$

where δ_{k0} is 1 or 0 according as $k = 0$ or $k \geq 1$ where $*$ and the superscript $(n-1)$ denote convolution in the usual manner. Since $\widetilde{h}(s) = \sum_{k=0}^{\infty} s^k h(k)$, the proof is complete. □

5. SPECIAL CASES OF $R = 0, 1$ AND $+\infty$

Consider first $r = 1$. For this case the states of the MRP are $R_{01}, R_{00}, R_{10}, L_{00}, L_{10}, L_{20}$, and L_{01}. In order to compute the transition probabilities of this 7-state MRP we need to describe the number of cars in S_{3L} and S_{3R} at each gap during an L phase. Let us think of the movement of cars between successive gaps during an L phase as being determined by successive repetitions of a game with three outcomes which we label as $+$, $-$, or 0 according as one R car, one L car, or two cars (an R and an L) merge during the last gap. Then for the next gap the $+$ game indicates that there is one R car in S_{3R}, the $-$ game indicates that there is one L car in S_{3L}, and the 0 game indicates that there are no cars in either S_{3L} or S_{3R}. Because of the Bernoulli model which has been assumed for the labeling of the cars, these games determine a Markov chain with the following transition matrix,

$$\mathbb{P} = \begin{pmatrix} & + & - & 0 \\ + & p & 0 & q \\ - & 0 & q & p \\ 0 & p^2 & q^2 & 2pq \end{pmatrix}$$

For example consider the transition from game 0 to game $+$. An outcome of 0 for the last game means that two cars merged, and hence, no information is available about the remaining cars on S_3 until the next gap appears. Thus at the next gap a single R car will merge with probability p^2, and so the outcome is $+$ with probability p^2.

Define $p_{ij}^{(n)}$ to be the ij^{th} element from the nth power of the matrix $\mathbb{P}$ where $i, j = +, -$, or 0. Recall that H_4 is the df for the length of time in state L and that F is the df for the length of time of a gap. The convolution $F^{(n)}(t)$ represents the probability that the sum of the lengths of n gaps is less than or equal to t. Let $P(A, B)$ denote the transition probability from state A to B.

The transition probabilities from the L states to the R states are all derived similarly so consider only $P(L_{00}, R_{00})$. If the length of time spent in state L is t, then the conditional probability of going from L_{00} to R_{00} is $\sum_{n=1}^{\infty} \{F^{(n-1)}(t) - F^{(n)}(t)\} p_{00}^{(n)}$. The result follows by taking the expectation of this quantity with respect to H_4, the exponential holding time df of state L. We now list these probabilities :

$$\begin{aligned} P(L_{00}, R_{00}) &= \int_0^\infty \sum_{n=1}^\infty [F^{(n-1)}(t) - F^{(n)}(t)] \\ &\quad p_{00}^{(n)} dH_4(t), \\ P(L_{20}, R_{00}) &= \int_0^\infty \sum_{n=2}^\infty [F^{(n-1)}(t) - F^{(n)}(t)] \\ &\quad p_{-0}^{(n-1)} dH_4(t), \\ P(L_{20}, R_{10}) &= \int_0^\infty \sum_{n=1}^\infty [F^{(n-1)}(t) - F^{(n)}(t)] \\ &\quad p_{--}^{(n-1)} dH_4(t). \end{aligned} \tag{18.6}$$

The quantities $P(L_{00}, R_{01})$, $P(L_{00}, R_{10})$, $P(L_{10}, R_{00})$,$P(L_{10}, R_{01})$, $P(L_{10}, R_{10})$, $P(L_{01}, R_{00})$, $P(L_{01}, R_{01})$, and $P(L_{01}, R_{10})$ are the same as $P(L_{00}, R_{00})$ with $p_{00}^{(n)}$ replaced by $p_{0+}^{(n)}$, $p_{0-}^{(n)}$, $p_{-0}^{(n)}$, $p_{-+}^{(n)}$, $p_{--}^{(n)}$, $p_{+0}^{(n)}$, $p_{++}^{(n)}$ and $p_{+-}^{(n)}$, respectively. $P(L_{20}, R_{01})$ is the same as $P(L_{20}, R_{00})$ with $p_{-+}^{(n-1)}$ in place of $p_{-0}^{(n-1)}$. Some formal simplifications are possible since H_4 is exponential; see Corollary 3. For example, one may show that

$$P(L_{00}, R_{00}) = \frac{1 - \tilde{f}(\lambda)}{\tilde{f}(\lambda)} \mathcal{P}_{00}(\tilde{f}(\lambda))$$

in which $\tilde{f}$ is the Laplace-Stieltjes transform of F and $\mathcal{P}_{ij}(s) = \sum_{n=1}^{\infty} s^n p_{ij}^{(n)}$ is the generating function of the transition probabilities.

Since the derivations of the transition probabilities from the R states to the L states are all similar, only the probability of going from R_{00} to L_{10} is derived. Given that $M_1 = j$, that is, the position of the first L car in the queue on S_3 is j, then the position of the second L car is $j + M_1^*$ where M_1^* is an independent rv identically distributed as M_1. Thus,

$$\begin{aligned} P(R_{00}, L_{10}) &= \sum_{j=1}^{\infty} P(R_{00}, L_{10}|M_1{=}j) P(M_1{=}j), \text{ and} \\ P(R_{00}, L_{10}|M_1 = j) &= P(j - 1 < \sum_{i=1}^{N} V_i \le M_1^* + j - 2) \\ &= \sum_{n=j}^{\infty} h(n) P(M_1^* \ge n - j + 2) = \sum_{n=j}^{\infty} h(n) p^{n-j+1} \end{aligned}$$

so that

$$P(R_{00}, L_{10}) = \sum_{j=1}^{\infty} p^{j-1} q \sum_{n=j}^{\infty} h(n) p^{n-j+1} = pq\widetilde{h}'(p).$$

We list all of these R-to-L probabilities as follows:

$$\begin{aligned}
P(R_{00}, L_{00}) &= \widetilde{h}(p), \\
P(R_{00}, L_{10}) &= pq\widetilde{h}'(p), \\
P(R_{01}, L_{00}) &= p^{-1}(\widetilde{h}(p) - h(0)), \\
P(R_{01}, L_{10}) &= q\widetilde{h}'(p) + qp^{-1}(h(0) - \widetilde{h}(p)), \\
P(R_{10}, L_{00}) &= 0, \\
P(R_{10}, L_{10}) &= \widetilde{h}(p) \\
P(R_{00}, L_{01}) &= 0, \\
P(R_{00}, L_{20}) &= 1 - \widetilde{h}(p) - pq\widetilde{h}'(p), \\
P(R_{01}, L_{01}) &= h(0), \\
P(R_{01}, L_{20}) &= 1 - \widetilde{h}(p) - q\widetilde{h}'(p), \\
P(R_{10}, L_{01}) &= 0, \\
P(R_{10}, L_{20}) &= 1 - \widetilde{h}(p).
\end{aligned} \tag{18.7}$$

The gap df, F, enters into these formulae for R-to-L transitions through g and g_1, and hence by Lemma 4, through h. The explicit dependence could be set out but the resulting formulae are messy.

Let $E(A, B)$ be the expected number of cars merging during state A under maximal traffic if the process goes from A next to B. Let $E(C)$ represent the expected number of cars merging during state C. Write

$$m_{ij}(k) = \sum_{\ell=2}^{k} p_{ij}^{(\ell-1)} p_{jj}^{(k-\ell+1)} (p_{ij}^{(k)})^{-1}$$

where $i, j = +,\ -$, or 0. Define $m_{ij}(1) = 0$. Also let $M(t) = \sum_{k=0}^{\infty} F^{(k)}(t)$ be the renewal function for the gap df. Then

$$\begin{aligned}
E(L_{00}, R_{00}) &= \int_0^{\infty} M(t) dH_4(t) + \sum_{k=1}^{\infty} (m_{00}(k) + 1) \\
&\quad \int_0^{\infty} [F^{(k-1)}(t) - F^{(k)}(t)] dH_4(t),
\end{aligned}$$

$$\begin{aligned}
E(L_{20}, R_{00}) &= \int_0^\infty [M(t) + F(t)]dH_4(t) + \sum_{k=2}^\infty m_{-0}(k-1) \\
&\quad \int_0^\infty [F^{(k-1)}(t) - F^{(k)}(t)]dH_4(t) \qquad (18.8) \\
E(L_{20}, R_{10}) &= \int_0^\infty M(t)dH_4(t) + \sum_{k=2}^\infty m_{--}(k-1) \\
&\quad \int_0^\infty [F^{(k-1)}(t) - F^{(k)}(t)]dH_4(t), \\
E(R_{00}) &= m_1(p),\ E(R_{10}) = m_0(p),
\end{aligned}$$

and $E(R_{01}) = p^{-1}m_1(p) - qp^{-1}m_0(p)$. The quantities $E(L_{00}, R_{10})$, $E(L_{00}, R_{01})$, $E(L_{10}, R_{00})$, $E(L_{10}, R_{10})$, $E(L_{10}, R_{01})$, $E(L_{01}, R_{00})$, $E(L_{01}R_{10})$, $E(L_{01}, R_{01})$ are the same as $E(L_{00}, R_{00})$ with $m_{00}(k) + 1$ replaced by $m_{0-}(k)$, $m_{0+}(k)$, $m_{-0}(k) + 1$, $m_{--}(k)$, $m_{-+}(k)$, $m_{+0}(k)+1$, $m_{+-}(k)$ and $m_{++}(k)$, respectively. The quantity $E(L_{20}, R_{01})$ is the same as $E(L_{20}, R_{10})$ with $m_{-+}(k-1)$ in place of $m_{--}(k-1)$.

Since the derivations of the quantities in (18.7) are all similar, only $E(L_{00}, R_{00})$ is derived. Let $P(k)$ be the probability of k gaps during an L state. Note that $P(k)$ is the same for L_{00}, L_{01}, L_{10}, and L_{20}. Let $E_T(L)$ be the expected number of gaps during an L state, and let $E(A, B|k)$ be the expected number of cars merging from A to B given that there were k gaps. If m gaps occur during an L state, let $\gamma_1, \ldots, \gamma_m$ represent the outcomes of the corresponding games played; thus $\gamma_i = +, -$, or 0. Also let $Z_i = 1$ or 0 according as $\gamma_i = 0$ or $\gamma_i \neq 0$. Finally define $E(0, k-1, \gamma_1, \gamma_{k+1})$ as the expected number of 0 games played in the $k-1$ gaps between γ_1 and γ_{k+1} if it is known what γ_1 and γ_{k+1} are. Thus,

$$\begin{aligned}
E(L_{00}, R_{00}) &= \sum_{k=1}^\infty P(k)E(L_{00}, R_{00}|k) \\
&= \sum_{k=1}^\infty P(k)[2\{E(0, k-1, 0, 0) + 1\} \\
&\quad + k - E(0, k-1, 0, 0) - 1] \\
&= \sum_{k=1}^\infty kP(k) + \sum_{k=1}^\infty P(k)[E(0, k-1, 0, 0) + 1].
\end{aligned}$$

Now

$$\begin{aligned}
E(0,k-1,0,0) &= E(\sum_{i=2}^{k} Z_i | J_1 = 0, J_{k+1} = 0) \\
&= \sum_{i=2}^{k} E(Z_i | J_1 = 0, J_{k+1} = 0) \\
&= \sum_{i=2}^{k} Pr(J_i = 0 | J_1 = 0, J_{k+1} = 0) \\
&= \sum_{i=2}^{k} Pr(J_i = 0, J_{k+1} = 0 | J_1 = 0)[Pr(J_{k+1} = 0 | J_1 = 0)] \\
&= \sum_{i=2}^{k} p_{00}^{(i-1)} p_{00}^{(k-i+1)} (p_{00}^{(k)})^{-1} = m_{00}(k).
\end{aligned}$$

Also

$$\begin{aligned}
P(k) &= \int_0^{\infty} [F^{(k-1)}(t) - F^{(k)}(t)] dH_4(t), \\
E_T(L) = \sum_{k=1}^{\infty} kP(k) &= \sum_{k=1}^{\infty} k \int_0^{\infty} [F^{(k-1)}(t) - F^{(k)}(t)] dH_4(t) \\
&= \int_0^{\infty} M(t) dH_4(t),
\end{aligned}$$

so that the formula for $E(L_{00}, R_{00})$ follows by substitution. By referring to the definition of $m_r(p)$ the results for $E(R_{00})$ and $E(R_{10})$ follow directly. If the conditional expectation of $E(R_{00})$ is taken given the label of the first car, then the resulting equation can be solved in order to get $E(R_{01})$.

For the MRP with states R_{00}, R_{01}, R_{10}, L_{00}, L_{10}, L_{20}, and L_{01} the transition matrix has been computed, but in order to calculate μ_{00}^N and μ_{00}^T, this process must be condensed into a smaller one. Our main interest is in the succession of L states. Let us denote by π_{ij} the probability that the next L state after L_i is L_j where $i, j = 00,\ 10,\ 20$, or 01. Then this transition matrix (π_{ij}) is derived from the previous one according to

$$\pi_{ij} = \sum_{k=00,01,10} P(L_i, R_k) P(R_k, L_j) \text{ where } i, j = 00, 10, 20, 01. \quad (18.9)$$

At this point all the quantities necessary to compute μ_{00}^N and μ_{00}^T are available. It only remains to set up the necessary notation and quote the appropriate equations. To this end let η_i^N denote the expected number of cars merging from the start of $L_i (i = 00, 01, 10, 20)$ to the start of the next L state when maximal traffic is assumed on S_3, and let η^T denote the expected length of time between successive L states.

The expected number of cars merging between successive L_{00} events is equal to

$$\mu_{00}^N = \mu_{00}^* \sum_{j=00,10,20,01} \eta_j^N (\mu_j^*)^{-1} \tag{18.10}$$

as follows from Lemma 4 of [11]. The μ_j^*'s are computable as stated in the section immediately following Corollary 3.

The expected number of cars arriving between successive L_{00} events is equal to

$$\lambda_3 \mu_{00}^T = \lambda_3 \eta^T \mu_{00}^* \tag{18.11}$$

where μ_j^* is the mean recurrence time of state j in the Markov Chain with transition matrix (π_{ik}), η_j^N is computable from equations (18.5), (18.6), and (18.7), and η^T is obtainable from Corollary 3 and Lemma 4. To justify (18.11) use [11] to obtain $\mu_{00}^T = \mu_{00}^* \sum_{j=00,10,20,01} \eta_j^T (\mu_j^*) \mu^{-1}$ where η_j^T denotes the expected length of time of state L_j, $j = 00, 10, 20,$ or 01. However, $\eta_j^T = \eta^T$ is independent of j, so the result follows since $\sum\{(\mu_j^*)^{-1} : j = 00, 01, 10, 20\} = 1$. For η^T, we have that $\eta^T = \eta_4 + \nu_1\eta_1 + \nu_2\eta_2 + \nu_3\eta_3$. η_j^N is obtained as follows,

$$\begin{aligned}
\eta_j^N &= \sum_{k=00,10,20,01} \pi_{jk} E(L_j, L_k) \\
&= \sum_k \pi_{jk} \sum_{i=00,01,10} P(L_j, R_i) P(R_i, L_k) \pi_{jk}^{-1} (E(L_j, R_i) + E(R_i, L_k)) \\
&= \sum_i P(L_j, R_i) E(L_j, R_i) + \sum_i P(L_j, R_i) \sum_k P(R_i, L_k) E(R_i, L_k) \\
&= \sum_i P(L_j, R_i) E(L_j, R_i) + \sum_i P(L_j, R_i) E(R_i). \qquad (18.12)
\end{aligned}$$

It remains to look at the two other cases; that of an infinite right turn lane $(r = \infty)$ and the case $r = 0$ where S_3 has only a single lane. For the latter, there are three states, L_1, L_0, and R_0 where L_1 indicates that the system entered an L phase when the previous $R - \emptyset$ phase was blocked by an L car. For this case, $P(L_0, R_0) = P(L_1, R_0) = 1$, $P(R_0, L_0) = \widetilde{h}(p)$ and $P(R_0, L_1) = 1 - \widetilde{h}(p)$. For the expectations,

$E(R_0) = m_0(p)$, $E(L_0, R_0) = E(L_1, R_0) = E_T(L)$. Thus $\pi_{11} = \pi_{01} = 1 - \widetilde{h}(p)$, $\pi_{10} = \pi_{00} = \widetilde{h}(p)$, and $\eta_i^N = E(L_i, R_0) + E(R_0)$, $i = 0, 1$. Thus all the quantities necessary to compute μ_{00}^N and μ_{00}^T are easily obtained when $r = 0$.

For an infinite right turn lane there is no distinction between L events on S_1 and S_2, and we have effectively two independent roads S_{3L} and S_{3R} with the average number of arrivals between L events on S_{3L} being $q\lambda_3\eta^T$ and on S_{3R} being $p\lambda_3\eta^T$. Then Theorem 1 implies that in order to have "freely flowing traffic", it would suffice to have $q\lambda_3\eta^T < E^T(L)$ and $p\lambda_3\eta^T < E^T(L) + E^T(R)$.

6. THE GENERAL CASE

During an L-phase, the numbers of cars in the two branches S_{3R} and S_{3L} on the secondary road may be described as an interesting Markov chain generalizing the Markov chain with states $+$, $-$, 0 used in Section 5 for the case of $r = 1$. With p denoting the common probability of picking label R for the sequence of cars on S_3, the transition probabilities between the states $L_{r0}, L_{r-1,0}, \ldots, L_{10}, L_{00}, L_{01}, \ldots, L_{01}, \ldots, L_{0r}$ are given in the transition matrix $\mathbb{P} = (p_{ij})$ for $-r \leq i, j \leq r$ in which the state L_{kl} is being identified with the integer $l - k$.. It can be shown that $p_{00} = 2pq$ and otherwise

$$p_{ij} = \begin{cases} p^{r+1-i} & \text{if } 0 \leq i \leq r \text{ and } j = r, \\ p^{j-i+1}q & \text{if } 0 \leq i \leq j+1 \text{ and } 1 \leq j < r, \\ q^{i-j+1}p & \text{if } j-1 \leq i \leq 0 \text{ and } -r < j \leq -1, \\ q^{r+1+i} & \text{if } -r \leq i \leq 0 \text{ and } j = -r. \end{cases} \tag{18.13}$$

It is helpful to display the matrix itself; observe that the upper left and lower right quadrants of $\mathbb{P}$ are lower and upper triangular, respectively, while the other quadrants are zero.

As in Section 5, the connection between this Markov chain and the traffic flow from S_3 during an L-phase is as follows: At a transition from state r (L_{r0}) or state $-r$ (L_{0r}) the tally of cars merged increases by one R car or one L-car, respectively. At transitions from all other states, the tally increases by one R-car *and* one L-car. If X_n (Y_n) denotes the number of R-cars (L-cars), respectively, that have merged during the first n transitions, and if $N_k(n)$ equals the number of visits to state k during the first n transitions, then one obtains

$$X_n = n - N_{-r}(n), \qquad Y_n = n - N_r(n), \tag{18.14}$$

since for example the only transitions at which an R-car does not merge is when the transition is into state L_{r0} which is state $-r$. Thus, traffic flow is completely describable in terms of the occupation probabilities of states r and $-r$. Write $\pi_k(n)$ for the probability of going from state 0 to state k in n steps and let $\{\pi_k : -r \leq k \leq r\}$ be their limits, the stationary probabilities. In particular, the stationary probabilities may be derived by standard methods to be

$$\pi_0 = (p-q)\left\{p^2\left(\frac{p}{q}\right)^r - q^2\left(\frac{q}{p}\right)^r\right\}^{-1},$$

$$\pi_k = p\left(\frac{p}{q}\right)^k \pi_0 \qquad \text{if } k = 1, 2, \ldots, r, \tag{18.15}$$

$$\pi_{-k} = q\left(\frac{q}{p}\right)^k \pi_0 \qquad \text{if } k = 1, 2, \ldots, r, \tag{18.16}$$

for $p \neq q$, and when $p = q = \frac{1}{2}$

$$\pi_0 = \frac{1}{r+1}, \qquad \pi_k = \pi_{-k} = \frac{1}{2(r+1)}; \qquad k = 1, 2, \ldots, r. \tag{18.17}$$

From (18.14), this implies for example that when cars are equally likely to turn left or right, the long run proportion of cars merging during L-phases is $2 - 1/(r+1)$ per gap, under maximal traffic, compared to 1 when there is no separate turn lane ($r = 0$).

As was done above for $r = 1$, quantities required for the application of Theorem 1 for general r may be determined using this Markov chain. For example, the extension of (18.5) would include the following, in which $\mathbf{P}^n \equiv (p_{ij}^{(n)})$:

$$P(L_{i0}, R_{j0}) = \int_0^\infty \sum_{n=1}^{\infty} [F^{(n-1)}(t) - F^{(n)}(t)] p_{-i,-j}^{(n)} dH_4(t)$$

for $0 \leq i, j \leq r$, and

$$P(L_{r+1,0}, R_{j0}) = \int_0^\infty \sum_{n=2}^{\infty} [F^{(n-1)}(t) - F^{(n)}(t)] p_{-r,-j}^{(n-1)} dH_4(t)$$

for $0 \leq j \leq r$. The expressions for $P(L_{i0}, R_{0j})$ and $P(L_{r+1,0}, R_{0j})$ are similar, requiring only that $-j$ on the right hand sides is changed to j.

References

[1] Adams, W. F., Road Traffic Considered as a Random Series, *J. Inst. Civil. Engrs. London*, 4 (1936), 121–130.

[2] Frechet, N., *Recherches Theoriques Modernes Sur Les Calcul des Probabilities*, Second Livre, Gauthier-Villars, Paris, 1938.

[3] Garwood, F., The Application of the Theory of Probability to the Operation of Vehicular Controlled Traffic Signals, *J. Roy. Statist. Soc. Suppl.*, **7** (1940), 65–77.

[4] Gaver, D. P., *Accommodation of Second Class Traffic*, Technical Report No. 74, Applied Math. and Stat. Lab., Stanford University, California.

[5] Haight, F. A., Mathematical Theories of Traffic Flow, *Mathematics in Science and Eng.*, **Vol. 7**, Academic Press, New York, 1963.

[6] Jewell, W. S., *Waiting for a Gap in Traffic*, Research Report No. 6, Operations Center, University of California–Berkeley (June 1961).

[7] Newell, G. F., The Effect of Left Turns on the Capacity of Traffic Intersection, *Quart. Appl. Math.*, **17**, No. 1 (1959), 67–76.

[8] Newell, G. F., Queues for a Fixed-cycle Traffic Light, *Ann. Math. Stat.*, **31** No. 3 (1960), 589–597.

[9] Oliver, R. M., Distribution of Gaps and Blocks in a Traffic Stream, *Operations Research*, **10** No. 2 (1962), 197–217.

[10] Pyke, R., Markov Renewal Processes: Definitions and Preliminary Properties, *Ann. Math. Stat.*, **32** (1961), 1231–1242.

[11] Pyke, R., Markov Renewal Processes with Finitely Many States, *Ann. Math. Stat.*, **32** (1961), 1243–1259.

[12] Pyke, R. and Schaufele, R., *Limit Theorems for Markov Renewal Processes*, Technical Report No. 6, University of Washington, June 1963.

[13] Raff, M. S., The Distribution of Blocks in an Uncongested Stream of Traffic, *J. Am. Stat. Assoc.*, **46** (1951), 114–123.

[14] Tanner, J.C., The Delay to Pedestrians Crossing a Road, *Biometrika*, **38** (1951), 383–392.

[15] Weiss, G. H. and Maradudin, A. A., Some Problems in Traffic Delay, *Operations Research*, **10** (1962), 74–104, No. 1.

V

FINANCIAL MODELS

Chapter 19

STOCHASTIC STABILITY AND OPTIMAL CONTROL OF SEMI-MARKOV RISK PROCESSES IN INSURANCE MATHEMATICS

Anatoly Swishchuk
National Academy of Sciences of Ukraine, Ukraine

Abstract We study semi-Markov risk processes which describe a dynamic of summary capital of an insurance company. The theorems on stability, asymptotic and exponential stability of zero state of the processes with probability 1 are proved. We also investigate the optimal stochastic control of the controlled semi-Markov processes. The Bellman equation for semi-Markov risk processes is derived. Analogue of Dynkin formulae and boundary value problem for semi-Markov random evolutions, and properties of the respected stochastic Liapunov functions are used.

Keywords: Insurance mathematics, semi-Markov risk process, discontinuous semi-Markov random evolutions, analogue of Dynkin's formula, boundary value problem, stochastic stability, optimal stochastic control, Bellman equation.

1. INTRODUCTION

Collective risk theory as a part of insurance mathematics deals with stochastic models of an insurance business. The traditional approach in the risk theory is to consider a model of the risk business of an insurance company, and to study the probability of ruin, i.e., the probability that the risk business ever will be below some specified value.

The simplest model-is so-called "classical risk model" by Cramer-Lundberg [1]. In the model only claim process were modelled as stochastic, namely, in the form of compound Poisson process. This classical risk process can be generalized in many ways.

One mainstream in present-day insurance mathematics undertakes to incorporate financial risk in the form of stochastic interest.

In our model of risk process, which is the summary capital of some insurance company, is described by the functional of semi-Markov process. It is concerned not only the claims process but also the interest process. The source of risk is the all insurance portfolio of the insurance company. Semi-Markov process is the mathematical model of external effects, for example, unexpected news or factors.

The semi-Markov risk process is the realization of discontinuous semi-Markov random evolutions [5].

This article is devoted to the study of stochastic stability and optimal control of semi-Markov risk process, applying analogue of Dynkin formula and boundary value problems for semi-Markov random evolutions [6], and super-martingale properties of the appropriate stochastic Lyapunov's functions.

We study stability and asymptotic stability in probability, and exponential stability of zero state of the semi-Markov risk process. These kinds of stabilities of zero state are described in the forms of estimations of probabilities for supremum of the semi-Markov risk process. The estimations of the probabilities are interested to study as they are closed to the ruin probabilities.

We investigate the optimal stochastic control of the controlled semi-Markov risk process, deriving Bellman equation. We also give constructive result for optimal stochastic control, and result on the comparing of the controls to separate optimal one.

2. SEMI-MARKOV RISK PROCESSES

We define the semi-Markov risk process $z(t)$ as the solution of the following equation

$$z(t) = z + \int_T^t v(z(s), x(s))ds - \sum_{k=1}^{\nu(t-T)} a(x_k) := z + B(t) - A(t); \; t > T; \tag{19.1}$$

Here, $x(t)$ is a regular semi-Markov process constructed by Markov renewal process (x_n, θ_n) [4]; $\nu(t) := \max\{n : \tau_n \geq t\}$ is a counting process, $\tau_n := \sum_{k=1}^n \theta_k$; θ_k are sojourn times, $x(T) = x \in X$, where (X, Ξ) is a phase space. Process $B(t)$ describes the sum of inputs to the insurance company up to the moment t; process $A(t)$ describes the sum of outputs (payments) of the insurance company to its clients up to the moment t; $\nu(t)$ is a number of payments; z is an initial capital of the insurance company, $z(T) = z$. Function $v(z, x)$ is the intensivity of insurance inputs (gross risk premium rate); function $a(x)$ is the value of one payment. Parameter x describes the outer effects, for example, unexpected news or factors. We note, if $v(z,x) \equiv v > 0$, $a(x_k)$ are independent identity distributed random variables, $\nu(t) \equiv N(t)$ is a Poisson process, then process $z(t)$ is a classical risk process [1].

3. SEMI-MARKOV RISK PROCESS AS DISCONTINUOUS SEMI-MARKOV RANDOM EVOLUTION

Semi-Markov Random Evolutions are defined by the following stochastic operator integral equation in separable Banach space B:

$$V(t)f = f + \int_T^t \Gamma(x(s))V(s)f ds + \sum_{k=1}^{\nu(t-T)} [\mathbf{D}(x_k) - I]V(\tau_k -)f; \quad (19.2)$$

Here, $\{\Gamma(x); x \in X\}$ are linear closed operators with common dense domain D: $\bar{D} = B$, $f \in B$; $\{\mathbf{D}(x); x \in X\}$ are bounded operators; $x(t)$ is a semi-Markov process. Let $\Gamma_x(t)$ be the semi-groups of operators generated by $\Gamma(x)$, $x \in X$-phase space of $x(t)$. Let us define these operators in the following way:

$$\Gamma_x(t) := f(q(t; z, x)),$$

$$f \in C^1(R),$$

where $q(t; z, x)$ is the solution of the following Cauchy problem:

$$dq/dt = v(q, x), \quad q(0; z, x) = z.$$

Here,

$$\Gamma(x)f(z) = v(z, x)df(z)/dz.$$

Also, let us define the operators $\mathbf{D}(x)$ in such a way:

$$\mathbf{D}(x)f(z) := f(z - a(x)).$$

Then functional equation for semi-Markov risk process $z(t)$ in (19.1) takes the form:

$$\begin{aligned} f(z(t)) &= f(z) + \int_T^t v(z(s), x(s))df(z(s))/dzds \\ &+ \sum_{k=1}^{\nu(t-T)} [f(z(\tau_k -) - a(x_k)) - f(z(\tau_k -))]; \end{aligned}$$

Hence,

$$V(t)f(z) = f(z(t)),$$

where $V(t)$ is defined in (19.2) and $z(t)$-in (19.1), and semi-Markov risk processes are the realizations of semi-Markov random evolutions.

4. ANALOGUE OF DYNKIN'S FORMULA FOR SEMI-MARKOV RANDOM EVOLUTIONS

Let τ be a stopping time such that $E_{x,T}\tau < +\infty$; $\tau > T$. Then the following formula is true [2], [6]:

$$E_{x,T}[V(\tau)f(y(\tau))] - f(y) = E_{x,T}\int_T^\tau [\hat{Q} + \Gamma(x)+$$

$$+\lambda(x)\int_X P(x(s), du)(\mathbf{D}(u) - I)]V(s)f(y(s))ds,$$

$$f(y) \in C^1(X \times R_+). \tag{19.3}$$

Here, $\hat{Q}$ is an infinitesimal operator of the Markov process $y(t) := (x(t), t - \tau_{\nu(t)})$ in the phase space $X \times R_+$, $y(T) := y := (x, T)$; $\lambda(x) := g_x(t)/\bar{G}_x(t)$, where $dG_x(t)/dt := g_x(t)$, $\bar{G}_x(t) := 1 - G_x(t)$, $P(x, dy)$ are transition probabilities for x_n; $G_x(t)$ is a distribution function for θ_n. If $\Gamma(x) \equiv 0$ and $\mathbf{D}(x) \equiv I$ in (19.3), and $x(t)$ is a Markov process, then this formula is well-known Dynkin formula. Realization of formula (19.3) for semi-Markov risk process in (19.1) is used for stochastic stability of semi-Markov risk process.

5. BOUNDARY VALUE PROBLEM FOR SEMI-MARKOV RANDOM EVOLUTIONS

Let we have the following boundary value problem for semi-Markov random evolutions [6]:

$$\hat{Q}C(y) + \Gamma(x)C(y) + \lambda(x)\int_X P(x, du)[\mathbf{D}(u) - I]C(u, T) = -k(y),$$

$$C(y) = b(y), \tag{19.4}$$

for all $y \in \partial G$, $G \in \Xi \times \mathbf{R}_+$, where $\mathbf{R}_+$ is a Borel sets in R_+.

Then the solution of this boundary value problem takes the following form:

$$C(y) = E_{x,T} \int_T^{\tau_G} V(s)k(y(s))ds + E_{x,T}[V(\tau_G)b(y(\tau_G))], \qquad (19.5)$$

where

$$\tau_G := \inf\{t : y(t) \notin G\}.$$

The solution (19.5) for (19.4) is used for stochastic optimal control of semi-Markov risk process.

6. STOCHASTIC STABILITY OF SEMI-MARKOV RISK PROCESSES

Let $z(t)$ be the semi-Markov risk process in (19.1). It is known ([6]) that process $(z(t), y(t))$ is a Markov process in phase space $R \times X \times R_+$ with infinitesimal operator:

$$Lf(z,y) = v(z,x)df(z,y)/dz + \lambda(x) \int_X [P(x,du)f(z - a(u), y) - f(z,y)] + \hat{Q}f(z,y). \qquad (19.6)$$

Let the following conditions be satisfied.

C1) There is nonnegative and continuous function $V(z,y)$ on the set

$$A_m := \{(z,y) : V(z,y) < m\},$$

for some $m > 0$; $V(0,y) = 0$; if $z > \epsilon$ then $V(z,y) > \epsilon$ for some positive ϵ;

C2) Let L_m be the infinitesimal operator of the stopped process

$$(z(t \wedge \tau_m), y(t \wedge \tau_m)),$$

where

$$\tau_m := \inf\{t : (z(t), y(t)) \notin A_m\},$$

and operator L is defined in (19.6);

C3)

$$V(z,y) \in Domain(L_m),$$

$$dV(z,y)/dz$$

is bounded and continuous.

Main Lemma : Let the conditions C1)-C3) be satisfied and

$$L_m V(z,y) \leq 0$$

on A_m.Then the process $V(z(t \wedge \tau_m), y(t \wedge \tau_m))$ is a nonnegative supermartingale and for $\lambda \geq m$:

$$P_{z,y}\{\sup_{T \leq t < +\infty} V(z(t \wedge \tau_m), y(t \wedge \tau_m)) \geq \lambda\} \leq V(z, y)/\lambda.$$

Proof We have the following analogue of Dynkin formula from (19.3) for function $V(z, y)$:

$$E_{z,y}[V(z(t \wedge \tau_m), y(t \wedge \tau_m))] - V(z, y) =$$

$$= E_{z,y} \int_T^{t \wedge \tau_m} [\hat{Q}V + v(z(s), x(s))dV/dz +$$

$$+\lambda(x(s)) \int_X P(x, du)(V(z(s) - a(u), y(s)) - V(z(s), y(s))]ds.$$

From this formula and Kolmogorov-Doob inequality for supermartingales we obtain the proof of the main lemma. □

The main result in the theory of stochastic stability of the risk processes is the following theorem - stochastic analogue of Lyapunov's theorem on stability.

Theorem 1 *Let the conditions of Main Lemma be satisfied and $v(0, x) = 0$, for all $x \in X$. Then zero state of semi-Markov risk process $z(t)$ in (19.1) is stable in probability.*

Proof Since $V(z, y)$ is continuous by z, for all $x \in X$, and $t \in R_+$, $V(0, t) = 0$, then

$$\{\sup_{T \leq t < +\infty} |z(t)| \geq \epsilon\} \subset \{\sup_{T \leq t < +\infty} V(z(t), y(t)) \geq \epsilon\}$$

because of the properties of the function $V(z, y)$. From here and the Main Lemma we obtain the proof of the Theorem 1. □

The next result is the stochastic analogue of Lyapunov's theorem on the asymptotic stability of risk processes.

Theorem 2 *Let the conditions C1)-C3) be satisfied, $v(0, x) = 0$ and*

$$L_m V(z, y) = -k(z, y)$$

on A_m with some positive bounded continuous function $k(z, y)$. Then zero state of semi-Markov risk process in (19.1) is asymptotical stable in probability.

Proof We note that if $k(z, y) = 0$ on A_m then Theorem 2 is a simple corollary of Theorem 1. By Main Lemma process $(z(t), y(t))$ belongs to the interior of the set A_m with probability no less then $1 - (V(z,y)/m)$. The further reasonings are analogous to the ones in the Theorem 2 in [8], p.57, but only for Markov process $(z(t), y(t))$. □

If the function $k(z,y)$ in the Theorem 2 is a linear of function $V(z,y)$ in A_m then we have the result analogous to the exponential stability.

Theorem 3 *Let the conditions C1)-C3) be satisfied, $v(0,x) = 0$ and*

$$L_m V(z,y) \leq -\alpha V(z,y)$$

on A_m, $\alpha > 0$. Then zero state of semi-Markov risk process in (19.1) is exponential stable in probability.

Proof Put $\xi_m := \inf\{t : (z(t), y(t)) \notin A_m\}$. Let's define the process $\hat{\zeta}(t) := (\hat{z}(t), \hat{y}(t))$ in such a way: $\hat{\zeta}(t) = \zeta(t) := (z(t), y(t))$ if $t < \xi_m$, and $\hat{\zeta}(t) = 0$, if $t \geq \xi_m$. Let $\hat{E}_\zeta$ respects to the process $\hat{\zeta}(t)$, which is a continuous from the right Markov process for $t < +\infty$. By analogue of Dynkin formula from (19.3) for function $V(\zeta)$ we obtain:

$$\hat{E}_\zeta V(\hat{\zeta}(t)) - V(\zeta) \leq E_y V(\zeta(t \wedge \xi_m)) - V(\zeta) = E_\zeta \int_T^{t\wedge\xi_m} L_m V(\zeta(s))ds \leq$$

$$-\alpha E_\zeta \int_T^{t\wedge\xi_m} V(\zeta(s))ds - \alpha \hat{E}_\zeta \int_T^t V(\hat{z}eta(s))ds,$$

where $V(\zeta) := V(z, x, T)$. From here and Gronwall-Bellman inequality we obtain:

$$\hat{E}_\zeta V(\zeta(T)) \leq V(\zeta) \exp\{-\alpha T\},$$

$$\hat{E}_\zeta V(\zeta(t)) \leq V(\zeta),$$

and

$$\hat{E}_\zeta V(\zeta(t)) \to V(\zeta),$$

if $t \to T$. In such a way, process $V(\hat{\zeta}(t))$ is nonnegative super-martingale. From Kolmogorov-Doob inequality and the above mentioned inequality we obtain:

$$P_{z,y}\{\sup_{T \leq t < +\infty} V(\zeta(t)) \geq \lambda\} \leq \frac{\hat{E}_\zeta V(\zeta(T))}{\lambda} \leq \frac{V(\zeta)\exp\{-\alpha T\}}{\lambda}.$$

It is obvious that

$$P_{z,y}\{\sup_{T \leq t < +\infty} |V(\zeta(t)) - V(\zeta(t))| > 0\} = P_\zeta\{\sup_{T \leq t < +\infty} V(\zeta(t)) \geq m\} \leq \frac{V(\zeta)}{m}.$$

From these estimations and from the inclusion of sets (see the proof of Theorem 2 we obtain the proof of Theorem 3. □

7. STOCHASTIC OPTIMAL CONTROL OF SEMI-MARKOV RISK PROCESSES

Let U be some control space and $u(t)$ be a control process, $u(t) \in U$. Controlled semi-Markov risk processes are defined by the solution of the following equation:

$$z(t) = z + \int_T^t v(z(s), x(s), u(s))ds - \sum_{k=1}^{\nu(t-T)} a(x_k). \tag{19.7}$$

The problem is to find such $u(t)$ that to minimize the following cost functional:

$$C(z, y, u) := E_{z,y}^u \int_T^{\tau} k(z(s), y(s), u(s))ds + E_{z,y}^u b(z(T), y(T)),$$

where τ is a some stopping time. Let

$$C_0(z, y) := \min_{u \in U} C(z, y, u),$$

Theorem 4 *(Bellman equation for semi-Markov risk process) Function $C_0(z, y)$ is the solution of the following equation*

$$\inf_{u \in U} [L^u C_0(z, y) + k(z, y, u)] = 0,$$

$$C_0(z, y) = b(z, y)$$

on the boundary ∂G, where $G \in \mathbf{R} \times \Xi \times \mathbf{R}_+$, and

$$L^u C_0(z, y) = \hat{Q} C_0 + v(z, x, u) dC_0/dz +$$

$$+\lambda(x)[PC_0(z - a(x), y) - C_0(z, y)].$$

Proof We apply the boundary value problem (19.4) to semi-Markov risk process $z(t)$ in (19.7) with operator L^u. □

Construction of stochastic optimal control for semi-Markov risk processes.

Theorem 5 *Let $V(z, y)$ be a nonnegative and continuous function on $R \times X \times R_+$ such that:*

$$V(z, y) = b(z, y)$$

on ∂G;

$$V(z, y) \in Domain(L^u);$$

$$L^u V(z,y) = -k(z,y,u)$$

with positive bounded continuous function k. Then there exists the sequence of stopping times τ_n such that $E^u_{z,y}\tau_n < +\infty$, τ_n converges to τ_u, and function $V(z,y)$ is defined as follows:

$$V(z,y) = C(z,y,u) = E^u_{z,y}\int_T^{\tau_u} k(z(s),y(s),u(s))ds + E^u_{z,y} b(z(\tau_u),y(\tau_u)).$$

Proof Let us put

$$\tau_n := \min\{n; \tau_u; \inf\{t : V(z(t),y(t)) \geq n\}$$

Then τ_n satisfy the conditions of Theorem 4 and

$$\lim_{n\to\infty} V(z(\tau_n),x(\tau_n),\gamma(\tau_n)) \to V(z(\tau_u),x(\tau_u),\gamma(\tau_u)).$$

From the analogue of Dynkin formula for semi-Markov risk processes with operator L^u and condition for function $V(z,y)$ we have:

$$V(z,x,t) = E^u_{z,x,t} V(z(\tau_n),x(\tau_n),\gamma(\tau_n)) - E^u_{z,x,t}\int_t^{\tau_u} k(z(s),x(s),\gamma(s))ds.$$

From here and boundary condition for function $V(z,y)$ we obtain under $n \to +\infty$:

$$V(z,x,t) = C(z,x,t,u) = E^u_{z,x,t} b(z(\tau_u),x(\tau_u),\gamma(\tau_u)) + E^u_{z,x,t}\int_t^{\tau_u} k(z(s),x(s),\gamma(s))ds,$$

and Theorem 5 is proved. □

Theorem 6 *Consider the control processes $u, v \in U$, and $\tau_w < +\infty$, where $w = u \vee v$. Let also the function $V(z,y)$ to be defined as in Theorem 4, and*

$$L^u V = -k(z,y,u)$$

and

$$L^v V \geq -k(z,y,v).$$

Then $C(z,y,u) \leq C(z,y,v)$.

Proof Let us define the following sequences:

$$\tau_n^\omega := \min\{n; \tau_\omega; \inf\{t : V(z(\tau_n^\omega),x(\tau_n^\omega),\gamma(\tau_n^\omega)) \geq n\}\}, \quad \omega = u \vee v.$$

They satisfy the conditions of Theorem 5 for every u and v. By the analogue of Dynkin formula for semi-Markov risk processes with operator L^ω we have:

$$V(z,x,t) = E^\omega_{z,x,t} V(z(\tau_n^\omega),x(\tau_n^\omega),\gamma(\tau_n^\omega)) - E^\omega_{z,x,t}\int_t^{\tau_n^\omega} L^\omega V(z(s),x(s),\gamma(s))ds,$$

where $\omega = u \wedge v$; τ_n^ω are defined above, $L^\omega = L^u \vee L^v$. From here and the condition of the Theorem 5 we obtain:

$$C_n(z,x,t,u) = E^\omega_{z,x,t} V(z(\tau_n^u), x(\tau_n^u), \gamma(\tau_n^u)) + E^u_{z,x,t} \int_t^{\tau_n^u} L^u V(z(s), x(s), \gamma(s))ds,$$

if $\omega = u$ and $L^\omega = L^u$. From another side,

$$C_n(z,x,t,v) \leq E^v_{z,x,t} V(z(\tau_n^v), x(\tau_n^v), \gamma(\tau_n^v)) + E^v_{z,x,t} \int_t^{\tau_n^v} L^v V(z(s), x(s), \gamma(s))ds,$$

if $\omega = v$ and $L^\omega = L^v$.

Passing to the limits in the last two expressions under $n \to +\infty$ we finally obtain:

$$\lim_{n\to\infty} C_n(z,x,t,u) = C(z,x,t,u) \leq C(z,x,t,v) := \lim_{n\to\infty} C_n(z,x,t,v),$$

and Theorem 6 is proved. □

Remarks Semi-Markov risk processes in series scheme and their approximations by averaged, diffusion and normal deviations processes, and ruin probabilities for later have been studied in [5]. Analogue of Dynkin formula and boundary value problems for multiplicative operator functionals and random evolutions of Markov and semi-Markov processes have been obtained in [2] and [6], respectively. Optimal stochastic control of Markov and semi-Markov random evolutions and respected evolutionary stochastic systems has been investigated in [3]. Some of the presented in the paper results are announced in [7].

References

[1] Embrechts, P. and Kluppelberg, C. (1993) Some aspects of Insurance Mathematics. *Theory Probability and its Applications*, **38**, No. 2, 374-416.

[2] Swishchuk, A.V. (1993) *The Analogue of Dynkin formula and boundary value problems for multiplicative operator functionals of Markov processes*, Institute of Mathematics, Kiev, Ukraine, Preprint 93.44, 17 pp.

[3] Swishchuk, A.V. (1994) Control of evolutionary stochastic systems, Proceed. In *Transactions of the 12th Prague Conference on information theory, statistical decision functions and random processes*, 235-238, Academy of Sciences of the Czech Republic, Prague.

[4] Korolyuk, V.S. and Swishchuk, A.V. (1995) *Evolution of Systems in Random Media*, CRC Press, Boca Raton, USA.

[5] Swishchuk, A.V. (1996) *Approximations and ruin probabilities for semi-Markov risk processes*, Institute of Mathematics, Kiev.

[6] Swishchuk, A.V. (1997) *Random Evolutions and Their Applications*, Kluwer Academic Publishers, Dordrecht.

[7] Swishchuk, A.V. and Goncharova, S.Y. (1998) Stability and control of semi-Markov risk processes in insurance mathematics, In *Abstracts of Communications of the 22nd European Meeting of Statisticians & 7th Vilnius Conference on Probability Theory & Mathematical Statistics*, 429, TEV, Vilnius.

[8] Kushner, H.J. (1967) *Stochastic stability and control*, Academic Press, New York-London.

Chapter 20

OPTION PRICING WITH SEMI-MARKOV VOLATILITY

Jacques Janssen*
Université Libre de Bruxelles, Belgium

Raimondo Manca
Universitá di Chieti, Italy

Ernesto Volpe
University "La Sapienza", Italy

Abstract We present a new extension of the fundamental Black-Scholes formula [1] in stochastic finance with the introduction of SM evolution for the volatility of an initial Black-Scholes model. The construction of this new model starts from the classical Cox-Rubinstein model [2] with one period.

Keywords: Black-Scholes formula, risk neutral measure, stochastic volatility, Markov chain, semi-Markov process, European call option pricing.

1. INTRODUCTION

Since the presentation of the Black-Scholes-Samuelson model [1] for the evolution of an asset, a lot of extensions of this model have been presented in the same framework of Itô calculus and the use of the absence of arbitrage possibility. This last assumption is equivalent to the existence of a new measure so that the present value, computed with the risk-less interest rate, is a martingale. Therefore, the pricing of

*Work performed under a MURST grant

derivative products involving such an underlying asset is equivalent to the computation of the expectation of the present value of the value of the option at exercise time.

One of the restrictions of the Black-Scholes model is the assumption of constant volatility upon time and there are a lot of attempts to relax this condition as for example in the model of [3] where the concept of stochastic volatility is introduced. Nevertheless, to our knowledge, in practice, no generalized model really supplants the classical Black-Scholes model.

Recently [8] develops a new model extending the Black and Scholes formula in presence of Markov or semi-Markov environments.

In this paper we develop another type of models. More precisely we present new semi-Markov (in short SM) models for the evolution of the volatility of the underlying asset.

In fact, the SM model presented here supposes a type of SM evolution for the volatility of an initial Black-Scholes model presented for the first time in an oral communication in the ETH Zurich (1995) by J. Janssen and in a different approach by E. Cinlar in an oral communication at the First Euro-Japanese Workshop on Stochastic Risk Modelling for Finance, Insurance, Production and Reliability, Brussels, September 1998. We show that such a type of model leads to a generalization of the classical Black- Scholes formula for the pricing of european calls.

For numerical applications, we refer to [9] and [7].

2. THE JMC SEMI-MARKOV MODEL IN STOCHASTIC FINANCE

The semi-Markov extension of the Black and Scholes model supposes a type of SM evolution for the volatility of an initial Black-Scholes model presented for the first time in an oral communication in the ETH Zurich (1995) and at the 8-th ASMDA conference in Capri, in June 1996 [5] and, more recently, in a different approach by E. Cinlar in an oral communication in Brussels (September 1998), at the First Euro-Japanese Workshop on Stochastic Risk Modelling for Finance, Insurance, Production and Reliability, Brussels, September 1998.

Hereby, we present our initial model of 1995 close to the presentation of Cinlar but he gives the formula for the pricing of a call option using the Markov renewal theory, we call this model the JMC (Janssen-Manca-Cinlar) semi-Markov model.

2.1 THE JMC SEMI-MARKOV MODEL (1995, 1998)

Let us consider a two-dimensional positive $(J-X)$ process [4] of kernel $\mathbf{Q}$ with as state space :

$$I = \{1, \ldots, m\}.$$

This means that on the probability space $(\Omega, \Im, P)$, we define the three-dimensional process

$$((J_n, (X_n, \sigma_n)), n \geq 0)$$

with :

$$J_n \in \mathbb{N}, \ (X_n, \sigma_n) \in \mathbb{R}^+ \times \mathbb{R}^+,$$

such that :

$$\begin{aligned} \mathrm{P}[X_n \leq x, \sigma_n \leq \sigma, J_n = j \mid (J_k, (X_k, \sigma_k)), k = 1, \ldots, n-1] = \\ = Q_{J_{n-1}j}(x, \sigma), \ p.s. \end{aligned}$$

We know that the $Q_{ij}(x, \sigma), \ i, j \in I$ can be factorized in such a way that :

$$Q_{ij}(x, \sigma) = p_{ij} F_{ij}(x, \sigma)$$

where :

$$p_{ij} = \mathrm{P}[J_n = j \mid (J_k, k \leq n-1, J_{n-1} = i]$$

$$F_{ij}(x, \sigma) = \mathrm{P}[X_n \leq x, \sigma_n \leq \sigma \mid (J_k, (X_k, \sigma_k)), k \leq n-1, Jn-1 = i].$$

We also introduce the following r.v. :

$$\begin{aligned} T_n &= X_1 + \cdots + X_n & n \geq 0 \\ N(t) &= \sup\{n : T_n \leq t\} & t \geq 0 \\ Z(t) &= I_{N(t)} & t \geq 0 \end{aligned}$$

The transition probability for the process $Z = (Z(t), t \geq 0)$ is designed by :

$$P_{ij}(t) = \mathrm{P}[Z(t) = j \mid Z(t) = i].$$

The stochastic processes $(N(t), t \in \mathbb{R}^+)$, $(Z(t), t \in \mathbb{R}^+)$ are respectively called the Markov renewal counting and the semi-Markov processes.

To give the financial interpretation of our model, let us define on the probability space $(\Omega, \Im, P)$, the following filtration $\Im = (\Im_t, t \in \mathbb{R}^+)$

$$\Im_t = \sigma((J_n, (X_n, \sigma_n)), n \leq N(t)).$$

Given $\Im_t$, let us consider the random time interval $[T_{N(t)}, T_{N(t)+1}]$ on which we define the new stochastic process $(S(t), t \in \mathbb{R}^+)$, representing the value of the considered financial asset, as solution of the stochastic differential equation :

$$\begin{aligned} &\frac{dS}{S(t')} = \mu_{J_{N(t)}J_{N(t)+1}} dt' + \sigma_{J_{N(t)}J_{N(t)+1}} dW_{J_{N(t)}J_{N(t)+1}}(t' - T_{N(t)}), \\ &t' \in [T_{N(t)}, T_{N(t)+1}], \; S(T_{N(t)}) = S(T_{N(t)}-), \end{aligned} \tag{20.1}$$

where the process $(W_{J_{N(t)}J_{N(t)+1}}(t'), t' \geq 0)$ is a standard Brownian motion on $[T_{N(t)}, T_{N(t)+1}]$ defined on the basic probability space stochastically independent of $(J_{N(t)}, X_{N(t)})$.

This model has the following financial meaning : at $t = 0$, the asset starts from the known initial value S_0, the known initial J-state J_0 representing the state of the initial economical and financial environment. On the time interval X_1, the asset has the random volatility σ_1 and has as stochastic dynamics the SDE (20.1) with $t = 0$; at time X_1, the J process has a transition to state J_1 and on the time interval $[T_1, T_2)$, the asset has the random volatility σ_2 and has as stochastic dynamics the SDE (20.1) with $N(t) = 1$ and so on ...

We always define $X_0 = 0$, a.s. So, it is now clear that we have in fact a perturbed Black-Scholes model due to this random change of volatility; Note that this model is quite general as, indeed, we have a random volatility on each time interval $[T_{N(t)}, T_{N(t)+1}]$. Of course for $m = 1$, we get back the classical Black-Scholes-Samuelson model for the description of an asset.

2.2 THE EXPLICIT EXPRESSION OF S(T)

Given $J_{N(t)}, J_{N(t)+1}$, the Itô calculus gives the solution of the SDE (20.1) :

$$\begin{aligned} &S(t') = S_{N(t)} e^{(\mu_{J_{N(t)}J_{N(t)+1}} - \frac{\sigma^2_{J_{N(t)}J_{N(t)+1}}}{2})t'} e^{\sigma_{J_{N(t)}J_{N(t)+1}} W(t' - T_{N(t)})}, \\ &t' \in [T_{N(t)}, T_{N(t)+1}] \end{aligned} \tag{20.2}$$

Starting from the state S_0 at time 0 and given a scenario for the economical and financial environment $(J_0, J_1, \ldots, J_n, \ldots)$, this expression gives the explicit form of the trajectories of the process $(S(t), t \geq 0)$. From relation (20.2), we get that for

$$(J_0, X_0, J_1, X_1, \ldots, J_{N(t)}, X_{N(t)}, J_{N(t)+1}, X_{N(t)+1}),$$

given :

$$\ln \frac{S(t')}{S_{N(t)}} = (\mu_{J_{N(t)}J_{N(t)+1}} - \frac{\sigma^2_{J_{N(t)}J_{N(t)+1}}}{2})t' + \sigma_{J_{N(t)}J_{N(t)+1}} W(t' - T_{N(t)}),$$
$$t' \in [T_{N(t)}, T_{N(t)+1}]$$

so that for $t' \in [T_{N(t)}, T_{N(t)+1}]$:

$$\ln \frac{S(t')}{S_{N(t)}} \prec N((\mu_{J_{N(t)}J_{N(t)+1}} - \frac{\sigma^2_{J_{N(t)}J_{N(t)+1}}}{2})(t' - T_{N(t)}), \sigma^2_{J_{N(t)}J_{N(t)+1}}(t' - T_{N(t)}))$$

$$E[\frac{S(t')}{S_{N(t)}} \mid \Im_t, J_{N(t)+1}] = e^{\mu_{J_{N(t)}J_{N(t)+1}}(t' - T_{N(t)})}$$

$$var\left[\frac{S(t')}{S_{N(t)}} \mid \Im_t, J_{N(t)+1}\right] = e^{2\mu_{J_{N(t)}J_{N(t)+1}}}(e^{\sigma^2_{J_{N(t)}J_{N(t)+1}}(t' - T_{N(t)})} - 1)$$

Let us suppose that the random variables

$$J_0, X_0, J_1, X_1, \ldots, J_{N(t)}, X_{N(t)}, J_{N(t)+1}, X_{N(t)+1}$$

are given; it follows that the conditional distribution function of $\frac{S(t')}{S_0}$ is a lognormal distribution, i.e. :

$$\begin{aligned} &\ln \tfrac{S(t)}{S_0} \prec N(\mu_{J_0 J_1} X_1 + \cdots + \mu_{J_{N(t)}J_{N(t)+1}}(t - T_{N(t)}), \\ &\sigma^2_{J_0 J_1} X_1 + \cdots + \sigma^2_{J_{N(t)}J_{N(t)+1}}(t - T_{N(t)})) \end{aligned} \tag{20.3}$$

3. CALL OPTION PRICING

Now to get a useful model, let us proceed as in [8] : for a fixed t, we assume that all the parameters μ, σ only depend on J_0, $J_{N(t)}$, $J_{N(t)+1}$ and t represented by

$$\mu_{J_0 J_{N(t)} J_{N(t)+1}}, \sigma_{J_0 J_{N(t)} J_{N(t)+1}}$$

so that from relation (20.3):

$$\ln \frac{S(t)}{S_0} \prec N((\mu_{J_0 J_{N(t)} J_{N(t)+1}} - \frac{\sigma^2_{J_0 J_{N(t)} J_{N(t)+1}}}{2})t, \sigma^2_{J_0 J_{N(t)} J_{N(t)+1}} t).$$

Of course, we can always simplify our basic assumption by suppressing the dependence with respect to $J_{N(t)+1}$ and even to $J_{N(t)}$. Nevertheless, we think that the dependence from the future environment state $J_{N(t)+1}$ is quite important as it gives the possibility for the first time to modelize the stochastic asset evolution taking into account of this anticipation of this next future state. Let us consider now an European call option with t as maturity time, K as exercise price that we must price at time 0. If we want to assume that there is no arbitrage possibility, we must impose that

$$\mu_{J_0 J_{N(t)} J_{N(t)+1}} = \delta_{J_0 J_{N(t)} J_{N(t)+1}}$$

where $\delta_{J_0 J_{N(t)} J_{N(t)+1}}$ represents the equivalent instantaneous riskless return on $[0, t]$ given J_0, $J_{N(t)}$, $J_{N(t)+1}$.

Doing so, we will use the risk neutral measure under which the forward value of the asset is a martingale; otherwise we work with the initial "physical" measure more appropriate for insurance than for finance. Knowing J_0, $J_{N(t)}$, $J_{N(t)+1}$ and working with the risk neutral measure, we can compute the value of the call at time 0 using the classical Black-Scholes formula :

$$\begin{aligned}
C_{J_0 J_{N(t)} J_{N(t)+1}}(S_0, t) &= S_0 \Phi(d_{J_0 J_{N(t)} J_{N(t)+1},1}) \\
&\quad - K r^{-t}_{J_0 J_{N(t)} J_{N(t)+1}} \Phi(d_{J_0 J_{N(t)} J_{N(t)+1},2}), \\
d_{J_0 J_{N(t)} J_{N(t)+1},1} &= \frac{\ln \frac{S_0}{K r^{-t}_{J_0 J_{N(t)} J_{N(t)+1}}}}{\sigma_{J_0 J_{N(t)} J_{N(t)+1}} \sqrt{t}} + \frac{1}{2} \sigma_{J_0 J_{N(t)} J_{N(t)+1}} \sqrt{t}, \\
d_{J_0 J_{N(t)} J_{N(t)+1},2} &= d_{J_0 J_{N(t)} J_{N(t)+1},1} - \sigma_{J_0 J_{N(t)} J_{N(t)+1}} \sqrt{t}, \\
r_{J_0 J_{N(t)} J_{N(t)+1}} &= e^{\delta_{J_0 J_{N(t)} J_{N(t)+1}}}.
\end{aligned}$$

To get the formula of the call only knowing S_0, J_0, we must use the following formula :

$$C_{J_0}(t) = E[C_{J_0 J_{N(t)} J_{N(t)+1}}(S_0, t) \mid J_0, S_0].$$

From the theory of semi-Markov processes, we get :

$$C_{J_0}(t) = \sum_{j \in I} \sum_{k \in I} P_{J_0 j}(t) p_{jk} C_{J_0 jk}(S_0 t) \tag{20.4}$$

If we have no information about the initial state J_0, we get of course the following formula :

$$C(t) = E[C_{J_0}(t)] = E[E[C_{J_0 J_{N(t)} J_{N(t)+1}}(S_0, t) \mid J_0, S_0]] \tag{20.5}$$

$$C(t) = \sum_{i \in I} a_i C_i(t)$$

Remark 1 *The case of non homogeneous semi-Markov processes ([6]) can also be considered.*

Remark 2 *Numerical treatments are possible (see [10]).*

4. STATIONARY OPTION PRICING FORMULA

In option pricing, it is a nonsense to let t tend towards $+\infty$; nevertheless, we can use the limit reasoning proposed by Janssen in a oral communication at the First Euro-Japanese Workshop on Stochastic models held in Brussels (1998) by supposing that on the time horizon $[0, t]$, the environment semi-Markov has more and more transitions in this finite time period. We can modelize this situation under the assumption that the conditional sojourn time means b_{ij}, $i, j \in I$ satisfy the following conditions

$$b_{ij} = \varepsilon \zeta_{ij},\ \varepsilon > 0$$

$$b_{ij} = E[X_n \mid J_{n-1} = i,\ J_n = j]$$

so that :

$$\eta_i = \sum_{j \in I} p_{ij} b_{ij} = \varepsilon \sum_{j \in I} p_{ij} \zeta_{ij} = \varepsilon \theta_i,\ i \in I$$

From the asymptotical theory of semi-Markov processes, we know that :

$$\lim_{\varepsilon \to 0} \mathrm{P}[J_{N(t)} = j,\ J_{N(t)+1} = k] = \frac{\pi_i p_{jk} \zeta_{jk}}{\sum_{l=1}^{m} \pi_l \theta_l},\ i, j \in I$$

where the vector $(\pi_1, \ldots, \pi_m)$ is the unique stationary distribution of the embedded Markov chain of matrix $\mathbf{P}$ supposed to be ergodic.

The new parameters ζ_{jk}, i, j, $k \in I$ represent factors expressing the proportionality of the sojourn in each environment state. Now the result (20.4) becomes :

$$C_{J_0}(t) = \sum_{j_i n I} \sum_{k \in I} \frac{\pi_j p_{jk} \zeta_{jk}}{\sum_{l=1}^{m} \pi_l \theta_l} C_{J_0 jk}(S_0, t)$$

Form (20.5), we get

$$C(t) = \sum_{i_i n I} a_i \sum_{j_i n I} \sum_{k \in I} \frac{\pi_j p_{jk} \zeta_{jk}}{\sum_{l=1}^{m} \pi_l \theta_l} C_{ijk}(S_0, t).$$

This last formula replaces the Black-Scholes formula without any a priori information at time 0 except of course the initial value of the asset S_0.

5. CONCLUSIONS

In conclusion, the new model proposed here extends the classical Black and Scholes formula in the case of the existence of an economical and financial environment modelized with a homogeneous semi-Markov process taking into account of this environment not only at time of pricing but also before and after the maturity date. This new family of Black-Scholes formulae seems to be more adapted to the reality particularly to take into account of the anticipations of the investor or of the consideration of stress scenario in the philosophy of the VaR approach.

References

[1] Black, F., Scholes, M. (1973), The pricing of options and corporate liabilities, *Journal of Political Economy*, 81, p. 637-654.

[2] Cox, J.C., and Rubinstein M. (1985), *Option Markets*, Prentice Hall, N.Y.

[3] Hull, J., White, A. (1987), The pricing of options on assets with stochastic volatilities. *Journal of Finance.* 42 p. 281-300.

[4] Janssen, J. (1969), Les Processus (J-X). *Cahiers du CERO.* 11 p. 151-161.

[5] Janssen, J. (1997), An extension of the mean semivariances portfolio models to include transaction cost and real characteristics of investiments. *Proceedings of the VIII Int. ASMDA Symposium, Anacapri.* Edited by J. Janssen and C. Lauro, vol.I p. 217-221.

[6] Janssen J., De Dominicis R. (1984), Finite non-homogeneous semi-Markov processes. *Insurance: Mathematics and Economics.* 3 p. 157-165.

[7] Janssen J., Manca R. (1999), Numerical solution for non-homogeneous semi-Markov processes in transient case, Proceedings of the second *Symposium on Semi-markov processes and their Applications* edited by J. Janssen and N. Limnios.

[8] Janssen J., Manca R. (1999), A new semi-Markov extension of the Black-Scholes formula in stochastic finance, to appear.

[9] Janssen J., Manca R., Acharki, R. (1999), Numerical solution for homogeneous and non- homogeneous semi-Markov processes extnesions of the Black-Scholes formula, to appear.

[10] Janssen J., Manca R., Acharki, R. (1999), Sensitivity analysis for the Markov and semi- markov extensions of the Black-Scholes formula, to appear.

[11] Janssen J., Manca R., De Medici G. (1996), Financial Operation Evaluation : a Semi-Markov Models for Pricing Derivative products, Proceedings of the *ASTIN 96 Conference.*

VI

CONTROLLED PROCESSES & MAINTENANCE

Chapter 21

APPLICATIONS OF SEMI-MARKOV PROCESSES IN RELIABILITY AND MAINTENANCE

Mohamed Abdel-Hameed
United Arab Emirates University, U.A.E.

Abstract In this paper we discuss applications of semi-Markov processes in reliability and maintenance. Concentrating, on optimal maintenance and replacement policies of devices subject to deterioration. Over time the deterioration process is assumed to be a semi-Markov process.

Keywords: Semi-Markov processes, optimal replacement policies, control-limit policies, optimal stopping, fixed point theorem.

1. INTRODUCTION AND SUMMARY

Feldman [6] considers the problem of determining the optimal replacement time for a system subject to a sequence of randomly occurring shocks; each shock causes some damage to the system. Any of the occurring shocks might cause the system to fail, and the probability of such a failure is a function of the sum of the magnitudes of the damage caused from all previous shocks. When the system fails it is immediately replaced by a similar system.

The accumulated damage is assumed to be an increasing semi-Markov process, with state space E. The semi-Markov presentation allows the use of Markov renewal equations in formulating the cost equation. Using standard results for solving Markov renewal equations, he obtains the optimal replacement level.

A control-limit policy is one that there is a fixed real value γ such that the system is replaced once the accumulated damage is greater than γ or when a failure occurs, whichever comes first. The cost of a replacement before failure is c_1, replacement at failure results in additional cost c_2. Among all such policies he obtains the optimal replacement time using the long-run average cost, that is, he finds the control-limit policy that minimizes the long-run average cost per unit of time.

Feldman [7] considers the same damage model described above with a different cost structure and optimality criterion. In this paper, the cost of replacement depends only on the state of the damage process at the time of replacement, the cost of replacing the system (before failure), when the damage level is x, is $f(x)$. The cost of replacing the system at failure is constant $f(\Delta)$, where Δ is a point not in the state space. He obtains the optimal replacement policy that minimizes the discounted cost per cycle.

Abdel-Hameed [2] treats the optimal maintenance of a system that is subject to deterioration, and the deterioration process is assumed to be non-stationary Lévy process. In [3] he treats the case where the deterioration problem is a pure jump Markov process, and the deterioration is monitored continuously. In [4] he treats the case where the deterioration is monitored periodically. Nakagawa [8] finds an optimal control-limit policy for a system subject to deterioration, the deterioration process is assumed to be a compound renewal process. Abdel-Hameed and Nakhi [5] consider the case where the system is subject to deterioration, the deterioration process is assumed to be an increasing right continuous semi-Markov process, withe state space $[0, \infty)$. The system has a random threshold and it fails once the deterioration exceeds the threshold. The system can be replaced before or at failure. Furthermore, the system is maintained at cost rate represented by a function g, defined on the state space of the deterioration process. The function g is assumed to be increasing and bounded. They find the optimal replacement and maintenance policy, among control-limit policies, using the total discounted cost and the long-run average criteria. Throughout we denote the set of non-negative real numbers by R_+ and the set of non-negative integers by N.

In section 2 we discuss some of the results mentioned above, in detail. In Section 3 we discuss the work of Abdel-Hameed and Nakhi [5], using the total discounted cost criterion. In Section 4 we discuss similar results using the long-run average cost criterion.

2. BASIC RESULTS

Feldman [7] defines $\hat{Z} = (\hat{Z}_t)_{t \geq 0}$ be the semi-Markov process describing the total accumulated damage. The embedded Markov renewal process is denoted by $(\hat{X}, \hat{T}) = (\hat{X}_n, \hat{T}_n)_{n \geq 0}$. Let $\hat{Q} = \{\hat{Q}(x, y, t); x, y \in$

E, $t \geq 0\}$ be the semi-Markov kernel associated with (X, T). That is, for $n \geq 1$,

$$\hat{Q}(x, y, t) = P_x\{\hat{X}_n \leq x, \hat{T}_n \leq t\}.$$

The system can fail only at times of occurrences of shocks. The random variable L denotes the n such that T_n is the failure time of the system. Its associated probability law is given by

$$P_x\{L > n \mid \hat{X}_0, \hat{X}_1, \ldots, \hat{X}_n\} = \prod_{i=1}^{n} g(\hat{X}_i),$$

for some non-increasing function g. It follows that $g(x)$ is the probability that the system does not fail at time $\hat{T}_n$ given that $\hat{X}_n = x$, and the system was working at time $\hat{T}_{n-1}$. In this model he assumes that the system cannot fail initially, even if a shock is realized during installation. The time of failure is defined by

$$\zeta = \hat{T}_L.$$

The killed process Z and its imbedded Markov renewal process (X_n, T_n) are defined by

$$\begin{aligned} Z_t &= \hat{Z}_t \text{ if } t < \zeta, \\ &= \Delta \text{ if } t \geq \zeta; \\ Z_n &= \hat{Z}_n \text{ if } n < L, \\ &= \Delta \text{ if } n \geq L; \\ T_n &= \hat{T}_n \text{ if } n \leq L, \\ &= \infty \text{ if } n > L. \end{aligned}$$

Definition 1 *For each n, let $\Im_n$ be the sigma algebra generated by T_n, and define $\Re = \{N \leq L : N$ is an $(\Im_n)$ stopping time$\}$. The optimal stopping time $N^* \in \Re$, is the one minimizing the discounted cost $E_x[\exp(-\alpha T_N) f(X_N)]$ over all stopping times in $\Re$. The minimal discounted cost, per a replacement cycle, $E_x[\exp(-\alpha T_{N^*}) f(X_{N^*})]$, is denoted by $v(x)$.*

Definition 2 *A function h defined on $\mathbf{E} \cup \Delta$ is called $\alpha -$ excessive if*

$$h(x) \geq E_x[\exp(-\alpha T_1) f(X_1)] \text{ for each } x \in E.$$

Using optimal stopping results, the following are the main two results in his paper :

Theorem 1 *The optimal function v in Definition 1, is the smallest $\alpha-$ excessive function that is smaller than or equal to the cost function f.*

Theorem 2 *The optimal replacement time N^* is given by*

$$N^* = \inf\{n : v(X_n) = f(X_n)\}.$$

To insure that the optimal replacement is a control-limit policy the following assumptions are imposed :

Assumption 1 *The cost function f is non-decreasing, and convex. The function $x \to P_x\{X_1 - x \geq y, T_1 < t\}$ is non-decreasing and right continuous for $x \in E$ and $t \geq 0$.*

From the above assumptions, the following theorem gives the form of optimal control limit policy.

Theorem 3 *Assume that Assumption 1 is satisfied, define*

$$\gamma = \inf\{x \in E : v(x) = f(x)\}.$$

Then the optimal replacement time N^* is given by

$$N^* = \inf\{n : X_n \geq \gamma\}.$$

Siedersleben, [10] generalizes the results obtained in Feldman [7], using a different approach. He considers the replacement problem where the devices subject to deterioration. The deterioration process is non-decreasing over time and has state space R_+. The device has a fixed threshold c, and it fails once the deterioration exceeds or equal to the threshold c. The system is observed at random times $\{T_n, n \geq 0, T_0 = 0\}$ and based on the state of the system X_n, if the state of the system is less than c, we decide either to replace the system and incur a cost $m(X_n)$ or leave the system alone and incur no cost. Of course if the system has failed at inspection, then replacement is mandatory. The sequence (T_n, X_n) is assumed to form a Markov renewal process. Using semi-Markov decision terminology we have : the action space $A = \{0, 1\}$, where action 0 stands for no replacement and 1 for replacement, respectively. By a replacement policy, π, we mean decision policy that specifies for each $x \in R_+$, an action $\pi(x) \in A$. The total discounted cost, over the infinite horizon, corresponding to such a policy is given by

$$V_\pi(x) = E_x[\sum_{n=0}^{\infty} e^{-\alpha T_n} C_\pi(X_n)]$$

where

$$C_\pi(x) = \begin{cases} 0 & \text{if } \pi(x) = 0, \\ m(x) & \text{if } \pi(x) = 1. \end{cases}$$

A replacement policy π^* is said to be optimal if, for all $x \in R_+$,

$$V_{\pi^*}(x) = \inf_\pi V_\pi(x).$$

Throughout we denote $V_{\pi^*}(x)$ by $V^*(x)$.

For any replacement policy π, we define

$$\tau_\pi = \inf\{n : \pi(X_n) = 1\}$$

and for $n \geq 2$, we let

$$\begin{aligned} \tau_\pi^n &= \inf\{m > \tau_\pi^{n-1} : \pi(X_m) = 1\}, \\ \tau_\pi^1 &= \tau_\pi. \end{aligned}$$

The sequence $(\tau_\pi^n, n \geq 1)$ represents the indices of the times of successive replacement, corresponding to the policy π. We note that, over the infinite horizon, the deterioration process is a regenerative process with regeneration points $(\tau_\pi^n, n \geq 1)$. It follows that

$$\begin{aligned} V_\pi(x) &= E_x[\sum_{n=1}^{\infty} \exp(-\alpha T_{\tau_\pi^n}) m(X_{\tau_\pi^n})] \\ &= E_x\{\exp(-\alpha T_{\tau_\pi})[m(X_{\tau_\pi}) + V_\pi(0)]\}. \end{aligned}$$

From the regeneration property, it follows that

$$V_\pi(0)] = \frac{E_0\{\exp(-\alpha T_{\tau_\pi})[m(X_{\tau_\pi})\}}{1 - E_0\{\exp(-\alpha T_{\tau_\pi})}.$$

Note that if $\pi(x) = 1$, then $\tau_\pi = 0$, and

$$V_\pi(x) = m(x) + V_\pi(0).$$

However, $\pi(x) = 0$, then $\tau_\pi > 0$, and it can be shown that (upon first conditioning on the sigma algebra generated by T_1)

$$V_\pi(x) = E_x[e^{-\alpha T_1} V_\pi(X_1)].$$

His method of obtaining the optimal replacement policy is based on the *"Fixed Point Theorem"* for "Contraction Mappings". For completeness we now discuss this matter, in some details following Ross [9] pp. 126-128. Let $B(R_+)$ be the normed linear space of all bounded function with domain R_+, equipped with the sup norm $\|\|$.

Definition 3 *A mapping $T : B(R_+) \to B(R_+)$ is said to be a contraction mapping if there exists a real number $\beta < 1$, such that for each $u, v \in B(R_+)$,*

$$\| Tu - Tv \| \leq \beta \| u - v \| .$$

The *"Fixed Point Theorem"* states that : If T is contraction mapping on $B(R_+)$, then there exists a unique point $v^* \in B(R_+)$ such that

$$Tv^* = v^*.$$

Furthermore, for all $u \in B(R_+)$,

$$\lim_{n\to\infty} T^n u = v^*.$$

Using the *"Fixed Point Theorem"*, the following theorem follows in a manner similar to the proof of Theorems 6.1, and 6.3, p.p 121-126 of [9].

Theorem 4 *On $B(R_+)$, define the operator*

$$TV(x) = \begin{cases} \min\{m(x) + V^*(0), E_x[e^{-\alpha T_1} V^*(X_1)]\} & x < c \\ m(x) + V^*(0) & x \geq c. \end{cases}$$

Let $V^*(x)$ be the total discounted cost associated with the optimal policy. Then V^* is the fixed point of T. Furthermore, the stationary policy, which, when the process is in state x, selects the action satisfying the right hand side of the above equation, is optimal.

To show that the optimal policy is a control-limit policy he imposes the following additional assumptions :

Assumption 2
A1. $V^*(0) \geq 0$,
A2. $x \to E_x e^{-\alpha T_1}$ *is a non-decreasing function in x,*
A3. $x \to E_x e^{-\alpha T_1} m(X_1) - m(x)$ *is a non-decreasing function in x.*

Define

$$B = \{x : m(x) + V^*(0) \leq E_x[e^{-\alpha T_1}\{m(X_1) + V^*(0)\}].$$

Using Assumption 2 above it follows that the set B is closed in the sense that, for all $n \geq 0$,

$$P\{X_{n+1} \notin B \mid X_n \in B\} = 0.$$

Let

$$\pi^*(x) = I_B(x).$$

The following is the main Theorem.

Theorem 5 *Let the replacement time $\pi^*(x)$ be defined as above, then $\pi^*(x)$ is the optimal limit control policy.*

We note that the difference $E_x[e^{-\alpha T_1}\{m(X_1)+V^*(0)\}]-[m(x)+V^*(0)]$ is the additional expected cost resulting from continuing one extra step. And the optimal control-limit policy is the first time it does not pay to continue further. This is a form of what is known as *One-step Look Ahead* Rule. For further examinations of this type of stopping rule the reader is referred to Abdel-Hameed [1].

Abdel-Hameed and Nakhi [5] consider the case where a system is subject to deterioration, the deterioration process is assumed to be an increasing right continuous semi-Markov process, with state space $[0,\infty)$. The system has a random threshold and it fails once the deterioration exceeds the threshold . The system can be replaced before or at failure, the cost of replacement before failure is c_1, while the cost of replacement at failure is c_1+c_2, c_1 and $c_2 \geq 0$. Furthermore, the system is maintained at cost rate represented by a function g, defined on the state space of the deterioration process. The function g is assumed to be increasing and bounded. As in Feldman[7] we let $\hat{Z} = \{\hat{Z}_t; t \in R_+\}$ be the semi-Markov process denoting the deterioration process, the process $\hat{Z}$ has state space R_+. Assume that the process $\hat{Z}$ is right continuous and non-decreasing, $\lim_{t\to\infty} \hat{Z}_t = \infty$, and has embedded Markov renewal process $(\hat{X},\hat{T}) = (\hat{X}_n, \hat{T}_n, n \in N)$, where for n$\in N, \hat{X}_n = \hat{Z}(\hat{T}_n)$. For $x, t \in R_+$ and $A \subset R+$, let $\hat{Q}(x,t,A)$ and $\hat{R}(x,t,A)$ be the semi-Markov kernel and Markov renewal kernel associated with $(\hat{X},\hat{T})$, respectively. Thus $\hat{Q}(x,t,A) = P\{\hat{X}_{n+1} \in A, \hat{T}_{n+1} - \hat{T}_n \leq t \mid \hat{X}_n = x\}$, and $\hat{R}(x,t,A) = \sum \hat{Q}^n(x,t,A)$, where the sum is taken over N. Assume that the device has a the device has a threshold Y, with right tail probability $\bar{G}$, and the device fails once the deterioration exceeds or equal to the threshold. Let ζ denote the failure time, that is $\zeta = \inf\{t : \hat{Z}_t \geq Y\}$. For a given deterioration level x and t$\geq$ 0, let $\bar{H}_z(t)$ be the survival probability given $\hat{Z}_0 = x$, that is $\bar{H}_x(t) = P_x\{\zeta > t\}$, and $H_x = 1 - \bar{H}_x$. Let $L = \{n : \hat{T}_n = \zeta\}$, denote the n such that T_n is the failure time of the system. Thus, the failure time of the system is given by $\zeta = \hat{T}_L$. Let Δ be a distinct point not in R_+ and define $Z_t = \hat{Z}_t, t < \zeta$ and Δ otherwise. Also, define $X_n^{(1)} = \hat{X}_n, n < L, \Delta$ otherwise. Furthermore, define $T^{(1)}{}_n = \hat{T}_n, n \leq L$, ∞ otherwise.

Let $\hat{Z} = \{\hat{Z}_t; t \in R_+\}$ be the semi-Markov process denoting the deterioration. Let $Z^{(1)} = \{Z_t^{(1)},\ t < \zeta\}$ be the cumulative deterioration before the first failure, and let $(X^{(1)}, T^{(1)})$ be the associated renewal

Markov process. If $Q^{(1)}$is the semi-Markov kernel of $(X^{(1)}, T^{(1)})$, then we have the following result (see Proposition 2.1 of [5]) :

Proposition 1 *For $x, t \geq 0$ and $A \subset R_+$*

$$Q^{(1)}(x, A, t) = \int_A \hat{Q}(x, A, t) \frac{\bar{G}(y)}{\bar{G}(x)}.$$

Let $R^{(1)}$ denote the Markov renewal kernel associated with $Q^{(1)}$ and $\hat{U}_\alpha, U_\alpha^{(1)}$ be the $\alpha-$potentials corresponding to $\hat{Z}$ and $Z^{(1)}$, respectively. Then we have the following proposition (see Proposition 2.2 of [5]) :

Proposition 2 *For $A \subset R_+$, we have*

$$\begin{aligned} R^{(1)}(x, A, t) &= \int_A \hat{R}(x, dy, t) \frac{\bar{G}(y)}{\bar{G}(x)}, \\ U^{(1)}(x, A, t) &= \int_A \hat{U}_\alpha(x, dy, t) \frac{\bar{G}(y)}{\bar{G}(x)}. \end{aligned}$$

Let $Z = (Z_t, t < \infty)$ be the process describing the cumulative damage over the infinite horizon, when replacements are made only at failure times, with imbedded Markov renewal (X, T) and associated semi-Markov kernel Q (see p. 144 [5]). For each $t \geq 0$, define $M(t) = \sum_{n=0}^{\infty} H_0^n(t)$, be the number of replacements in $[0, t)$. Let $P_t(x, A), P^{(1)}{}_t(x, A)$ be the transition kernels corresponding to Z and $Z^{(1)}$ respectively. Then we have the following proposition whose proof is found in page 145 of [5].

Proposition 3 *Let $P_t(x, A), P^{(1)}{}_t(x, A)$ be the transition kernels defined above. Then for each $t \geq 0, A \subset R_+$*

$$P_t(x, A) = P^{(1)}{}_t(x, A) + M * H_x * P^{(1)}{}_t(0, A).$$

Let $\hat{h}(z, t) = P_z\{\hat{T}_1 > t), h_\alpha(z)$ be the Laplace transform of $\hat{h}(z, t)$, $\hat{Q}_\alpha(z, A), Q_\alpha(z, A), Q_\alpha^{(1)}(z, A)$ and $R_\alpha^{(1)}(z, A)$ be the Laplace transform of $\hat{Q}(z, A, t), Q(z, A, t), Q^{(1)}(z, A, t)$ and $R^{(1)}(z, A, t)$ with respect to t, respectively. Then we have

Proposition 4 *For each z, t and $\alpha \geq 0$, we have*

$$\begin{aligned} \hat{h}(z, t) &= 1 - \hat{Q}(z, R_+, t), \\ \hat{h}_\alpha(z) &= \frac{1}{\alpha}[1 - \hat{Q}_\alpha(z, R_+)]. \end{aligned}$$

Let $h(z,t) = 1 - Q(z, R_+, t) = 1 - P_z(X_1^{(1)} \neq \Delta, T_1^{(1)} \leq t)$ and $h_\alpha(z)$ be its Laplace transform. The proof of the following proposition can be found in page 146 of [5].

Proposition 5 *Let* $T_y = \inf\{t : Z_t^{(1)} > y\}$, *then*

$$E_x(\exp(-\alpha T_y)) = 1 - \alpha \int_x^y R_\alpha^{(1)}(x, dz)\, \hat{h}_\alpha(z).$$

Furthermore, we have the following proposition (see Proposition 2.7 of [5]).

Proposition 6

$$P_x(T_y = \zeta) = \int_x^y R^{(1)}(x, dz, \infty)\, h(z, \infty).$$

Let $m(z) = E_z(\hat{T}_1)$, then we have the following

Proposition 7

$$E_z T_y = \int_x^y R^{(1)}(x, dz, \infty)\, m(z).$$

3. OPTIMAL REPLACEMENT UNDER THE EXPECTED TOTAL DISCOUNTED COST CRITERION

Consider the maintenance cost rate $g(z)$ and as the before, we consider replacement times of the form

$$T_y = \inf\{t : Z_t^{(1)} > y\} = \{T_n^{(1)} : X_n^{(1)} \geq y\}.$$

The following proposition gives the total discounted cost of maintenance until the first replacement (see Proposition 2.9 of [5], for a proof).

Proposition 8 *Assume that maintenance rate cost g is bounded. Then*

$$E_z \int_0^{T_y} e^{-\alpha t} g(Z_t) dt = \int_x^y R_\alpha^{(1)}(x, dz)\, \hat{h}_\alpha(z) g(z).$$

The following proposition gives the total maintenance cost (non-discounted), until the first replacement occurs.

Proposition 9 *Assume that the maintenance rate function g is bounded, and* $R^{(1)}(x, R_+, \infty) < \infty$. *Then we have*

$$\begin{aligned} E_x \int_0^{T_y} g(Z_t) dt &= \int_x^y R_\alpha^{(1)}(x, dz, \infty)\, m(z)\, g(z), \\ E_x \int_0^{\zeta} g(Z_t) dt &= \int_x^\infty R_\alpha^{(1)}(x, dz, \infty)\, m(z)\, g(z). \end{aligned}$$

Let the cumulative deterioration process over the infinite horizon Z, as defined before. Define

$$\tau^0 = \inf\{t : Z_t \geq y\}$$

and for $n \geq 1$,

$$\begin{aligned} \tau^n &= \inf\{t > \tau^{n-1} : Z_t \geq y\}, \\ W_n &= \tau^n - \tau^{n-1}. \end{aligned}$$

For the maintenance rate function g, define

$$g_\alpha^*(0) = \int_0^{\tau} e^{-\alpha t} g(Z_t) dt$$

and for $n \geq 1$,

$$\begin{aligned} g_\alpha^*(n) &= \int_{\tau^{n-1}}^{\tau^n} e^{-\alpha t} g(Z_t) dt, \\ g^*(n) &= \int_{\tau^{n-1}}^{\tau^n} g(Z_t) dt. \end{aligned}$$

We define the total discounted cost of maintenance, over the infinite horizon and the long run average cost of maintenance per a unit of time as

$$g_\alpha^*(x) = E_x \int_0^{\infty} e^{-\alpha t} g(Z_t) dt, \quad g^* = \lim_{t \to \infty} \frac{1}{t} \int_0^t g(Z_t) dt$$

Then we have the following theorem, whose proof can be found in page 151 of [5].

Theorem 6

$$\begin{aligned} g_\alpha^*(x) &= E_x[g_\alpha^*(0)] + \frac{E_x(e^{-\alpha\tau^0})\, E_0 g_\alpha^*(1)}{1 - E_0(e^{-\alpha W_1})}, \\ g^* &= \frac{E_0 g^*(1)}{E_0(W_1)}. \end{aligned}$$

From the above, we have the following main result of this section. The proof of this theorem can be found in page 154 of [4].

Theorem 7 *Assume that the measure $A \to R^{(1)}(x, A, \infty)$ is absolutely continuous with respect to the Lebesgue measure on R_+. Suppose that the following hold :*

1. *$z \to \hat{h}_\alpha(z)$ is non-decreasing and bounded,*

2. $z \to g(z)$ *is non-decreasing and bounded,*

3. $z \to [\hat{Q}_\alpha(z, R_+) - Q_\alpha^{(1)}(z, R_+)]$ *is non-decreasing.*

Then the optimal replacement level y^* *is the unique solution to*

$$c_1 = \int_0^y R^{(1)}(0, dz)[c_2(\frac{h_\alpha(z)}{h_\alpha(y)}\{\hat{Q}_\alpha(y, R_+) - Q_\alpha^{(1)}(y, R_+)\} - \{\hat{Q}_\alpha(z, R_+) - Q_\alpha^{(1)}(z, R_+)\}) + \hat{h}_\alpha(z)\,(g(y) - g(z))].$$

If no solution exists, then the optimal policy is to wait until failure for replacement.

4. OPTIMAL REPLACEMENT UNDER THE LONG RUN AVERAGE COST CRITERION.

Assume the same maintenance and replacements structure described before. In this section we discuss the form of the optimal maintenance and replacement using the long run average cost criterion. In this case, for a given control-limit policy, we denote the cost incurred in the interval $[0, T_y)$ by $C_0(y)$. It follows that

$$C_0(y) = \begin{cases} c_1 + \int_0^{T_y} g(Z_t)dt, & T_y < \zeta \\ (c_1 + c_2) + \int_0^{T_y} g(Z_t)dt & T_y = \zeta. \end{cases}$$

It follows that the long-run average cost of maintenance and replacement per a unit of time is given by

$$\psi(y) = \frac{E_0[C_0(y)]}{E_0(T_y)}.$$

From Proposition 8 we know that $E_z T_y = \int_x^y R^{(1)}(x, dz, \infty)\, m(z)$, and the following theorem follows (see p. 155 of [5]).

Theorem 8

$$\psi(y) = \frac{c_1 + \int_0^y R_\alpha^{(1)}(x, dz, \infty)\,(c_2 h(z, \infty) + m(z)g(z))}{\int_0^y R_\alpha^{(1)}(x, dz, \infty)\, m(z)}.$$

From the above we have the following main theorem in this section.

Theorem 9 *Assume that the measure* $A \to R^{(1)}(x, A, \infty)$ *is absolutely continuous with respect to the Lebesgue measure on* R_+. *Assume that the following hold :*

1. $z \to m(z)$ *is non-increasing,*

2. $z \to h(z, \infty)$ *is non-decreasing,*

3. $z \to g(z)$*is non-decreasing and bounded.*

Then the optimal replacement level z^* *is the unique solution of the integral equation*

$$c_1 = \int_0^y R^{(1)}(0, dz, \infty)[c_2\{\frac{m(u)}{m(z)}h(y, \infty) - h(z, \infty)\} + m(z)(g(y) - g(z))].$$

If no solution exists, then the optimal policy is to wait until failure for replacement.

References

[1] Abdel-Hameed, M.S. (1977) Optimally of the one-step look-ahead stopping times. *J. Appl. Prob.*, 14, 162-169.

[2] Abdel-Hameed, M.S. (1984) Life distribution properties of devices subject to a Levy wear process. *Math. Operat. Res.*, 9, 606-614.

[3] Abdel-Hameed, M.S. (1984) Life distribution properties of devices subject to a pure jump damage process. *J. Appl. Prob.*, 21, 816-825.

[4] Abdel-Hameed, M.S. (1987) Inspection and maintenance of systems subject to deterioration. *Advances in Applied Probability*, 19, 917-931.

[5] Abdel-Hameed, M.S, and Nakhi, Y. (1991) Optimal replacement and maintenance of systems subject to semi-Markov damage. *Stoch. Proc. Appl. 37*, 141-160.

[6] Feldman, R.M. (1976) Optimal replacement with semi-Markov shock models. *J. Appl. Prob.* 13, 108-117.

[7] Feldman, R.M. (1977) Optimal replacement with semi-Markov shock mod using discounted costs. *Math. Operat. Res.* 2, 78-90.

[8] Nakagawa, T. (1976) On a replacement problem of a cummulatrive damage model. *Operat. Res. Quart.* 27, 895-900

[9] Ross, S.M. (1970) *Applied Probability Models with Optimization Applications.* Holden Day, San Francisco.

[10] Siedersleben, J. (1981) Dynamically optimized replacement with a Markovian renewal process. *J. Appl. Prob.* 18, 641-651.

Chapter 22

CONTROLLED QUEUEING SYSTEMS WITH RECOVERY FUNCTIONS

Tadashi Dohi
Hiroshima University, Japan

Shunji Osaki
Hiroshima University, Japan

Naoto Kaio
Hiroshima Shudo University, Japan

Abstract Consider the $M/G/1$ queueing system with removable server subject to failure interruptions, where the failure occurs only during operating busy period. More precisely, jobs arrive according to the identical and independent Poisson stream and are processed with an identical and independent general service distribution. Two kinds of recovery functions from failures are analyzed in this paper. One is the retry function, the other is the repair function. In the retry model, after a system failure the system undergoes instantly the corrective repair and the interrupted job is reprocessed again after the system becomes as good as new. On the other hand, in the repair model, the minimal repair is executed for each failure and at the termination of the process of a job, the preventive repair is carried out again. In the framework of the T-policy for the $M/G/1$ queue, we formulate the relevant expected costs under two recovery functions, respectively, and derive the optimal timing strategies minimizing them. In numerical examples, we compare two strategies under respective recovery functions in terms of cost effectiveness and examine the dependence of model parameters in the recovery functions.

Keywords: $M/G/1$ queue, failure interruption, corrective maintenance, semi-Markov processes, performance comparison.

1. INTRODUCTION

Consider the $M/G/1$ queueing system with removable server. Suppose that jobs arrive according to the identical and independent Poisson stream and are processed with an identical and independent general service distribution. Since the server is removable, the timing to start processing jobs has to be determined by the decision maker. Yadin and Naor[1], Heyman[2], Sobel[3] and Bell[4] considered the operating strategy called *the N-policy.* In particular, Sobel[3] proved the optimality of the N-policy by applying the theory of Markov decision processes. In the N-policy, the decision maker turns the server *off* until the number of jobs in the buffer becomes the specified threshold N. When the number becomes N, then the server is turned *on* and jobs stored in the buffer are started processing. When the number of jobs in the buffer becomes zero, *i.e.* at the time when all jobs arrived are completed, the server is turned off and the same cycle is repeated again and again. Kimura[5] applied the diffusion approximation approach to represent the discounted cost function over an infinite time horizon and provided a tractable approximated N-policy. In this respect, Balachandran[6] and Balachandran and Tijms[7] proposed *the D-policy* in which the server is turned on when the cumulative service time on jobs stored in the buffer exceeds the specified threshold D. If the state of buffer can not be observed, however, both N-policy and D-policy will lose their effectiveness and such a situation can be often seen in practice. In other words, it seems to be rather strong that the time required for processing is known in advance. For this problem, Heyman[8] proposed *the T-policy* in which the server is turned on at the prespecified time T. Okamura, Dohi and Osaki[9] also derived approximately the optimal T-policy minimizing the discounted total expected cost and further investigated the combined (N, T)-policy. For an excellent survey on queueing control with removable server, see Teghem[10].

In this article, we consider the $M/G/1$ queueing system with removable server subject to failure interruptions, where the failure occurs only during operating busy period. In fact, it is of great importance to inve- stigate effects of failures in the controlled queueing system quantitatively. If the failure occurs, the system has to be maintained. Here, two kinds of recovery functions from failures are studied. One is the retry function, the other is the repair function. In the retry model, after a system failure the system undergoes instantly the corrective repair and the interrupted job is reprocessed again after the system becomes as good as new. On the other hand, in the repair model, the minimal repair (see *e,g.* Barlow and Proschan[11]) is executed for each failure and at the termination of the process of a job the preventive repair is carried out. Such recovery mechanizms for system failures can be seen in the context of the database management system and the bufferd production system. In the framework of the T-policy for the $M/G/1$ queue, we formulate the relevant expected costs under two recovery functions, respectively, and derive the optimal timing strategies minimizing them. The theory of semi-Markov processes will be applied to analyze the local behaviour of the system consisted of three states; failure, maintenance/retry and operation. More concretely, the state-space method is used to obtain the Laplace transform of the effective service time for a job. This is the similar technique used by Gaver[12], Keilson[13] and Sumita, Kaio and Goes[14]. In numerical examples, we compare two recovery functions in terms of cost effectiveness and also examine the dependence of model parameters in the recovery functions.

2. MODEL DESCRIPTION

We consider an *M/G/1* queueing system with a removable server. It is assumed that k-th job ($k = 1, 2, \cdots$) arrives according to a Poisson process $\{N(t), t \geq 0\}$ with $N(0) = 0$ and parameter $\lambda\,(> 0)$, and its processing time S_k is nonnegative i.i.d. random variable having a distribution function $A(t) = \Pr\{S_k \leq t\}$ with finite mean $1/\mu\,(> 0)$ and finite variance $\sigma^2\,(> 0)$. Define the Laplace-Stieltjes transform (L.S.T.) of the distribution function $A(t)$ as $\tilde{A}(\beta)$, where β is a complex number. Without loss of generality, the heavy traffic condition is assumed, *i.e.* the traffic intensity $\rho = \lambda/\mu$ is less than unity, and the number of jobs in the system is zero at time $t = 0$.

Following Heyman[8], we consider the $T-$ policy. Under this policy, the server is kept to be off from $t = 0$ to time $t = T$ (≥ 0) and jobs arriving at the system during this period are cumulated in the buffer. At time T, the sever is turned on, the service operation is started and continues until all jobs including those arrived later are completed in the system. Figure 22.1 illustrates the configuration of the queueing system under consideration. In this paper, we call this kind of control policy *the timing strategy* instead of the T-policy. Since the arrival process is the Poisson, the system is regenerative at the end of busy period. To this end, the queueing process under the timing strategy is an ordinary renewal process, and one cycle is defined as the time interval from the end of busy period to the next one.

Define the following cost parameters.

h (> 0) : holding cost per unit job per unit time

R (> 0) : switching cost to turn the system on and off

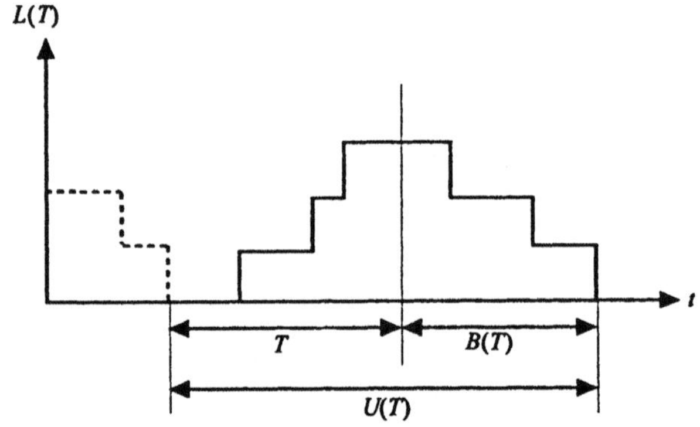

Figure 22.1 $M/G/1$ queueing system under the timing strategy.

Then, if the failure does not occur with probability one, the expected cost per unit time in the steady-state is

$$C(T) = hL(T) + R/U(T), \tag{22.1}$$

where $L(T)$ is the mean number of jobs in the system in the steady-state and $U(T)$ is the mean time length of one cycle (see Fig. 22.1). From the elementary queueing results, the probability that k ($k = 1, 2, \cdots$) jobs arrive at the system until $t = T$ is

$$p_k(T) = \frac{e^{-\lambda T}(\lambda T)^k}{k!}, \tag{22.2}$$

since the arrival process is the Poisson with parameter λ (> 0). Also, the mean time required to process all jobs on condition that k jobs exist until $t = T$ queue and the mean mumber of all jobs on condition that k jobs arrived until $t = T$ are processed in the $M/G/1$, are

$$B_k = \frac{k}{\mu(1-\rho)} \tag{22.3}$$

and

$$N_k = \frac{k}{(1-\rho)}, \tag{22.4}$$

respectively. As depicted in Fig. 22.1, since $B(T)$ and $U(T)$ are the mean load time and the mean time length of one cycle, respectively, then we have

$$B(T) = \sum_{k=0}^{\infty} B_k p_k(T) = \rho T/(1-\rho) \tag{22.5}$$

and

$$U(T) = T + B(T) = T/(1-\rho). \tag{22.6}$$

Let L_k and W_k denote the mean number of jobs processed when k jobs arrive at the system until $t = T$ and the mean waiting time, respectively. Then, from the result by Heyman[8], it can be seen that

$$L_k = L + \frac{k-1}{2} \tag{22.7}$$

and

$$W_k = \frac{L_k}{\lambda}, \tag{22.8}$$

where

$$L = \rho + \frac{\rho^2(\sigma^2\mu^2 + 1)}{2(1-\rho)} \tag{22.9}$$

is due to the well-known Pollaczec-Khinchine's formula (see *e.g.* Tijms[15]).

Hence, under the timing strategy, the mean waiting time in the steady-state is

$$W(T) = \frac{\sum_{k=0}^{\infty} W_k N_k p_k(T)}{\sum_{k=0}^{\infty} N_k p_k(T)} = \frac{1}{\lambda}\left(L + \frac{\lambda T}{2}\right). \tag{22.10}$$

Using the Little's formula, the mean number of jobs in the systems in the steady-state becomes

$$L(T) = \lambda W(T) = L + \lambda T/2. \tag{22.11}$$

Finally, from Eqs. (22.1), (22.6) and (22.11), the expected cost per unit time in the steady-state is given by

$$C(T;\mu,\sigma) = hL + h\lambda T/2 + R(1-\rho)/T. \tag{22.12}$$

From the simple algebraic manipulation, the optimal timing strategy T^* which minimizes $C(T;\mu,\sigma)$ is

$$T^* = \sqrt{\frac{2R(1-\rho)}{h\lambda}}. \tag{22.13}$$

From Eq. (22.13), it is found that the optimal timing strategy is increasing in μ and decreasing in λ.

3. RETRY MODEL

In this section, the retry function is considered for the system subject to failures. One job has a random service time S (> 0) measured by the cumulative operating time with $A(x) = \Pr\{S \leq x\}$, which is absolutely continuous having density $a(x) = dA(x)/dx$, finite mean $1/\mu$ (≥ 0), variance σ^2 (> 0) and hazard function $r_a(y) = a(y)/\overline{A}(y)$, where in general $\overline{\varphi}(\cdot) = 1 - \varphi(\cdot)$ (hereafter, the subscript k is dropped to simplify the notation). Also, the system is subject to failure interruptions which occur according to an ordinary renewal process with inter-occurence time distribution $G(x)$ which is absolutely continuous having density function $g(x) = dG(x)/dx$ and failure rate $r_g(x) = g(x)/\overline{G}(x)$. The stochastic process $\{X_p(t), t \geq 0\}$ is defined as the cumulative operating time required to process a job during the time interval $(0, t]$. Define the stochastic process $\{I(t), t \geq 0\}$ as follows.

$$I(t) = \begin{cases} 0, & \text{if the process of a job has been completed by time } t \\ 1, & \text{if a job is processed at time } t \\ 2, & \text{if the system is in a system down period at time } t. \end{cases} \tag{22.14}$$

If we pay our attention to the local behaviour of one job and the failure occurs until the cumulative operating time attains to S, *i.e.* during the processing of a job, then it is detected instantly and the major repair is started. More precisely, if the failure occurs at time t with $X_p(t) = x \in [0, S]$, then the corrective repair time is given by a non-negative random variable V, and after completing to repair the system can be recovered as good as new. We assume that $B(y) = \Pr\{V \leq y\}$ is absolutely continuous having density $b(y) = dB(y)/dy$, finite mean m_1 (> 0), variance v_1^2 (> 0) and hazard function $r_b(y) = b(y)/\overline{B}(y)$. To simplify the discussion, we assume $X_p(t) = 0$ if $I(t) = 0$.

On the other hand, when the cumulative operating time attains to S without any failure, the preventive repair is started. At the moment, suppose the preventive repair time is a nonnegative randome variable W with absolutely continuous probability distribution $D(x) = \Pr\{W \leq x\}$, density $d(x) = dD(x)/dx$, finite mean m_w (> 0), variance v_w^2 (> 0) and $r_d(x) = d(x)/\overline{D}(x)$ (this assumption is later simplified). After completing the preventive maintenance the system can be recovered as good as new. A typical sample path of the retry function is depicted in Fig. 22.2. Further, we define the stochastic process $\{Y_p(t), t \geq 0\}$ as the elapsed time since a failure interruption, given $I(t) = 2$. In other words, $Y_p(t)$ is defined to be zero if $I(t) \neq 2$. Now we are in the position to characterize the stochastic system by the trivariate process $Z_p\{X_p(t), Y_p(t), I(t), t \geq 0\}$. Especially, of our interest is the effective

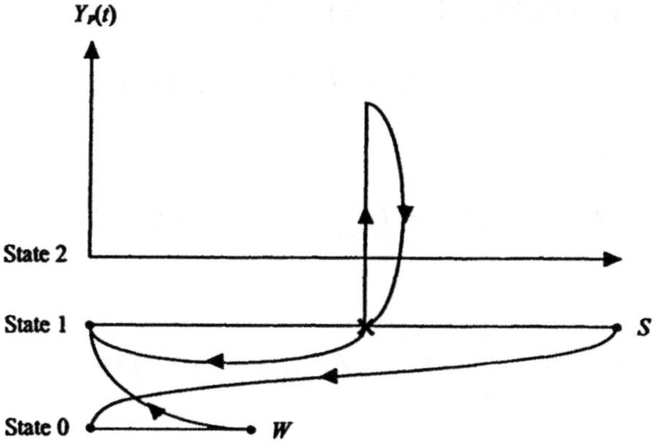

Figure 22.2 Configuration of the retry model.

operating time $S_{\text{eff-p}}$ describing the cumulative elapsed time until completing the preventive/corrective maintenance for one job. Define

$$F_0(x,w,t) = \Pr\{X_p(t) \le x,\ W \le w,\ I(t) = 0\};$$

$$f_0(x,\ w,t) = \frac{\partial^2}{\partial x \partial w} F_0(x,w,t), \tag{22.15}$$

$$F_1(x,t) = \Pr\{X_p(t) \le x, I(t) = 1\}; \quad f_1(x,t) = \frac{\partial}{\partial x} F_1(x,t), \tag{22.16}$$

$$F_2(x,y,t) = \Pr\{X_p(t) \le x, Y_p(t) \le y, I(t) = 2\};$$

$$f_2(x,y,t) = \frac{\partial^2}{\partial x \partial y} F_2(x,y,t) \tag{22.17}$$

and

$$F_3(t) = \Pr\{I(t) = 3\} = \Pr\{S_{\text{eff-p}} \le t\}; \quad f_3(t) = \frac{\partial}{\partial t} F_3(t), \tag{22.18}$$

where state 3 is the termination of one job.

Since the trivariate stochastic process $Z_p\{X_p(t), Y_p(t), I(t), t \ge 0\}$ is a semi-Markov process, we have

$$f_2(x,y,t) = f_2(x,0,t-y)\overline{B}(y) \tag{22.19}$$

and

$$f_2(x,0,t) = r_g(x) f_1(x,t). \tag{22.20}$$

From (22.19) and (22.20), it is seen that

$$f_2(x,y,t) = r_g(x) f_1(x, t-y) \overline{B}(y). \tag{22.21}$$

Similarly, we have

$$f_0(x,w,t) = r_a(x) f_1(x, t-w) \overline{D}(w) \tag{22.22}$$

and

$$\begin{aligned} f_3(t) &= \int_0^\infty \int_0^\infty f_2(x,y,t) r_b(y) dy dx \\ &+ \int_0^\infty \int_0^\infty f_0(x,w,t) r_d(w) dw dx \\ &= \int_0^\infty r_g(x) \int_0^\infty f_1(x,t-y) b(y) dy dx \\ &+ \int_0^\infty r_a(x) \int_0^\infty f_1(x,t-w) d(w) dw dx. \end{aligned} \tag{22.23}$$

Taking the Laplace transform (L.T.) of $f_3(t)$ and solving the corresponding differential equation, we have

Theorem 1 *The L.T. of the effective operating time $S_{\text{eff}-\text{p}}$ is*

$$\begin{aligned} \mathrm{E}\Big[\exp(-sS_{\text{eff}-\text{p}})\Big] &= \int_0^\infty \exp(-sx) \overline{A}(x) \hat{b}(s) dG(x) \\ &+ \int_0^\infty \exp(-sx) \overline{G}(x) \hat{d}(s) dA(x), \end{aligned} \tag{22.24}$$

where

$$\hat{b}(s) = \int_0^\infty \exp(-sy) b(y) dy \tag{22.25}$$

and

$$\hat{d}(s) = \int_0^\infty \exp(-sy) d(y) dy. \tag{22.26}$$

From 1, it is straightforward to derive the following result.

Lemma 1 *(i) The mean effective operating time under the retry function is*

$$\mathrm{E}[S_{\text{eff}-\text{p}}] = \int_0^\infty \overline{G}(x) \overline{A}(x) dx$$

$$+\int_0^\infty g(x)m_1\overline{A}(x)dx + m_w\int_0^\infty g(x)A(x)dx. \tag{22.27}$$

(ii) The variance of the effective operating time under the retry function is

$$\mathrm{Var}[S_{\mathrm{eff-p}}] = \mathrm{E}[S^2_{\mathrm{eff-p}}] - \mathrm{E}[S_{\mathrm{eff-p}}]^2, \tag{22.28}$$

where

$$\begin{aligned}\mathrm{E}[S^2_{\mathrm{eff-p}}] &= 2\int_0^\infty x\overline{G}(x)\overline{A}(x)dx + \int_0^\infty g(x)\{2xm_1 + v_1^2 + m_1^2\}\overline{A}(x)dx \\ &+ \int_0^\infty \overline{G}(x)\{2xm_w + v_w^2 + m_w^2\}a(x)dx. \end{aligned}\tag{22.29}$$

Next, we formulate the expected cost under the retry function. Define $1/\mu_p \equiv \mathrm{E}[S_{\mathrm{eff-p}}]$ and $\sigma_p^2 \equiv \mathrm{Var}[S_{\mathrm{eff-p}}]$. To simplify the discussion, suppose that the preventive repair is made instantaneously. This leads to $W = m_w = v_w^2 = 0$ and is often used in the maintenance theory (see Barlow and Proschan[11]). Further, define the major repair cost per unit time and the preventive repair cost per unit job as c_l (> 0) and c_p (> 0), respectively. Then the expected cost under the retry function becomes

$$V_p(T) = C(T;\mu_p,\sigma_p) + c_l\lambda\int_0^\infty m_1\overline{A}(x)dG(x) + c_p\lambda. \tag{22.30}$$

Consequently, we have the optimal timing strategy

$$T_p^* = \sqrt{\frac{2R(1-\rho_p)}{h\lambda}}, \tag{22.31}$$

where $\rho_p = \lambda/\mu_p$.

4. REPAIR MODEL

Next, consider the optimal timing strategy under the repair function. From the well-known result, the number of failures under the minimal repair follows the non-homogeneous Poisson process with mean value function $\Lambda(t) = -\log\overline{G}(t)$ and intensity $r(t) = d\Lambda(t)/dt$. If the failure occurs, the system undergoes a minimal repair where the repair time V is required for undisturbing the system age. Similar to Section 3, suppose that the minimal-repairing time for the system obeys an arbitrary general distribution with $B(y) = \Pr\{V \le y\}$, density function $b(y) = dB(y)/dy$, hazard rate function $r_b(y) = b(y)/\overline{B}(y)$, finite mean

m_2 (> 0) and variance v_2^2 (> 0). After completing the service, the preventive repair is made at time S instantaneously. Figure 22.3 shows the configuration of the repair model. In a fashion similar to the retry model, the stochastic process $\{X_m(t), t \geq 0\}$ is defined as the cumulative operating time required to process a job during the time interval $(0, t]$.

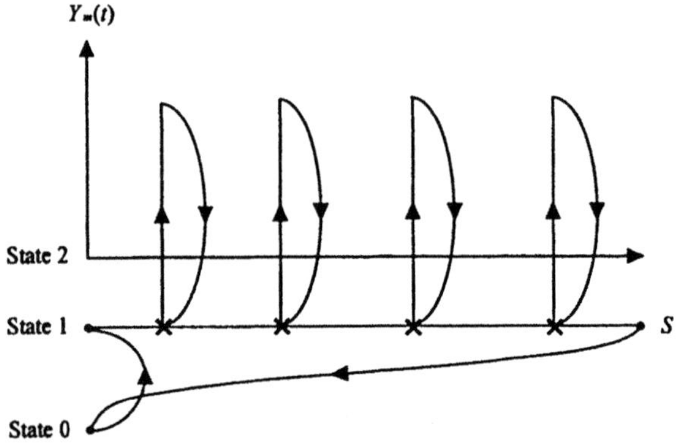

Figure 22.3 **Configuration of the repair model.**

Using (22.14), we define the stochastic process $\{I(t), t \geq 0\}$ where $X_m(t) = 0$ if $I(t) = 0$ and $\{Y_m(t), t \geq 0\}$ as the elapsed time since the last failure interruption, given $I(t) = 2$. Then from the trivariate process $Z_m\{X_m(t), Y_m(t), I(t), t \geq 0\}$, we can evaluate the effective operating time $S_{\text{eff}-\text{m}}$ describing the cumulative elapsed time until the preventive maintenance for one job. Define

$$F_0(t) = \Pr\{I(t) = 0\}; \quad f_0(t) = \frac{\partial}{\partial t} F_0(t), \tag{22.32}$$

$$F_1(x, t) = \Pr\{X_m(t) \leq x, I(t) = 1\}; \quad f_1(x, t) = \frac{\partial}{\partial x} F_1(x, t) \tag{22.33}$$

and

$$\begin{aligned} F_2(x, y, t) &= \Pr\{X_m(t) \leq x, Y_m(t) \leq y, I(t) = 2\}; \\ f_2(x, y, t) &= \frac{\partial^2}{\partial x \partial y} F_2(x, y, t). \end{aligned} \tag{22.34}$$

The following results are due to Sumita, Kaio and Goes[14].

Theorem 2 *The L.T. of the effective operating time $S_{\text{eff}-\text{m}}$ is*

$$\mathrm{E}[\exp(-sS_{\text{eff}-\text{m}})] = \int_0^\infty \exp\left\{-\left[sx + \int_0^x r(\tau)(1 - \hat{b}(s))d\tau\right]\right\} dA(x).$$

Lemma 2 *(i) The mean effective operating time under the repair function is*

$$\mathrm{E}[S_{\mathrm{eff-m}}] = \frac{1}{\mu} + \int_0^\infty dA(x) \int_0^x r(\tau) m_2 d\tau. \qquad (22.35)$$

(ii) The variance of the effective operating time under the repair function is

$$\mathrm{Var}[S_{\mathrm{eff-m}}] = \mathrm{E}[S^2_{\mathrm{eff-m}}] - \mathrm{E}[S_{\mathrm{eff-m}}]^2, \qquad (22.36)$$

where

$$\begin{aligned}\mathrm{E}[S^2_{\mathrm{eff-m}}] &= \sigma^2 + 1/\mu^2 + \int_0^\infty dA(x) \Big\{ \int_0^x r(\tau)(v_2^2 + m_2^2) d\tau \\ &\quad + 2x \int_0^x r(\tau) m_2 d\tau + \Big[\int_0^x r(\tau) m_2 d\tau\Big]^2 \Big\}. \qquad (22.37)\end{aligned}$$

Let us define $1/\mu_m \equiv \mathrm{E}[S_{\mathrm{eff-m}}]$ and $\sigma_m^2 \equiv \mathrm{Var}[S_{\mathrm{eff-m}}]$. In addition, let c_m (> 0) and c_p (> 0) be the minimal repair cost per unit time and the cost for the preventive maintenance per unit job, respectively. Then the expected cost becomes

$$C_m(T) = C(T; \mu_m, \sigma_m) + c_m \lambda \int_0^\infty dA(x) \int_0^x r(\tau) m_2 d\tau + c_p \lambda \qquad (22.38)$$

and the optimal timing strategy is

$$T_m^* = \sqrt{\frac{2R(1-\rho_m)}{h\lambda}}, \qquad (22.39)$$

where $\rho_m = \lambda/\mu_m$.

5. NUMERICAL EXAMPLES AND SOME REMARKS

In what follows, we assume that the failure occurs following the homogeneous Poisson process with $r_g(t) = r_g$ (constant) and compare two recovery functions in terms of cost effectiveness. Figure 22.4 presents the dependence of the parameter r_g in the optimal timing strategy and its associated expected cost. As increasing the failure rate, the optimal timing strategy under the repair function is drastically decreasing. This result explains that the optimal timing strategy under the repair function is more pessimistic and behaves to avoid the system down for an unreliable service facility. Also, both the expected costs tends to increase

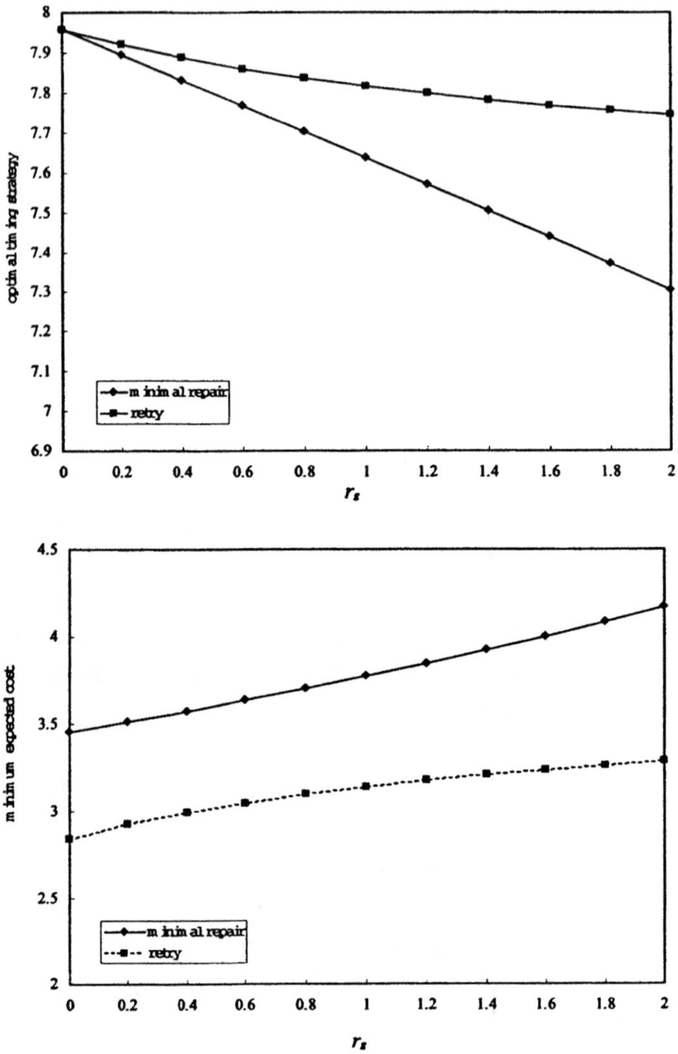

Figure 22.4 Dependence of parameter r_g in the optimal timing strategy (exponential failure): $h = 3$, $R = 10$, $c_m = c_l = 4$, $c_p = 3$, $\lambda = 0.1$ $\mu = 2$, $m_1 = m_2 = 1.5$, $v_1 = v_2 = 2$.

monotonically for increasing r_g, but the retry function under-estimates the corresponding expected cost.

In Fig. 22.5, the dependence of the parameter μ in the optimal timing strategy and its associated expected cost is presented. Since the mean processing time becomes small as μ increases, the optimal timing strategies and their minimum expected costs increase and decrease, respectively. These results are intuitively understandable, but it is re-

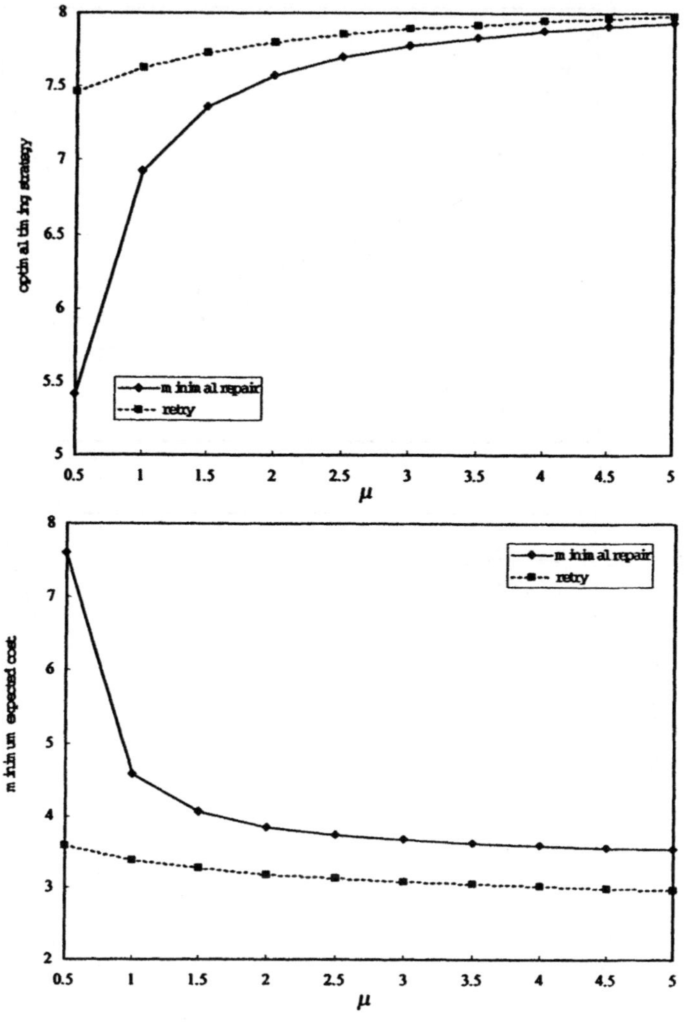

Figure 22.5 Dependence of parameter μ in the optimal timing strategy (exponential failure): $h = 3$, $R = 10$, $c_m = c_l = 4$, $c_p = 3$, $\lambda = 0.1$ $r_g = 1.2$, $m_1 = m_2 = 1.5$, $v_1 = v_2 = 2$.

markable that the repair function reduces the optimal timing strategy for a small μ. Further, it is observed that the retry function is not sensitive for varying μ comparing the repair case, and this result may be useful to design the retry model in the queueing system with removable server. Finally, the relation between the optimal timing strategy and the arrival parameter λ is illustrated in Fig. 22.6. Since an increasing λ implies the shorter arrival time interval in the sense of expectation, in the more idle situation the retry function reduces the expected cost.

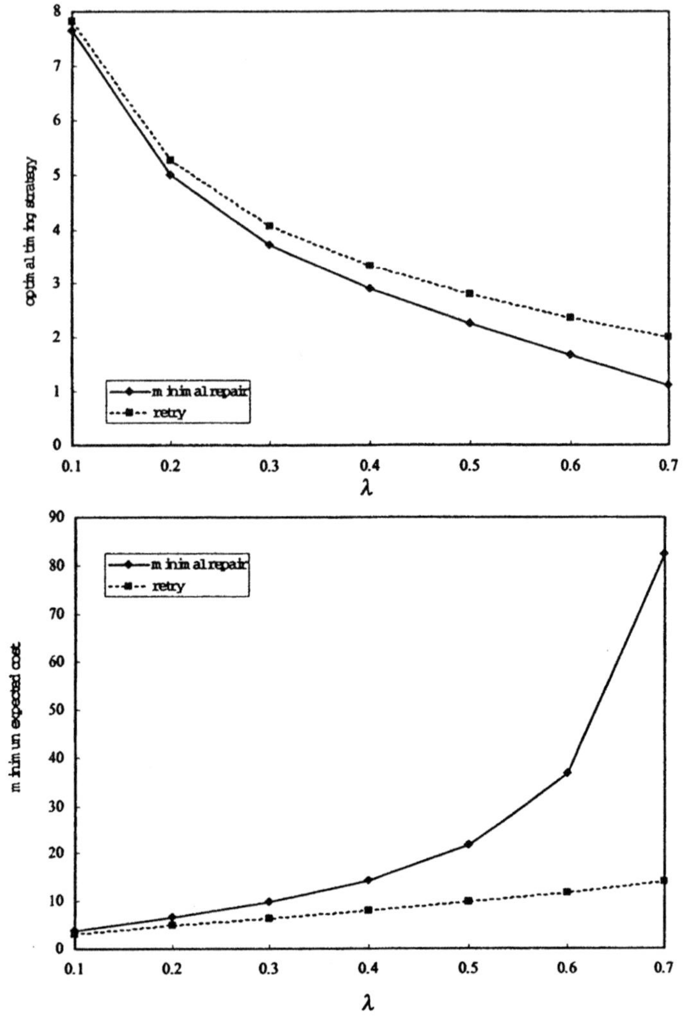

Figure 22.6 Dependence of parameter λ in the optimal timing strategy (exponential failure): $h = 3$, $R = 10$, $c_m = c_l = 4$, $c_p = 3$, $r_g = 1$ $\mu = 2$, $m_1 = m_2 = 1.5$, $v_1 = v_2 = 2$.

In this article we have considered two recovery functions for the $M/G/1$ queueing system with removable server subject to failure interruptions. We have compared two timing strategies under respective recovery functions in terms of cost effectiveness and have examined the dependence of model parameters in the recovery functions. The results by the sensitivity analysis have denoted that the retry function over-estimates the optimal timing strategy and under-estimates the ecpected cost for the same corrective maintenance cost $c_m = c_l$. In future, the remaining problem will be to analyse the different control methods from the timing strategy, *e.g.* such as the N-policy and the D-policy. Also, multi-channel queue with removable servers subject to failures should be discussed in the similar framework to this paper.

Acknowledgments

The authors would like to thank Mr. K. Kitou, Department of Industrial and Systems Engineering, Hiroshima University, for his numerical assistance. This work was partially supported by a Grant-in-Aid for Scientific Research from the Ministry of Education, Sports, Science and Culture of Japan under Grant No. 09780411 and No. 09680426.

References

[1] Yadin, M. and Naor, P. (1963). Queuing systems with a removable service station, *Opl Res. Q.*, **14**, 393-405.

[2] Heyman, D. P. (1968). Optimal operating policies for *M/G/1* queuing systems, *Opns. Res.*, **16**, 362-382.

[3] Sobel, M. J. (1969). Optimal average-cost policy for a queue with start-up and shut-down costs, *Opns. Res.*, **17**, 145-162.

[4] Bell, C. E. (1971). Characterization and computation of optimal policies for operating an *M/G/1* queuing system with removable server, *Opns. Res.*, **19**, 208-218.

[5] Kimura, T. (1981). Optimal control of an *M/G/1* queuing system with removable server via diffusion approximation, *Euro. J. Opl. Res.*, **8**, 390-398.

[6] Balachandran, K. R. (1973). Control policies for a single server system, *Management Sci.*, **19**, 1013-1018.

[7] Balachandran and Tijms, H. (1975). On the D-policy for the $M/G/1$ queue, *Management Sci.*, **21**, 1073-1076.

[8] Heyman, D. P. (1977). The T-policy for the *M/G/1* queue, *Management Sci.*, **23**, 775-778.

[9] Okamura, H., Dohi, T. and Osaki, S. (1997). Optimal timing strategies for controlled $M/G/1$ queueing system via diffusion approximation approach, *Proc. 28th ISCIE Int. Symp. Stochastic Systems Theory and Its Applications*, 143-148.

[10] Teghem, J. Jr. (1986). Control of the service process in a queueing system, *Euro. J. Opl. Res.*, **23**, 141-158.

[11] Barlow, R. E. and Proschan, F. (1975). *Statistical Theory of Reliability and Life Testing: Probability Models*, NY: Holt, Rinehart and Winston.

[12] Gaver, D. P. (1962). A waiting line with interrupted service including priorities, *J. Roy. Statist. Soc.*, **B-24**, 73-90.

[13] Keilson, J. (1982). Queues subject to service interruptions, *Ann. Math. Statist.*, **33**, 1314-1322.

[14] Sumita, U, Kaio, N. and Goes, P. B. (1989). Analysis of effective service time with age dependent interruptions and its application to optimal rollbac k policy for database management, *Queueing Syst.*, **4**, 193-212.

[15] Tijms, H. C. (1994). *Stochastic Models; An Algorithmic Approach*, NY: John Wiley & Son.

VII

CHROMATOGRAPHY & FLUID MECHANICS

Chapter 23

CONTINUOUS SEMI-MARKOV MODELS FOR CHROMATOGRAPHY

Boris P. Harlamov

Russian Academy of Sciences, Russia

Abstract Continuous semi-Markov processes on a metric space are considered. Some properties of such a process and related functions are discussed, e.g. semi-Markov transition functions, characteristic operators, conditional distribution given sequence of states, connection to Markov processes, differential equations for diffusion type semi-Markov process. The semi-Markov property to have trajectories with intervals of constancy can be used for modeling in chromatography. The model takes into account a special character of movement of a particle through a filter. Formulae for some chromatograph characteristics are derived. The well-known semi-empirical formula of van Deemter is deduced theoretically.

Keywords: Stopping time, first exit, Markov property, transition function, generating function, time change, independent increments, Lévy representation, curvilinear integral, inverse process, interval of constancy, random delay, eluent, detector, absorption, theoretical plate, chromatogram.

1. INTRODUCTION

We consider the application of continuous semi-Markov processes in order to study the transfer of matter with a solvent through a porous medium. This transfer and separation of mixtures of substances, connected to them, is the important constituent of such natural phenomena as metasomatism (Korzhinski (1982)), filtration of oil and gas (Lejbenzon (1947)) and such engineering processes, as percolation of liquids and gases through constructional materials, that determines their reliability. In its most pure aspect, this process is exhibited in chromatography (Giddings (1965), Gut and Ahlberg (1981), Hofmann (1972)). The theoretical treatment of transfer processes is conducted within the framework of the appropriate applied disciplines and frequently reduces to overlapped results. For example, there is an explicit parallelism in outcomes of the theory of metasomatic zonality (Korzhinski (1982)) and the theory of chromatography (Hofmann (1972)). There are common regularities of transfer process, which can be studied independently from concrete engineering applications.

The stochastic model begins with the assumption that the observable measure of distribution of the substance is the expectation of some random measure which is a random point field in the states space. Each point of this field can be interpreted as a molecule or other particle keeping a wholeness when transferring with a solvent. The exact interpretation is not important for a model, as only the averaged magnitudes are observed. The point field and its average, generally speaking, depend on time. This dependence is related to the movement of particles. The stochastic model assumes moving particles under the rule of some random process. The essential hypothesis of a model is the assumption of independence of random processes, corresponding to various particles, from each other. With these suppositions two laws determine a measure of the matter we are interested in: 1) expectations of initial and (or) boundary point field and 2) individual rules of moving of a separate particle . In the present paper the basic attention is given to a substantiation and application of the semi-Markov law of moving of separate particles and corollaries from this supposition.

The semi-Markov processes are an approximate model for processes of transfer of substance with a solution through a porous medium due to practical absence of inertia of moving particles permitting to apply a premises of the theory of semi-Markov processes, and also due to the character of trajectories of particles of transferable substance. The experimental fact is, that with transfer of matter through a porous medium in each instant one can find out particles both in a solution, and in substance of a filter in absorbed condition. Only by discontinuous moving with stoppings on a rigid phase such distribution of particles between phases is possible to explain. In other words, the continuous trajectories of moving should contain intervals of a constancy. And this is also one of the peculiarities of the trajectories of (non-Markov) semi-Markov processes.

2. CONTINUOUS SEMI-MARKOV PROCESS

1. Semi-Markov process. We consider continuous semi-Markov processes within the framework of the theory of semi-Markov processes of a general view, which describes, for example, stepped semi-Markov processes, and also Markov diffusion processes. Let $\mathbb{X}$ be a complete σ-compact separable metric space, $\mathfrak{A}$ the class of all open subsets of space $\mathbb{X}$, $\mathcal{D}$ the class of all functions, continuous on the right and having limits at the left, $\xi : \mathbb{R}_+ \to \mathbb{X}$. The semi-Markov process is determined by the family of probability measures $(P_x)_{x\in\mathbb{X}}$ (where $P_x(\xi(0) = x) = 1$) on the Borel σ-algebra $\mathcal{F} \equiv \mathcal{B}(\mathcal{D})$ of subsets of a set $\mathcal{D}$. The family (P_x) refers to as *semi-Markov*, if it has the (homogeneous in time) Markov property concerning the first exit time σ_G from any set $G \in \mathfrak{A}$, $\sigma_G(\xi) = \inf\{t \geq 0 : \xi(t) \notin G\}$.

2. Semi-Markov transition function. We shall designate $\pi_G(\xi) = \xi(\sigma_G)$ a point of the first exit of a trajectory ξ from a set G, which is defined only with $\sigma_G(\xi) < \infty$. The following kernel refers to as a *semi-Markov transition function* of the process

$$F_G(dt \times dx_1 \,|x) = P_x(\sigma_G \in dt,\ \pi_G \in dx_1).$$

For a semi-Markov process with any open sets G_1, G_2 such, that $G_1 \subset G_2$, the following equation holds

$$F_{G_2}([0,t] \times S \,|x_0) = \int_0^t \int_{\mathbb{X}} F_{G_2}((0,t-t_1] \times S|x_1)\, F_{G_1}(dt_1 \times dx_1 \,|x_0) \quad (23.1)$$

On the other hand, from (23.1) the semi-Markov condition of the coordination in a family (P_x) follows.

3. Continuous semi-Markov process. The continuity of the trajectories of the process is equivalent to the condition: for all $G \in \mathfrak{A}$ and $x \in G$ P_x-almost sure the point of the first exit $\xi(\sigma_G)$ belongs to the boundary of a set G i.e. $\xi(\sigma_G) \in \partial G$:

$$F_G([0,t] \times S|x) = F_G([0,t] \times (S \cap \partial G)|x).$$

The *continuous semi-Markov process* can be given by the family of measures (P_x) on the Borel σ-algebra $\mathcal{B}(\mathcal{C})$ of subsets of space $\mathcal{C}$ of continuous functions.

4. Semi-Markov transitional generating function. In the theory of semi-Markov processes it is convenient to use a Laplacian image on t of the transition function:

$$f_G(\lambda, dx_1 \,|x) = \int_0^\infty e^{-\lambda t} F_G(dt \times dx_1 \,|x), \quad (\lambda \geq 0),$$

that is called a *semi-Markov transitional generating function.* From the equation (23.1) follows, that for a transitional generating function the following equation holds

$$f_{G_2}(\lambda, S\,|x) = \int_{\mathbb{X}} f_{G_2}(\lambda, S\,|x_1) f_{G_1}(\lambda, dx_1\,|x). \qquad (23.2)$$

5. λ-characteristic operator. Let $x \in G \in \mathfrak{A}$, φ be some measurable function on $\mathbb{X}$. Let us consider a ratio

$$A_\lambda^G \varphi(x) \equiv \frac{P_x(e^{-\lambda\sigma_G}\varphi(\pi_G)) - \varphi(x)}{P_x(\sigma_G,\, \sigma_G < \infty)} \qquad (\lambda \geq 0)$$

(here and further we designate an integral of function $\varphi : \mathcal{D} \to \mathbb{R}$ on a set $B \in \mathcal{F}$ by a measure M as $M(\varphi, B)$, omitting B, if $M(B) = M(\mathcal{D})$). The limit $A_\lambda\varphi(x)$ of this ratio when $G \downarrow \{x\}$, defines the so-called λ-*characteristic operator* A_λ. The domain of definition of this operator consists of the functions φ, for which this limit exists for any point $x \in \mathbb{X}$ and for any sequence of neighborhoods $G \in \mathfrak{A}$ of x converging to this point.

In terms of an operator A_λ many properties of semi-Markov process are described. In particular, the semi-Markov process without fixed intervals of a constancy is Markov process if and only if the function $A_\lambda\mathbb{I}$ is linear on λ ($\mathbb{I}(x) \equiv 1$).

6. One-dimensional diffusion process. Let $\mathbb{X} = \mathbb{R}^1$, $G = (a, b)$, $a \leq x \leq b$. In this case $f_G(\lambda, dx_1\,|x)$ decomposes to two functions $f_G(\lambda, \{a\}\,|x) = g_G(\lambda, x)$, $f_G(\lambda, \{b\}\,|x) = h_G(\lambda, x)$. From the equation (23.2) follows, that these functions satisfy the set of equations

$$g_{G_2}(\lambda, x) = g_{G_1}(\lambda, x) g_{G_2}(\lambda, c) + h_{G_1}(\lambda, x) g_{G_2}(\lambda, d), \qquad (23.3)$$

$$h_{G_2}(\lambda, x) = g_{G_1}(\lambda, x) h_{G_2}(\lambda, c) + h_{G_1}(\lambda, x) h_{G_2}(\lambda, d), \qquad (23.4)$$

where $G_2 = (a, b), G_1 = (c, d) \subset G_2$. Let $G = G(r) = \{x_1 : |x - x_1| < r\}$. Conditions

$$g_{G(r)}(\lambda, x) = \frac{1}{2} - A(x)\, r + B(\lambda, x)\, \frac{r^2}{2} + o(r^2), \qquad (23.5)$$

$$h_{G(r)}(\lambda, x) = \frac{1}{2} + A(x)\, r + B(\lambda, x)\, \frac{r^2}{2} + o(r^2), \qquad (23.6)$$

($r \to 0$), where A', $B(\lambda, \cdot)$ are some continuous on x functions, are necessary and sufficient for the functions g_{G_2}, h_{G_2} to be solutions of the differential equation

$$f'' + 4A(x) f' + 2B(\lambda, x) f = 0 \qquad (23.7)$$

with boundary conditions $g_{G_2}(\lambda, a) = h_{G_2}(\lambda, b) = 1$, $h_{G_2}(\lambda, a) = g_{G_2}(\lambda, b) = 0$ (see Harlamov (1983a)). From the properties of Laplace transform it follows, that $(\forall \lambda > 0)\ B(\lambda, \cdot) < 0$, $B(0, \cdot) = 0$, $B(\infty, \cdot) = -\infty$, and $(\forall x \in \mathbb{X})\ (-B'_\lambda(\lambda, x))$ is a completely monotone function of λ. We call processes satisfying equation (23.7) as *semi-Markov processes of a diffusion type*. Markov diffusion process homogeneous in time with Kolmogorov parameters $v(x)$ (velocity of a drift) and $d(x)$ (the growth rate of a variance) has coefficients of the equation (23.7) as

$$A(x) = \frac{2v(x)}{d(x)}, \quad B(\lambda, x) = -\frac{\lambda}{d(x)}. \tag{23.8}$$

The λ-characteristic operator of semi-Markov process of a diffusion type has an aspect

$$A_\lambda \varphi(x) = \frac{1}{2\gamma(x)}(\varphi'' + 4A(x)\varphi' + 2B(\lambda, x)\varphi),$$

where $\gamma(x) = \lim_{r \to 0} P_x(\sigma_{G(r)}, \sigma_{G(r)} < \infty)/r^2$. In particular, $A_\lambda \mathbb{I}(x) = B(\lambda, x)/\gamma(x)$.

7. Time change. Let Φ be the set of all continuous functions $\varphi : \mathbb{R}_+ \to \mathbb{R}_+$, $\varphi(0) = 0$, $\varphi(t) \to \infty\ (t \to \infty)$; the φ^* function is such, that $(\forall t \in \mathbb{R}_+)\ \varphi^*(t) = \inf \varphi^{-1}[t, \infty)$. It is not difficult to prove, that $(\forall G \in \mathfrak{A})\ (\forall \varphi \in \Phi)\ (\forall \xi \in \mathcal{D})\ \sigma_G(\xi \circ \varphi) = \varphi^*(\sigma_G(\xi))$. The same property holds for finite iterations of the first exit time and some limits of such times (see Harlamov (1976a)). In relation to ξ (initial function) and $\xi \circ \varphi$ (the transformed function) φ refers to as an inverse time change and φ^* as a direct one. The *random time change* can be given as a probability measure on the measurable space $\Phi \times \mathcal{D}$.

8. Sequence of states (curve). Let $\mathcal{M}$ be the set of all equivalence classes of a set $\mathcal{D}$ with respect to equivalence relation:

$$\xi_1 \overset{\ell}{\sim} \xi_2 \iff (\exists \varphi_1, \varphi_2 \in \Phi) \quad \xi_1 \circ \varphi_1 = \xi_2 \circ \varphi_2.$$

We call such an equivalence class a *sequence of states* (or curve) in metric space. Let ℓ be a map $\mathcal{D} \to \mathcal{M}$, relating to each function $\xi \in \mathcal{D}$ its equivalence class $\xi_1 \overset{\ell}{\sim} \xi_2 \iff \ell\xi_1 = \ell\xi_2$ (see Kolmogorov and Fomin (1972)).

For a study of *a course of time concerning a sequence of states* the little bit more rough partition of a set $\mathcal{D}$, is more convenient than $\mathcal{M}$. The sequence of *essential states* is defined with the help of the values of a trajectory on some countable set $\mathcal{T}_1$ of the so called intrinsic stopping times such, as moment of the first exit and their finite iterations. The set of sequences of essential states $\mathcal{M}_1$ determines the σ-algebra $\widehat{\mathcal{F}}$ of subsets of a set $\mathcal{D}$ and the appropriate conditional distribution $P_x(\cdot|\widehat{\mathcal{F}})$ named distribution of a course of time with respect to a sequence of states.

There are continuous semi-Markov processes, for which the sequence of condition has an especially simple view, for example, monotone processes, for which the sequence of states makes an interval on the line, and the distribution of process is reduced to the distribution of a course of time concerning a sequence of states, which is the inverse process with independent positive increments. The class of such processes we consider in the following chapter. On the other hand, for a Markov diffusion process the sequence of states, actually, completely determines the process: for this process the distribution of a course of time concerning a sequence of states is degenerate (Harlamov (1984)). One of the principal problems of the theory of semi-Markov processes is to prove a possibility of representation of any such process as Markov process transformed by a random time change (see Harlamov (1983b)).

9. Curvilinear integral. Let $\ell_1(\xi) \in \mathcal{M}_1$ be a sequence of essential states of a function $\xi \in \mathcal{D}$. Let us consider $\mathcal{T}_1(\xi)$ a set of contractions of all stopping times $\tau \in \mathcal{T}_1$ to a set $\ell^{-1}\ell(\xi)$. It is easy to show, that the set $\mathcal{T}_1(\xi)$ is linearly ordered. The function of two arguments $A(\tau_1, \tau_2)$ ($\tau_i \in \mathcal{T}_1(\xi)$) refers to as an additive functional, if $A(\tau_1, \tau_2) = 0$ with $\tau_1 \geq \tau_2$, and, if $\tau_1 \leq \tau_2 \leq \tau_3$, $A(\tau_1, \tau_3) = A(\tau_1, \tau_2) + A(\tau_2, \tau_3)$. It is interesting to consider the *curvilinear integral* of function $f(\pi_\tau)$ ($\pi_\tau(\xi) = \xi(\tau)$) on a segment of a sequence of states by an additive functional $A(\tau_1, \tau_2)$, which is determined in a traditional mode with the help of a converging sequence of integral sums appropriate to a sequence of finite partitions of this segment (Harlamov (1990)). In our denotation for the integral

$$(\ell) \int_{\tau_1}^{\tau_2} f(\pi_\tau)\, dA(\tau)$$

the index (ℓ) means left, that is connected to a choice of values of function $f(\pi_\tau)$ in the appropriate integral sums, which are calculated with τ equal to the left boundary points of appropriate intervals of partitions.

10. Weak convergence of semi-Markov processes. Time change with the help of process with independent positive increments, produced in strictly Markov process, is the most important source for semi-Markov processes of a general view, in particular, continuous, to be derived.

Other source is the passage to the limit, when a sequence of stepped semi-Markov processes converges weakly to a continuous semi-Markov process. The language of the first exit times and their iterations has appeared convenient for necessary and sufficient conditions of weak compactness and weak convergence in Skorokhod's space to be formulated (see. Harlamov (1976b)).

3. SEMI-MARKOV MODEL OF CHROMATOGRAPHY

3.1 CHROMATOGRAPHY

11. Method. Chromatography is a physics-chemical method of separation based on distribution of divided components between two phases: motionless (filter) and mobile (eluent). A mobile phase flows continuously through a motionless one. The purpose of separation is the preparative selection of substances in the pure state or physics-chemical measurements. Separation of complicated mixtures when moving along the surface of sorbent happens at the expense of distinctions of intermolecular interactions (various sorbableness) and consequent definition of divided components on an exit of a device with the help of special transmitters — detectors. Chromatography was open by the Polish botanician M.S. Tswett in 1901-1903 during a study of the structure of chlorofill and the mechanisms of photo-synthesis (see Giddings (1965), Yashin (1976), Golbert and Vigderhaus (1974) et al.).

12. Device. A central part of a classical chromatograph is a column, i.e. narrow cylindrical vessel, completed with sorbent. It solves the main problem: separation of the components of a mixture. All remaining devices in the chromatograph are intended either for filing divided components on an exit from a column, or for creation of stables conditions for work of a column. In analytical chromatograph in overwhelming majority of cases an elution variant of chromatography is used, in which inert liquid or the gas-carrier continuously passes through a column with a constant velocity. Before a column puts a device for input of an analyzable test in a column. The test injected in an eluent stream begins to move to an exit from a column. The components of a mixture, which weakly sorbable, pass through a column with a greater velocity and go out a column earlier. As a rule for chromatographic separation they use convertible physical adsorption, where adsorbing substances can be desorbed with a stream of liquid or the gas-carrier.

Column chromatography is possible to be classified as one-dimensional. As another example we mark planar chromatography, in which a layer of a paper or other porous material, on which the liquid moves in space of two dimensions and the filing is carried out by tracks of components on a surface of a material, plays the role of a column.

13. Chromatogram. The stream of eluent, including desorbing components, passes through a feeler of the detector, whose signal is registered. The signal curve of the detector that depends on time refers to as a *chromatogram*. Typical chromatogram of one absorbing substance is represented by a curve of the bell-shaped asymmetric form, which maxima is displaced with respect to the origin of coordinates (instant of introduction of the test). With small asymmetry, which is usually connected to a large displacement, this curve is well approximated by a Gaussian curve. Thus, basic characteristics of this curve are displacement, height of curve peak (maxima), width of a bell-shaped part and asymmetry. The area limited by this curve, is directly proportional to a quantity of the analyzed substance in test. Therefore it is natural to consider as performance of chromatographic process the parameters of a normalized curve, i.e. the result of division of it on magnitude of this area. The normalized curve can be interpreted as a density function of some random variable. In statistical terms the displacement is the first initial moment of this distribution, width is mean-square deviation (radical square of a variance), while asymmetry is the third central moment. The height of peak in this case is derivative magnitude from the previous parameters and actual form of a curve. Derivative magnitude is also such popular performance of chromatogram as a height equivalent to a theoretical plate (HETP), connecting variance and displacement. Below we shall give the exact definition of an HETP in terms of a mathematical model and a formula of its approximate value used for an evaluation.

The rather large difference in displacements with rather small variances of distributions appropriate to two various adsorbing substances is necessary for separation of these substances. Therefore efforts of investigators were directed on the study of the dependence of parameters of chromatogram on physics-chemical properties of a filter and eluent, chromatograph design and mode of operations, in particular, a velocity of a course eluent through a filter. One of such associations, formula of van Deemter, is deduced below as a corollary from the offered semi-Markov model of chromatography.

3.2 MODEL OF LIQUID COLUMN CHROMATOGRAPHY

14. Monotone process. For the beginning we consider the simplest but practically important case of process $\xi(t)$, when the adsorbing particle moves in one-dimensional space $\mathbb{X} = \mathbb{R}^1$ monotonously and without terminal absorption in a filter. This case is realized, for example, in long and narrow chromatography columns with a liquid mobile phase without chemical interaction of analyzable substance with a material of a filter. In this case the semi-Markov property turns into the requirement for a particle to have independent sojourn times in not intersected intervals: if $\tau_x(\xi)$ is a hitting time of the level x and $0 < a < b < c$, then $\tau_b - \tau_a$, $\tau_c - \tau_b$ are independent. From here follows, that the random process $(\tau(x))_{x \geq 0}$ is a process with independent increments (here and further $\tau(x) \equiv \tau_x$), and the process $\xi(t)$ is so called *inverse process with independent positive increments.*

15. The formula of Lévy. Let (P_x) be a family of distributions of this process. For any $\lambda > 0$ the *formula of Lévy* holds (see Ito and McKean (1965)):

$$P_x(\exp(-\lambda(\tau(x_2) - \tau(x_1)))) = \exp(-b(\lambda, x_1, x_2)),$$

where

$$\begin{aligned} b(\lambda, x_1, x_2) &= \lambda\, a([x_1, x_2)) + \int_{0+}^{\infty} (1 - e^{-\lambda u})\, n(du \times [x_1, x_2)) - \\ &- \sum_{x_1 \leq y_i < x_2} \log P_{y_i}(e^{-\lambda \sigma_0}), \end{aligned}$$

where $a(dx)$ is locally finite absolutely continuous measure on an axes $\mathbb{X}$, and $n(du \times dx)$ is a measure on a half-plane $(0, \infty) \times (-\infty, \infty)$, called a measure of Lévy, (y_i) is a countable set of points on an axes $\mathbb{X}$, σ_0 is the first exit time from an initial point, where $P_{y_i}(\sigma_0 > 0) > 0$.

16. Representation of Lévy and its interpretation. The most interesting corollary from the formula Lévy is the *representation of Lévy*: P_x-almost sure

$$\tau(x_2) - \tau(x_1) = a([x_1, x_2)) + \int_{0+}^{\infty} u\, \mathcal{P}(du \times [x_1, x_2)) + \sum_{x_1 \leq y_i < x_2} c_i,$$

where $\mathcal{P}(du \times dx)$ is a Poisson measure (random point field of a special view) on a half-plane $(0, \infty) \times (-\infty, \infty)$ with an intensity measure $n(du \times$

dx), c_i is a random variable, and the random variables in a set (c_i) are mutually independent and independent of a Poisson field $\mathcal{P}$.

The interpretation of Lévy representation is an interesting and the most attractive feature of a semi-Markov model chromatography. The deterministic part $a((x_1, x_2))$ can be interpreted as a result of driving of particles with a liquid and can be connected to a velocity $v(x)$ of moving of a liquid-carrier :

$$a([x_1, x_2)) = \int_{x_1}^{x_2} \frac{1}{v(x)}\, dx,$$

where $1/v(x)$ is a density of the measure $a(dx)$ with respect to Lebesgue measure on the axes $\mathbb{X}$. In a spatially-homogeneous case $v(x) \equiv v$ and $a([0, x)) = x/v$. The random part consists of intervals of a constancy of two sorts. At first, it is a sum of random delays in random points, whose duration and the position on a segment $[x_1, x_2)$ is determined by a Poisson measure $\mathcal{P}$. Secondly, it is a sum of independent random delays in fixed points from a countable set (y_i). For an homogeneous in space process $\xi(t)$ the delay of the second sort is absent. Further we shall be limited to processes without this sort of delays. As we see, the emerging of a random delay follows from the sole and very natural supposition: independence of sojourn times in not intersected intervals.

We note that the magnitudes of intervals of a constancy can be infinitely small (carrier of a measure $\mathcal{P}(du \times [x_1, x_2))$ begins from zero), however summarized length of these small intervals can be comparable with summary length of large intervals. It will be in case when for any $\varepsilon > 0$ $n((0, \varepsilon] \times [x_1, x_2)) = \infty$. Nowadays experimental data about magnitudes of intervals of a constancy are not known and consequently there are no reasons to neglect a case of infinite intensity.

17. Density of Lévy measure. The measure of Lévy can be represented as an integral of some conditional measure $\nu(du|x)$:

$$n(B \times [x_1, x_2)) = \int_{x_1}^{x_2} \nu(B|x)\, dx,$$

which in the homogeneous case turns into the equality $n(B \times [0, x)) = \nu(B)\, x$, where $\nu(B) = \nu(B|0)$. From here in a homogeneous case we have

$$b(\lambda, 0, x) = x \left(\frac{\lambda}{v} + \int_{0+}^{\infty} (1 - e^{-\lambda u}) \nu(du) \right).$$

18. Moment of distribution of delay. A residual $\tau(x) - a([0, x))$ in a monotone case refers to as a random delay, where $a([0, x))$ is a passage time of eluent through chromatographic column. Thus, it is a random part of Lévy representation (in this case, poissonian).

$$\tau(x) - a([0,x)) = \int_{0+}^{\infty} u\,\mathcal{P}(du \times [0,x)).$$

Moment of distribution of a random variable $\tau(x)$ can be obtained from the formula Lévy by differentiation on λ with $\lambda = 0$:

$$\begin{aligned} P_0\tau(x) &\equiv \int_0^{\infty} tP_0(\tau(x) \in dt) = -\frac{\partial}{\partial\lambda} P_0 e^{-\lambda\tau(x)}\Big|_{\lambda=0} = \\ &= a([0,x)) + \int_{0+}^{\infty} u\,n(du \times [0,x)), \\ \sigma^2(\tau(x)) &\equiv P_0(\tau(x) - P_0\tau(x))^2 = \int_{0+}^{\infty} u^2 n(du \times [0,x)). \end{aligned}$$

We also obtain 3-rd and 4-th central moment:

$$\begin{aligned} P_0(\tau(x) - P_0\tau(x))^3 &= \int_{0+}^{\infty} u^3 n(du \times [0,x)), \\ P_0(\tau(x) - P_0\tau(x))^4 &= \int_{0+}^{\infty} u^4 n(du \times [0,x)) + 3\sigma^4(\tau(x)). \end{aligned}$$

It is not difficult to count up other moment of distribution of the random variable $\tau(x)$. Since the fourth order the central moment are non-linear in x.

Let us remark, that in the considered case of monotone moving the important characteristic of a filter the magnitude $H_\tau \equiv \sigma^2(\tau(x))/x$, which in homogeneous case is equal to $\int_{0+}^{\infty} u^2\nu(du)$, depends on a velocity v of a carrying liquid only in the case, when the conditional measure Lévy $\nu(du)$ depends on v. The form of this dependence does not follow from the semi-Markov model itself and serves an additional assumption, which can be justified only with the help of experiment. It seems to be reasonable to suppose for the measure ν to be deformed in inverse proportion to a velocity of eluent:

$$\nu(B) = \frac{\nu_1(B)}{v}. \tag{23.9}$$

In this case, the relation between determinate and random parts of a delay does not depend on the velocity eluent: the longer particle is in moving, the longer on the average it is in a motionless condition on sorbent. The velocity of eluent plays a role of a scale parameter for process.

19. Height equivalent to a theoretical plate. Together with the variance $\sigma^2(\tau(x))$ the variance of an random variable $\xi(t)$ is of interest. Exactly this value in terms of a measure Lévy can be expressed with the awkward formula following from an inversion of Laplace transformation. To estimate this variance one can use the approximate equality: $\sigma(\xi(t_x)) \cong \sigma(\tau(x))\, v_c$, where t_x is obtained by a condition $P_0\xi(t_x) = x$ and $v_c = x/P_0\tau(x)$ (Giddings (1965)). In the homogeneous case we have

$$v_c = \left(\frac{1}{v} + \int_{0+}^{\infty} u\,\nu(du)\right)^{-1} = vR,$$

where R is a factor of a diminution of a velocity of transferable substance comparatively to the velocity of the carrier. In particular, if condition (23.9) is satisfied, R does not depend on the velocity.

The magnitude $\sigma^2(\xi(t_x))/x$ in work Giddings (1965) is called a *height equivalent to a theoretical plate* (HETP). This title is borrowed from practice of distillation, in which the separation of substances happens as a result of multiple transfusions of solutions from each plate of a long series of plates in following. In an outcome of these transfusions the test located in the first plate of a series, is spreaded on all plates, and, the maxima of concentration also moves along a series. We explicitly shall not describe this procedure, and also substantiation of the definition HETP for chromatograph. For practical purposes, in particular, for an evaluation of quality chromatograph they use approached HETP, namely

$$H \equiv (\sigma(\tau(x))\, v_c)^2/x. \tag{23.10}$$

The above quality chromatograph is less than H. In an one-dimensional monotone case we have $H = H_\tau R^2 v^2$ and, if condition (23.9) is satisfied

$$H = R^2 v \int_{0+}^{\infty} u^2 \nu_1(du),$$

i.e. it is directly proportional to an eluent velocity.

Let us remark, that the formula (23.10) is deduced without the supposition of a monotonicity of process $\xi(t)$. We use it below when analyzing continuous semi-Markov process of a diffusion type.

3.3 SOME MONOTONE SEMI-MARKOV PROCESSES

20. Model of Gut and Ahlberg. The set of measures $(\nu(du|x))_{x\in\mathbb{X}}$ of monotone process determines a so-called field of a delay. This field in a common case is infinite-dimensional, i.e. the measure $\nu(\cdot|x)$ is characterized by an infinite set of parameters. For practical applications it is preferable to use measures with a finite number of parameters. The example of a field with finite intensity is considered in work of Gut and Ahlberg (1981). In this work, devoted to a problem of summation of series with a random number of summands, the stochastic model of chromatography was offered. In this model a particle moves with stops in such a way that intervals of uniform moving with a velocity v alternate with intervals of a constancy. The supposition about an exponential distribution of length of an interval of uniform moving (with a parameter α), which is accepted in this work, refers these processes to a class of homogeneous monotone continuous semi-Markov processes. The considered process has finite intensity of a random point field of jumps of function $\tau(x)$ and exponential distribution of magnitudes of jumps, which are intervals of a constancy of function $\xi(t)$ (with a parameter β). The index of an exponential function in Lévy expansion for this process is of the form

$$b(\lambda,0,x) = \frac{x\lambda}{v} + x\alpha \int_0^\infty (1-e^{-\lambda u})e^{-\beta u}\,du.$$

The parameters v, α and β of this process are functions of the first three moment of distribution of the random variable $\tau(x)$:

$$P_0\tau(x) = x\left(\frac{1}{v}+\frac{\alpha}{\beta^2}\right), \quad \sigma^2(\tau(x)) = \frac{2\alpha x}{\beta^3}, \quad P_0(\tau(x)-P_0\tau(x))^3 = \frac{6\alpha x}{\beta^4},$$

which can be evaluated with a chromatogram.

21. Inverse gamma-process. The inverse gamma-process (without drift) is a monotone semi-Markov process with independent positive increments of a random function $\tau(x)$, distributed according to a gamma-distribution:

$$\frac{1}{\Gamma(x\alpha)} h^{x\alpha} t^{x\alpha-1} e^{-ht} \quad (x>0),$$

where $\Gamma(x)$ is the gamma-function, $\alpha > 0$ is a scale parameter and $h > 0$ is a form parameter. The index of an exponential function of Lévy of this process has the form

$$x\alpha \ln\frac{h+\lambda}{h} = x\alpha \int_0^\infty (1-e^{-\lambda u})\frac{e^{-hu}}{u}\,du$$

(see Prudnikov et al (1981) formula 2.3.19.28). For practical purposes, it is expedient to supplement this index with a term $x\lambda/v$, where v is velocity of eluent. This supplemented index

$$b(\lambda,0,x) = \frac{x\lambda}{v} + x\alpha \int_0^\infty (1-e^{-\lambda u})\frac{e^{-hu}}{u}\,du$$

corresponds to an inverse gamma-process with drift, for which

$$P_0\tau(x) = \frac{x}{v} + \frac{x\alpha}{h}, \quad \sigma^2(\tau(x)) = \frac{x\alpha}{h^2}, \quad P_0(\tau(x) - P_0\tau(x))^3 = \frac{2x\alpha}{h^3}.$$

The principal advantage of an inverse gamma-process with drift on a comparison with other monotone semi-Markov models is its simplicity, sufficient flexibility due to its three parameters, and also its well-known properties. One can find tables of its values, for example, in the book Pagurova (1963).

22. Process of maximum values. The large subclass of inverse processes with independent positive increments is constituted with processes of record values connected to semi-Markov non-monotone processes. For continuous one-dimensional process $\xi(t)$ with an initial point $x = \xi(0)$ the process $\eta(t) = \max(\xi(s) : 0 \le s \le t)$ is called a *process of maximum values*. Thus, if $\xi(t)$ is a semi-Markov process, $\eta(t)$ is semi-Markov too. Besides it is monotone process and thereof inverse process with independent positive increments.

Let us consider, for example, such continuous semi-Markov process as a homogeneous Brownian motion with drift. The density of its one-dimensional distribution is given by

$$p_t(x) = \frac{1}{\sqrt{2\pi d\, t}} \exp(-\frac{(x-\mu t)^2}{2d\, t}).$$

The transitional semi-Markov generating function of this process $h_G(\lambda, x)$ (where $G = (a, b)$ and $a < x < b$) is expressed by the formula

$$h_G(\lambda, x) = e^{2A(b-x)} \frac{\sinh(x-a)\sqrt{4A^2 - 2B(\lambda)}}{\sinh(b-a)\sqrt{4A^2 - 2B(\lambda)}},$$

where $A = \mu/(2d)$ and $B(\lambda) = -\lambda/d$, (see item 6). The generating function for time of the first exit of the appropriate process of maximum values from an interval $(-\infty, b)$ can be obtained from this formula by taking the limit as *goes* to $-\infty$. Taking into account values of factors, we obtain

$$P_0(e^{-\lambda\tau(x)}) = \exp\Big(-x\sqrt{\frac{2}{d}}\Big(\sqrt{\frac{\mu^2}{2d} + \lambda} - \frac{\mu}{\sqrt{2d}}\Big)\Big). \tag{23.11}$$

Let us take apart a case $\mu \ge 0$. In this case the process $\eta(t)$ does not have an infinite interval of a constancy: $P_0(\tau(x) = \infty) = 0$. We need to find the parameters in the formula of Lévy, i.e. to find the unknown measures $a(\cdot)$ and ν in the equation

$$\lambda a([0,x)) + x\int_{0+}^{\infty}(1-e^{-\lambda u})\nu(du) = x\sqrt{\frac{2}{d}}\Big(\sqrt{\frac{\mu^2}{2d}+\lambda}-\frac{\mu}{\sqrt{2d}}\Big). \quad (23.12)$$

By dividing both parts of the equation by λ and letting λ go to infinity, we obtain $a([0,x)) = 0$. It is easy to check up, that the function

$$\frac{1}{\sqrt{2\pi d\, u^3}}\exp\big(-u\frac{\mu^2}{2d}\big). \quad (23.13)$$

is a density of a measure of Lévy in the equation (23.12). Adding to an exponent in the formula of Lévy a term $x\lambda/v$, where v is the velocity of the eluent, we obtain the process of maximum values with drift. The first moment of the distribution $\tau(x)$ for this process looks like

$$P_0\tau(x) = \frac{x}{v}+\frac{x}{\mu}, \quad \sigma^2(\tau(x)) = \frac{x\,d}{\mu^3}, \quad P_0(\tau(x)-P_0\tau(x))^3 = \frac{3x\,d^2}{\mu^5}.$$

For modeling of real processes it is possible to add a scale parameter, for which the evaluation of the fourth central moment of distribution submitted chromatogram would be necessary. With $\mu = 0$ for a process without drift we obtain monotone, slowly growing process: the average first passage time of any level is equal to infinity, though the probability of its reaching is equal to 1. To the case of a process of maximum values without drift with a condition $\mu < 0$ we shall return below.

3.4 PROCESSES OF TRANSFER WITH TERMINAL ABSORPTION

23. Probability of final stopping and conditional distribution of the first exit time. We shall consider continuous monotone semi-Markov process, for which the trajectory has an infinite interval of a constancy, i.e. in some instant it stops for ever (we distinguish cases of a final stopping and breakage of a trajectory). In substantial terms this process is connected to the terminal absorption of a particle by a material of a filter. The final (terminal) stopping can be interpreted as a sedimentation of a particle in a filter, for example, in an outcome of a chemical transformation. Such processes are interesting from a point of view of geological applications (see Golubev (1981), Korzhinski (1982)).

From a semi-Markov property follows, that with $0 < x_1 < x$,

$$P_0 e^{-\lambda\tau(x)} = P_0 e^{-\lambda\tau(x_1)} P_{x_1} e^{-\lambda\tau(x)},$$

whence $P_0(\tau(x) < \infty) = P_0(\tau(x_1) < \infty)P_{x_1}(\tau(x) < \infty)$, and, hence, at first,

$$P_0(e^{-\lambda\tau(x)}|\tau(x) < \infty) = P_0(e^{-\lambda\tau(x_1)}|\tau(x_1) < \infty)P_{x_1}(e^{-\lambda\tau(x)}|\tau(x) < \infty) =$$

$$= \exp(-b(\lambda, 0, x)),$$

(see item 15) and, secondly, $P_0(\tau(x) < \infty) = \exp(-c([0, x))$, where $c(dx)$ — measure on an axes $\mathbb{X}$. We shall assume, that this measure (as well as the measure $a(dx)$) is absolutely continuous concerning Lebesgue measure. Hence, in this case Lévy's formula is supplemented by one more term (not depending on λ)

$$P_{x_1}(e^{-\lambda\tau(x_2)}) = \exp(-c([x_1, x_2)) - \lambda a([x_1, x_2)) - \\ - \int_{0+}^{\infty} (1 - e^{\lambda u}) n(du \times [x_1, x_2))). \tag{23.14}$$

The distribution of $\tau(x)$ is sub-probability, and the moment, obtained in item 18, are moment concerning a measure $P_0(\cdot|\tau(x) < \infty)$, i.e. conditional moment. However, distribution of $\xi(t)$ with any $t \geq 0$ is probability:

$$P_0(\xi(t) < x) = P_0(\tau(x) = \infty) + P_0(\tau(x) < \infty) P_0(\xi(t) < x|\tau(x) < \infty).$$

The limiting distribution of a position of an absorbed particle as $t \to \infty$ for a spatially homogeneous case is exponential. It is characteristic of one-dimensional channel with an absorption, when the substance is carried out by an eluent from a point source and the velocity of the eluent along the channel is constant.

24. Process of maximum values of one-dimensional diffusion with negative drift. An example of homogeneous process with terminal absorption is the process, taken apart in item 22, of maximum values of a Brownian motion with negative drift. Let us consider expression (23.11) under condition $\mu < 0$. This process finally stops in a random instant and at a random level. From the formula (23.11) with $\lambda = 0$ we obtain probability of an exit on a level x

$$P_0(\tau(x) < \infty) = \exp(-2x|\mu|) < 1.$$

According to item 22 the formula (23.14) for this process turns into the expression

$$P_0(\exp(-\lambda\tau(x))) = \exp\left(-2x|\mu| - x \int_{0+}^{\infty} (1 - e^{\lambda u}) \frac{1}{\sqrt{2\pi d\, u^3}} \exp\left(-u\frac{\mu^2}{2d}\right) du\right).$$

From the equality

$$P_0(\exp(-\lambda\tau(x-\varepsilon)), \tau(x) = \infty) = P_0(\exp(-\lambda\tau(x-\varepsilon)))\, P_{x-\varepsilon}(\tau(\varepsilon) = \infty)$$

($\varepsilon > 0$) we obtain a conditional generating function of a moment of a beginning of an infinite interval of a constancy:

$$P_0(\exp(-\lambda\tau(x-0))|\tau(x)=\infty) = \exp(-x(\sqrt{2\frac{\lambda}{d}+\mu^2} - |\mu|)),$$

though this moment is not a stopping time in the sense of Ito and McKean (1965).

25. Two-dimensional and three-dimensional radial streams with terminal absorption. In case of spreading of substance from a point source in all directions on a plane (two-dimensional space) or in volume (three-dimensional space) the scalar velocity eluent decreases with the distance from a source: $|v(x)| \sim 1/|x|$ $(d=2)$ and $|v(x)| \sim 1/|x|^2$ $(d=3)$, where the source is located in an origin of coordinates.

Let eluent flows be uniform on all directions of a plane. In this case we shall consider the transfer of substance along a ray from the origin, where it goes with a velocity $|v(x))| = v_1/r$ (here x and $v(x)$ are vectors, $|x| = r$ is length of a vector x). Thus, we deal again with one-dimensional monotone semi-Markov process. Let $r \geq 0$ be its spatial (one-dimensional) coordinate. We shall assume, as well as in a case with a measure Lévy (1.9), that the density $c'(r)$ of the measure $c(dr)$ is inversely proportional to the velocity of a stream:

$$c'(r) = \frac{c_1}{|v(r)|}. \tag{1.15}$$

In this case

$$P_0(\tau(r)<\infty) = \exp(-\int_0^r \frac{c_1}{v_1} r\,dr) = \exp(-\frac{c_1 r^2}{2v_1}).$$

From here follows, that the bivariate distribution of the absorbed particle is normal and centrally symmetric with a density

$$f_2(x) = \frac{1}{2v_1\pi}\exp(-\frac{c_1 r^2}{2v_1}).$$

In a three-dimensional case a velocity of the eluent decreases as $v(r) = v_1/(r^2)$. Doing the supposition (1.15) about the influence of the velocity on probability of a sedimentation, we obtain a three-dimensional density

$$f_3(x) = \frac{c_1}{4v_1\pi}\exp(-\frac{c_1 r^3}{3v_1})$$

($r = |x|$), which decreases faster, than normal, with $r \to \infty$.

In the two previous examples we assumed, that eluent flows uniformly from a point source and is spreaded in the appropriate space (plane or volume) lost-free. However, in a plane case the loss of eluent (evaporation) is possible. That means that the velocity reduction is greater than in the case of lack of evaporation. Let $v(x) \doteq v_2/(r^3)$. From here $P_0(|\xi(\infty)| \geq r) = \exp(-c_1 r^3/(3v_2))$. To find the density of two-dimensional distribution of the absorbed particle , we have to differentiate on r this function with a sign minus and then to divide on $2\pi r$. In an outcome we obtain

$$f_4(x) = \frac{c_1 r}{2v_2\pi} \exp(-\frac{c_1 r^3}{3v_2}),$$

i.e. in contrast to two previous cases the measure of the absorbed particles has a minimum in a place of inflow of the eluent. The maxima of a measure forms a circle with a center in a point of the source.

3.5 PROCESS OF TRANSFER WITH DIFFUSION

26. Moment of distribution of a random delay. Taking apart of a diffusion is necessary with the gaseous carrier, for example, in gas chromatography. In this case they use one-dimensional column chromatograph. Let us consider continuous semi-Markov process of a diffusion type $\xi(t)$ in space $\mathbb{X} = \mathbb{R}^1$. For the sake of simplicity we shall assume, that the trajectories of process have not infinite intervals of a constancy. The transitional semi-Markov generating function $f = f_G(\lambda, x)$ of this process satisfies the differential equation (23.7) with the appropriate boundary conditions. In particular, if $G = (-\infty, b)$, then $f(-\infty) = 0$, $f(b) = 1$. In a homogeneous case ($A(x) \equiv A$, $B(\lambda, x) \equiv B(\lambda)$) the solution of the equation with these boundary conditions looks like

$$f_G(\lambda, x) = \exp\Big(-2(b-x)\Big(\sqrt{A^2 - B(\lambda)/2} - A\Big)\Big)$$

(see (23.11)). Substituting in this formula $b - x$ on x and differentiating with respect to λ, we obtain moment of distribution of an random variable $\tau(x)$ (time of the first reaching of a level x):

$$P_0\tau(x) = -xB'(0)/(2A),$$

$$\sigma^2(\tau(x)) = x(B''(0)/(2A) + (B'(0))^2/(2A)^3).$$

27. Determination of velocity of eluent. Using one-dimensional column chromatograph they usually connect one extremity of a column to a vessel with condensed gas of sufficient volume permitting to consider a column as semi-infinite. On the other extremity (usually open for going out gas) the detector is established to record the first emerging time of the substance of the test. That means that the detector registers the process of a maximum value of another process, which is a continuous semi-Markov process of a diffusion type with positive drift (moving of individual adsorbing particle of the substance of the test). These suppositions enable to express basic moment of distribution of a random variable $\tau(x)$ in terms of factors of the appropriate differential equation. However, with the help of this monotone process the velocity of eluent cannot be determined, because the transformation of passage to maximum values adds to intervals of a constancy of initial process (intervals of adsorption) intervals of a constancy associate with negative evolutions of initial process. To receive the formula of the dependence of H on v, representing the theoretical variant of the formula of Van Deemter, and to determine the factor of a delays associate with intervals of an adsorption, it is necessary to compare the initial process with the appropriate Markov process without intervals of a constancy, which determines a velocity of eluent.

It is accepted to consider, that the velocity of gas-carrier is a velocity of a maxima of a density of spatial distribution of least adsorbing component of a mixture. It is natural to accept as the law of moving of this component the law of moving of adsorbing component without a random delay. Thus we make the assumption, that, as well as in case of monotone moving, the laws of moving of various substances of a mixture differ only by random delay. In this case the velocity of gas-carrier (determinate component) can be evaluated from the outcome of the analysis of chromatogram of adsorbing substance.

To separate the time of real moving from the time of delay for an adsorption, we consider the conditional distribution of a course of time with respect to a sequence of states. According to the outcomes of work Harlamov (1985) (with a lack of fixed points of an adsorption) P_0-a.s.

$$P_0(\exp(-\lambda\tau)|\widehat{\mathcal{F}}) = \exp\Big(-\lambda a_\tau - \int_{0+}^{\infty}(1-e^{-\lambda u})n_\tau(du)\Big),$$

where a_τ, $n_\tau(du)$ are $\widehat{\mathcal{F}}$-measurable additive functionals. In this Lévy expansion the component n_τ determines intensity of a Poisson field of jumps of time on a sequence of states, which make intervals of a constancy of all trajectories with the given sequence of states, i.e. intervals of delay of a transferable particle on a filter. The appropriate process $(\overline{P}_x)$ with the same distribution on a set of sequences of states, but without intervals of an adsorption, has a conditional generating function who takes the following form

$$\overline{P}_0(\exp(-\lambda\tau)|\widehat{\mathcal{F}}) = \exp(-\lambda a_\tau).$$

It is known (Harlamov (1984)), that this process is a diffusion Markov process. The semi-Markov generating function of this process satisfies the differential equation

$$f'' + 4A(x)\, f' + \overline{B}(\lambda, x)\, f = 0.$$

Let us define the parameter $\overline{B}(\lambda, x)$ of this equation. From Harlamov (1989) and (1983a) the representation with the help of left curvilinear integral is fair

$$P_0(\exp(-\lambda\tau)|\widehat{\mathcal{F}}) = \exp\Big((\ell)\int_0^\tau \frac{B(\lambda, \pi_{\tau_1})}{\gamma(\pi_{\tau_1})}\, d\mu(\tau_1)\Big),$$

$$\overline{P}_0(\exp(-\lambda\tau)|\widehat{\mathcal{F}}) = \exp\Big((\ell)\int_0^\tau \frac{\overline{B}(\lambda, \pi_{\tau_1})}{\overline{\gamma}(\pi_{\tau_1})}\, d\overline{\mu}(\tau_1)\Big),$$

where $\gamma(x) = \lim_{r\to 0} P_x(\sigma_r)/r^2$, $\overline{\gamma}(x) = \lim_{r\to 0} \overline{P}_x(\sigma_r)/r^2$, and the additive functionals are defined as $\mu(0,\tau) = P_0(\tau|\widehat{\mathcal{F}})$, $\overline{\mu}(0,\tau) = \overline{P}_0(\tau|\widehat{\mathcal{F}})$.

On the other hand, as the function $-B'(\lambda, x)$ is completely monotone (see Harlamov (1983a)), under the theorem of Bernstein (see Feller (1966)) the expansion is fair: $(\exists a_1(x),\, n_1(du|x))$

$$-B(\lambda, x) = a_1(x)\lambda + \int_{0+}^\infty (1 - e^{-\lambda u}) n_1(du|x).$$

From here P_0-almost sure

$$(\ell)\int_0^\tau \frac{\overline{B}(\lambda, \pi_{\tau_1})}{\overline{\gamma}(\pi_{\tau_1})}\, d\overline{\mu}(\tau_1) = (\ell)\int_0^\tau \frac{-a_1(\pi_{\tau_1})\lambda}{\gamma(\pi_{\tau_1})}\, d\mu(\tau_1),$$

Here τ is an arbitrary interior stopping time, for example, $\tau = \sigma_{G(r)}$. From the outcomes of work Harlamov (1989) follows, that $(\forall x \in \mathbb{X})$ with respect to a measure P_x weak convergence $\mu(\sigma_r)/r^2 \to \gamma(x)$ and $\overline{\mu}(\sigma_r)/r^2 \to \overline{\gamma}(x)$ is fair, while $\overline{B}(\lambda, x) = -a_1(x)\lambda$. Hence, the appropriate Markov process has a parameter of a diffusion $d(x))1/a_1(x)$. The parameter of drift of diffusion Markov process $(\overline{P}_x)$ is equal to $v(x) \equiv 2A/a_1(x)$.

28. Formula of van Deemter. From the formula (23.10) and from item 26 we obtain

$$H = \frac{x\,\sigma^2(\tau(x))}{(P_0\tau(x))^2} = \frac{B''(0)}{(B'(0))^2}\, 2A + \frac{1}{2A} = \frac{B''(0)}{(B'(0))^2}\, a_1 v + \frac{1}{a_1 v} =$$

$$= \frac{\int_{0+}^{\infty} u^2 n_1(du)}{(a_1 + \int_{0+}^{\infty} u\, n_1(du))^2}\, a_1 v + \frac{1}{a_1 v}.$$

In this formula it is not difficult to find out a known relation of van Deemter for dependence of a height equivalent to a theoretical plate on a velocity of gas-carrier (Giddings (1965)). The term of inverse dependence a velocity is generated by a diffusion of process. It is present even with $n_1(du) = 0$. The term with a direct dependence on a velocity occurs only for adsorbing substance.

The deduced formula is obtained for a fixed length of chromatograph column. For a porous material of a filter, what a column is filled with, this supposition is not fulfilled, as the point of an injection of test is connected to the point of installation of the detector with a set of possible paths, on which the absorbing particle can move. To take into account this situation it is necessary to consider the length of a column as a random variable with some distribution, which can be estimated experimentally. Variance of this variable brings in an additional term to the formula of HETP, which does not depend on a velocity of the eluent. Thus, the full formula of dependence on a velocity of eluent looks like

$$H = H_0 + a_1 v \int_{0+}^{\infty} u^2\, n_1(du)/(a_1 + \int_{0+}^{\infty} u\, n_1(du))^2 + (a_1 v)^{-1}.$$

In contrast to monotone process for diffusion type process there exists an optimum velocity of eluent, ensuring a minimum of function $H = H(v)$.

29. Notes. For monotone process following the equality holds

$$P_0(\xi(t) > x) = P_0(\tau(x) < t),$$

that enables to restore spatial distribution of process from distributions of the first exit time. For non-monotone process, what the considered semi-Markov process of a diffusion type is, the following inequality holds

$$P_0(\xi(t) > x) \le P_0(\tau(x) < t).$$

However, if $P_0(\xi(t) > a) > 1 - \varepsilon$ for all $t > t_0$, for homogeneous process $P_0(\tau(x) < t) \le \varepsilon + P_0(\xi(t + t_0) > x + a)$, whence the connection between spatial and time performances follows. In particular, the velocity of distribution of expectation of magnitude $\xi(t)$ is approximately equal to $v_c \equiv x/P_0\tau_x$.

The generalization of the obtained formulae on a case of inhomogeneous in space processes does not meet principal difficulties. In particular, for an evaluation of such local performance, as approached HETP, it is enough in the last formula to replace constants factors with some depending on x functions.

The supposition that a deterministic component of moving for different substances of a mixture is the same, requires experimental check. It seems rather probable, that in a lack of an adsorption the drift for various substances of the same mixture, because the separation of a mixture does not happen. However substances can differ by a parameter of a variance. Hence, the form of an association of function H from v for various substances of a mixture will be the same, but with various factors with v and $1/v$ accordingly. These factors depend on the parameters a_1 and n_1, each of which can accept different values for various substances.

References

[1] Feller, W. (1966). *An introduction to probability theory and its applications*, **II**, NY.

[2] Giddings, J.C. (1965). *Dynamics of chromatography*, **1**, NY.

[3] Golbert, K.A. and Vigderhaus, M.S. (1974). *Course of gas chromatography*, M. (in Russian).

[4] Golubev, V.S. (1981). *Dynamics of geochemical processes*, M. (in Russian).

[5] Gut, A. and Ahlberg, P. (1981). On the theory of chromatography based upon renewal theory and a central limit theorem for randomly indexed partial sums of random variables. *Cheica Scripta*, **v.18, 5**, 248-255.

[6] Harlamov, B.P. (1976a). On a connection between random curves, time change and regenerative times of random processes. *Zapiski nauch. semin. LOMI*, **55**, 128–194 (in Russian).

[7] Harlamov, B.P. (1976b). On convergence of semi-Markov walks to continuous semi-Markov process. *Theory of probability and its applications*, **XXI, 3**, 497–511 (in Russian).

[8] Harlamov, B.P. (1983a). Transition functions of a continuous semi-Markov process on the line. *Zapiski nauch. semin. LOMI*, **130**, 190–205 (in Russian).

[9] Harlamov, B.P. (1983b). Representation of semi-Markov process as Markov process transformed with time change. *Theory of probability and its applications*, **XXVIII, 4**, 653-667 (in Russian).

[10] Harlamov, B.P. (1984). Continuous semi-Markov processes and an extremal property of Markov processes. In *Proceedings of the 7-th Conference on Probability Theory, aug.29–sept.4, 1982, Braşov*, 423–430, România, Bucureşti.

[11] Harlamov, B.P. (1985). Distribution of course of time with respect to sequence of states in a semi-Markov process. *Zapiski nauch. semin. LOMI*, **142**, 167-173 (in Russian).

[12] Harlamov, B.P. (1989). Characteristic operator and curvilinear integral for a semi-Markov process. *Zapiski nauch. semin. LOMI*, **177**, 170–180 (in Russian).

[13] Harlamov, B.P. (1990). Random curvilinear integrals and their application. *Theory of probability and its applications*, **35**, **1**, 118–130 (in Russian).

[14] Hofmann, A. (1972). Chromatographic theory of infiltration metasomatism and its application to feldspars. *Amer. J. of Sci.*, **272**, 69–90.

[15] Ito, K. and McKean, G. (1965). *Diffusion processes and their sample paths*, NY.

[16] Kolmogorov, A.N. and Fomin, S.V. (1972). *Elements of the theory of function and functional analysis*, M. (in Russian)

[17] Korzhinskii, D.S. (1982). *Theory of metasomatic zonality*, M. (in Russian)

[18] Lejbenzon, L.S. (1947). *Moving of natural liquids and gases through porous medium*, M. (in Russian)

[19] Pagurova, V.I. (1963). *Tables of incomplete gamma-function*, M. (in Russian)

[20] Prudnikov, A.P., Brychkov, Yu.A. and Marichev, O.I. (1981). *Integrals and series*, M. (in Russian)

[21] Yashin, Ya.I. (1976). *Physics-chemical basis of chromatographic separation*, M. (in Russian)

Chapter 24

THE STRESS TENSOR OF THE CLOSED SEMI-MARKOV SYSTEM. ENERGY AND ENTROPY

George M. Tsaklidis

Aristotle University of Thessaloniki, Greece

Abstract The set of the attainable structures of a closed continuous time Homogeneous Semi-Markov System (HSMS) with n states, is considered as a continuum and the evolution of the HSMS in the Euclidean space E^n corresponds to its motion. A suitable model of a continuum –for which the stress tensor depends on the acceleration gradient and the density-is proposed in order to explain the motion of the system. The adoption of this model enables us to establish the concept of the energy and the entropy of the HSMS.

Keywords: Stochastic population systems, continuous time homogeneous semi-Markov system, continuum mechanics.

1. INTRODUCTION

The continuous-time semi-Markov model can be considered in an analogous way as the discrete-time semi-Markov model. The aspects which have been studied in discrete and also in continuous time models are the asymptotic behaviour, stability, variability, attainability etc. Continuous-time semi Markov models arise in physics, actuarial work, biometry and manpower planning. The basic results can be found in the work of McClean (1976, 1978, 1980 1986), Mehlman (1979), Bartholomew (1982, 1986), Howard (1971) , Janssen (1986) and Bartholomew, Forbes and McClean (1991). The concept of the non homogeneous semi Markov system (NHMS) was introduced by Vassiliou and Papadopoulou (1992, 1998) in order to provide a general framework for Markov chain models in manpower systems.

2. THE CLOSED CONTINUOUS-TIME SEMI-MARKOV MODEL

In what follows denote by $S = \{1, 2, \ldots, n\}$ the state space of the model. The states $1, 2, \ldots, n$ are exclusive and exhaustive, i.e. every member of the system may be in one and only one state at some time point t. Also denote by p_{ij}, $i, j = 1, 2, \ldots, n$ the probability that a member who entered state i on its last transition will enter state j on its next transition, by τ_{ij}, $i, j = 1, 2, \ldots, n$, the holding time in state i before moving to state j (after j have been selected) and by $h_{ij}(t)$ the probability density function of the random variable τ_{ij} $(i, j = 1, 2, \ldots, n)$. Symbolize by τ_i the waiting time in i providing the successor state chosen is not known, and by $w_i(t)$ its probability density function. Thus, we have

$$prob\{\tau_{ij} \leq t\} = \int_0^t h_{ij}(t)dt \quad \text{and} \quad w_i(t) = \sum_{j=1}^{n} p_{ij} h_{ij}(t).$$

Also, let ${}^{\leq}h_{ij}(t)$ be the cumulative probability distribution of τ_{ij} and ${}^{>}h_{ij}(t)$ the complementary cumulative probability distribution of τ_{ij}. Hence,

$${}^{\leq}h_{ij}(t) = \int_0^t h_{ij}(t)dt \quad \text{and} \quad {}^{>}h_{ij}(t) = prob\{\tau_{ij} > t\} = \int_t^\infty h_{ij}(t)dt.$$

In the same way let

$${}^{\leq}w_i(t) = prob\{\tau_i \leq t\} = \int_0^t w_i(t)dt = \sum_{j=1}^{n} p_{ij} {}^{\leq}h_{ij}(t)$$

and

$${}^{>}w_i(t) = prob\{\tau_i > t\} = \int_t^\infty w_i(t)dt = \sum_{j=1}^{n} p_{ij} {}^{>}h_{ij}(t).$$

Moreover, let the interval transition probabilities $\varphi_{ij}(t)$ and the counting transition probabilities $\varphi_{ij}(k/t)$ be defined as follows :

$\varphi_{ij}(t)$ = *prob* {a member of the system which entered state i at some time point is in state j after time t}, and

$\varphi_{ij}(k/t)$ = *prob*{ a member of the system is in state j at time t having made k transitions in the time interval $(0, t]$, given that it entered state i at time zero}

In order to describe the random variables for all possible states we will use the matrix notation

$$\begin{array}{lll} \mathbf{P} = (p_{ij}), & \mathbf{H}(t) = (h_{ij}(t)), & {}^{\leq}\mathbf{H}(t) = ({}^{\leq}h_{ij}(t)), \\ {}^{>}\mathbf{H}(t) = ({}^{>}h_{ij}(t)), & \mathbf{\Phi}(t) = (\varphi_{ij}(t)), & \mathbf{\Phi}(k/t) = (\varphi_{ij}(k/t)), \end{array}$$

and

$$\begin{array}{ll} {}^{\leq}\mathbf{W}(t) = diag({}^{\leq}w_i(t)), & {}^{>}\mathbf{W}(t) = diag({}^{>}w_i(t)), \\ \mathbf{C}(t) = \mathbf{P} \diamond \mathbf{H}(t) = (p_{ij}h_{ij}(t)). & \end{array}$$

If $a(t)$ is a function, then we symbolize by $a^G(t)$ the geometric transform and by $a^L(s)$ the Laplace transform of $a(t)$. Extending the notation for the matrices, we define the geometric and the Laplace transform of $\mathbf{A}(t) = (a_{ij}(t))$, to be $= \mathbf{A}^G(s) = (a^G_{ij}(s))$ and $\mathbf{A}^L(s) = (a^L_{ij}(s))$ respectively.

Now, by the same logic used for the discrete time model a useful formula for the interval transition probabilities can be derived. Thus (by [5]),

$$\varphi_{ij}(t) = \delta_{ij}{}^{>}w_i(t) + \sum_{k=1}^{n} p_{ik} \int_0^t h_{ik}(\tau)\varphi_{kj}(t-\tau)d\tau,$$

(where δ_{ij} is the Kronecker delta) and in matrix notation

$$\mathbf{\Phi}(t) = {}^{>}\mathbf{W}(t) + \int_0^t (\mathbf{P} \diamond \mathbf{H}(\tau))\mathbf{\Phi}(t-\tau)d\tau = {}^{>}\mathbf{W}(t) + \int_0^t \mathbf{C}(\tau)\mathbf{\Phi}(t-\tau)d\tau.$$

Taking the Laplace transform of the last equation we get $\mathbf{\Phi}^L(s) = {}^{>}\mathbf{W}^L(s) + \mathbf{C}^L(s)\mathbf{\Phi}^L(s)$, thus

$$\mathbf{\Phi}^L(s) = (\mathbf{I} - \mathbf{C}^L(s))^{-1} . {}^{>}\mathbf{W}^L(s) \tag{24.1}$$

(while the matrix $(\mathbf{I} - \mathbf{C}^L(s))^{-1}$ always exists). In the same way using the counting transitions we have (by [5])

$$\mathbf{\Phi}^{GL}(y/s) = (\mathbf{I} - y\mathbf{C}^L(s))^{-1} . {}^{>}\mathbf{W}^L(s), \tag{24.2}$$

where

$$\mathbf{\Phi}^{GL}(y/s) = (\varphi^{GL}_{ij}(y/s)) \quad \text{and} \quad \varphi^{GL}_{ij}(y/s) = \sum_{k=0}^{\infty} y^k \int_0^{\infty} \varphi_{ij}(k/t)e^{-st}dt.$$

Now, the possibility of deriving $\mathbf{\Phi}(t)$ from (24.1) or from (24.2) (taking into account that $\varphi_{ij}(t) = \sum_{k=0}^{\infty} \varphi_{ij}(k/t)$) depends on the form of the transformed $\mathbf{\Phi}^L(s)$ or $\mathbf{\Phi}^{GL}(y/s)$, and thus on the functions $h_{ij}(t)$).

Let $\mathbf{p}(0)$ denote the starting vector of a closed continuous-time semi-Markov system and $\mathbf{x}(t) = (x_i(t))_{i\in S}$ the probability vector, the i-th component of which is the probability for a member which entered the system at time 0 to be in state i at time t. Then,

$$\mathbf{x}'(t) = \mathbf{x}'(0)\Phi(t), \tag{24.3}$$

and

$$\dot{\mathbf{x}}'(t) = \mathbf{x}'(0) \cdot \dot{\Phi}(t), \tag{24.4}$$

where $\dot{\mathbf{x}}(t) = (\frac{d}{dt}x_i(t))$, $\dot{\Phi}(t) = (\frac{d}{dt}\varphi_{ij}(t))$ and the prime means transposition of the respective vectors (or -generally- of a matrix). The matrix $\Phi(t)$ turns out to be stochastic. From (24.3) and (24.4) we get that the velocity field $\dot{\mathbf{x}}(t)$ of the HSMS at time t in some point $\mathbf{x}$ is given by

$$\dot{\mathbf{x}}'(t) = \mathbf{x}'(t) \cdot \{(\Phi(t))^{-1}\dot{\Phi}(t)\}, \tag{24.5}$$

(providing that $det(\Phi(t)) \neq 0$).

3. THE CONTINUUM VIEWPOINT

Equation (24.3) determines the motion (the trajectory) of a stochastic structure in E^n. If we consider every structure of the HSMS as a 'particle' of the n-dimensional space E^n, then we assign material behaviour to E^n. Now, by adopting the viewpoint that

"the motion of the particle at every time point is due to the interaction of it with its surrounding",

E^n can be regarded as a continuum. Symbolize by A_t the set of the attainable structures of a HSMS with fixed size, i.e. the set of the vectors $\mathbf{x}(t)$ which we get from (24.3) as the initial state vector $\mathbf{x}(0)$ runs over all stochastic n-tuples, and denote by $A_n(t)$ the region of E^n defined by A_t. In what follows we are interested in the motion-evolution of the continuum possessing the region $A_n(0) \subset E^n$ at time $t = 0$ in the velocity field defined by (24.3). Every part of the "material continuum" $A_n(t)$, $t \geq 0$, is supposed to be subjected to surface forces. The material is supposed to be homogeneous and isotropic. The $n \times 1$ stress vector $\mathbf{t}^{(\mathbf{n})}$ at some point P enclosed by the infinitesimal surface S, is defined (in the same way as in E^2 and E^3) by

$$\mathbf{t}^{(\mathbf{n})} = \lim_{\Delta S \to 0}(\Delta \mathbf{f}/\Delta S) = d\mathbf{f}/dS, \tag{24.6}$$

where $\mathbf{n}$ is the $n \times 1$ outward unit normal of the surface element ΔS of S (i.e. ΔS lies on the hyper-plane of E^n which is perpendicular to $\mathbf{n}$) and

$\Delta\mathbf{f}$ the resultant force exerted across ΔS upon the material within S by the material outside of S. The state of stress at the point P is given by the set of $\mathbf{t}^{(\mathbf{n})}$ generating from all the unit vectors $\mathbf{n}$, according to the formula

$$\mathbf{t}^{(\mathbf{n})} = \mathbf{T}\cdot\mathbf{n},$$

where $\mathbf{T}$ is the (symmetric) $n \times n$ stress tensor ([4]).

4. THE STRESS TENSOR OF THE CLOSED CONTINUOUS TIME HSMS

Equation (24.3) represents a system of n equations and (equivalently) equation (24.5) represents a system of n differential equations; since $x_1(t) + x_2(t) + ... + x_n(t) = 1$, for every $t \geq 0$, the variables $x_i(t)$, $i = 1, 2, \dots, n$ are dependent, and the motion takes place on the hyper-plane (Π) $: x_1 + x_2 + ... + x_n = 1$. Now, assume - without loss of generality - that the matrix of the transition probabilities $\Phi(t)$ is irreducible for every t. Then, it can be easily verified ([1],[6]) that (2.5) has a (stochastic) stability point, which will be denoted by π. Consider at π an orthogonal coordinate system $\{\mathbf{f}_1, \mathbf{f}_2, \dots, \mathbf{f}_n\}$, where $\mathbf{f}_1, \mathbf{f}_2, \dots, \mathbf{f}_{n-1}$ belong to (Π) and $\mathbf{f}_n \perp (\Pi)$, and let $\mathbf{F} = [\mathbf{f}_1, \mathbf{f}_2, \dots, \mathbf{f}_n] = [\mathbf{F}_1|\mathbf{f}_n]$, where $\mathbf{F}_1 = [\mathbf{f}_1|\mathbf{f}_2|...|\mathbf{f}_{n-1}]$. Denote by $z_1, z_2, \dots, z_n$ the components of some point $\mathbf{z} \in= E^n$ with respect to the coordinate system $\{\mathbf{f}_1, \mathbf{f}_2, \dots, \mathbf{f}_n\}$. Since we are interested in the case $\mathbf{z} \in (\Pi)$ then $z_n = 0$, and Equation (24.5) could be considered for the hyper-plane (Π), with respect to the coordinate system $\{\mathbf{f}_1, \mathbf{f}_2, \dots, \mathbf{f}_{n-1}\}$, in the form

$$\dot{\mathbf{z}}'(t) = \mathbf{z}'(t)\cdot\mathbf{S}^{(1)}(t) + \mathbf{f}_1'(t), \tag{24.7}$$

where $\mathbf{z}(t) = (z_1(t), z_2(t), \dots, z_{n-1}(t))'$ and $\mathbf{S}^{(1)}(t)$ and $\mathbf{f}_1(t)$ depend on $\dot{\Phi}(t)$.

Now, the acceleration $\mathbf{a}(t)$ can be derived from (24.7):

$$\mathbf{a}(t) = \ddot{\mathbf{z}}(t) = \mathbf{z}'(t)\mathbf{S}^{(2)}(t) + \mathbf{f}'(t),$$

where $\mathbf{S}^{(2)}(t)$ and $\mathbf{f}(t)$ depend on $\ddot{\Phi}(t)$.

In conclusion, we have reduced (24.5) to the form (24.7) and we can use (24.7) instead of the (equivalent) initial formula (24.3). Under the assumptions of section 3 we regard (24.7) as expressing the velocity field of a continuum. For this purpose we shall find out an appropriate stress tensor $\mathbf{T} = (T_{ij})$, $i, j = 1, 2, \dots, n-1$, which would satisfy Cauchy's equation of motion at every point P (infinitesimal region at P) of the continuum, i.e.

$$\rho(t)\mathbf{a}(t) = div\mathbf{T}, \tag{24.8}$$

where $\mathbf{a}(t)$ is the acceleration at the point P at time t with respect to the coordinate system $\{\mathbf{f}_1, \mathbf{f}_2, \ldots, \mathbf{f}_{n-1}\}$, $\rho(t)$ is the density of the region at P at time t, and

$$div\mathbf{T} = (\sum_{j=1,2,\ldots,n-1} (\partial \mathbf{T}_{ij}/\partial z_j)), \quad i = 1, 2, \ldots, n-1.$$

As regards the density of the (homogeneous) continuum, let us assume that at time $t = 0$ it is 1. Then,

$$\rho(t) = \{vol(A_n(t))\}^{-1} = \{vol(A_n(0))det(\Phi = (t))\}^{-1}, \tag{24.9}$$

where $vol(A_n(t))$ is the volume of the region $A_n(t)$. According to [13] we have that $vol\{A_n(0)\} = \frac{\sqrt{n}}{(n-1)!}$, thus relation (24.9) becomes

$$\rho(t) = \frac{\sqrt{n}\,(n-1)!}{n}\{det(\Phi(t))\}^{-1}.$$

As regards the acceleration $\mathbf{a}(t)$ in (4.2) , note that $\mathbf{a}(t) = \partial \mathbf{v}/\partial t + \nabla \mathbf{v} \cdot \mathbf{v}$, where the (i, j) entry of the $(n-1) \times (n-1)$ matrix $\nabla \mathbf{v}$ is equal to $\partial \mathbf{v}_i / \partial z_j$ ([4]).

Following the same arguments made in connection with isotropic linear elastic materials and Newtonian fluids, we will seek a stress tensor of the form

$$\mathbf{T} = -p\mathbf{I} + \lambda tr\mathbf{M} \cdot \mathbf{I} + 2\mu \mathbf{M}, \tag{24.10}$$

where p, λ, μ are scalars, $\mathbf{I}$ is the $(n-1) \times (n-1)$ identity matrix and $\mathbf{M}$ is an $(n-1) \times (n-1)$ second-order tensor. It can be verified that neither the model of the elastic solid, nor the model of the Newtonian fluid, (for which the tensor $\mathbf{M}$ depends on the *displacement gradient* and the *velocity gradient* respectively), can in the general case explain the motion (24.7). Then, let us examine the case where $\mathbf{M}$ depends on the *acceleration* gradient. For this purpose, let $\mathbf{S}^{(2)} = (s_{ij}^{(2)})$, $\mathbf{f} = (\mathbf{f}_i)$, and $\mathbf{M} = (M_{ij})$, $i, j = 1, 2, \ldots, n-1$, where

$$M_{ij} = \begin{cases} \frac{1}{4}(z_i^2 s_{ij}^{(2)} + z_j^2 s_{ji}^{(2)}), & \text{if} \quad i \neq j \\ \\ \frac{1}{4} \cdot z_i^2 s_{ii}^{(2)} + \frac{1}{2} z_i f_i, & \text{if} \quad i = j \end{cases} \tag{24.11}$$

It can be easily verified that the matrix (24.11) satisfies the equation of motion (24.8) with $\lambda = 0$ and $\mu = \rho(t)$. It is apparent that more complicated functions of $\mathbf{M}$ could also be used instead of $\mathbf{M}$ itself (given in (24.11)). Then, while referring to the stress tensor of the system, we will mean the tensor $\mathbf{T} = 2\mu\mathbf{M}$ (letting $p = 0$ in (24.10)) generated by the matrix $\mathbf{M}$ of the simple, symmetric form (24.11).

5. THE ENERGY EQUATION

By adopting the assumption $p = 0$, $\lambda = 0$ and since $\mu = \rho(t)$ the stress tensor becomes

$$\mathbf{T} = 2\rho(t)\mathbf{M}. \tag{24.12}$$

Every structure-particle of the HSMS-continuum moves on the hyperplane (Π) according to (24.8), having at some time point kinetic energy K and internal energy U. The rate of change of the kinetic energy can be easily evaluated by taking (24.7) into account.

The rate of change of the internal energy U of the particle at time t can be evaluated according to the formula ([4])

$$\rho(t)\frac{dU}{dt} = tr(\mathbf{T}\cdot\nabla\mathbf{v}) \tag{24.13}$$

Then, taking into account (24.8) and (24.12), (and since $\nabla\mathbf{v} = (\mathbf{S}^{(1)})'$) the rate of change of the internal energy of a structure could be evaluated by (24.13). Thus,

$$\rho(t)\frac{dU}{dt} = 2\rho(t)tr(\mathbf{S}^{(1)}\mathbf{M}),$$

and

$$\frac{dU}{dt} = 2tr(\mathbf{S}^{(1)}\mathbf{M}) \tag{24.14}$$

In the same way as for the homogeneous Markov systems, the quantity $2tr(\mathbf{S}^{(1)}\mathbf{M})$ is negative at every point $\mathbf{z}$, and dissipation of energy takes place [15].

6. THE ENTROPY OF THE SYSTEM

For real materials the internal energy U is in general related to the concept of the temperature T. Thus, in order to draw a parallel between an HSMS and a (non-Newtonian) fluid, we could consider that the variation of the internal energy reflects the variation of the "temperature" of the system. It is apparent that the character of the internal energy and the temperature depends on the character of the system.

Now, the decrease of the internal energy implies loss of energy (heat-transfer) of the HSMS towards the outer environment. Thus, the system becomes "colder". This means that the mobility of the structures-particles becomes smaller in the course of time, or - in other words - the trajectories of the structures are restricted after some time into a small area of E^n (i.e. the neighborhood of the stability point π). In connection with real fluids it seems reasonable to consider the function $T = T(U)$ as an increasing one. Nevertheless, it seems rather inconvenient and arbitrary to define here some formula for this function, and to assign some temperature-value to the HSMS. As a consequence, an exact estimation of the entropy of the HSMS becomes useless, while from the point of view of thermodynamics the variation $\mathcal{E}$ of the entropy is more useful.

If we do not want to pay special attention to the temperature T of the HSMS (it depends on the character of the system) we could assume that it remains constant during the process. Then, the variation $\Delta = \mathcal{E}$ of the entropy can be evaluated as a function of the heat variation ΔQ and - consequently - of the internal energy variation according to the equation

$$\Delta\mathcal{E} = \frac{\Delta Q}{T} = \frac{\Delta U}{T}.$$

Under this consideration, the entropy decrease is proportional to the internal energy decrease. The HSMS cannot be considered with respect to the energy as a closed system, because of the loss of energy to the environment. The loss of internal energy is a necessary condition for the entropy decrease.

References

[1] D.J. Bartholomew, *Stochastic models for social processes*, 3rd edn., Wiley, New York, 1982.

[2] D. J. Bartholomew, A. F. Forbes and S. I. McClean, (1991). *Statistical Techniques for Manpower Planning*, second ed. Wiley, Chichester.

[3] G. S. Davies, (1978). Attainable and maintainable regions in Markov chain control, *Recent Theoret. Developm. in Control*, 371-381.

[4] C. A. Eringen, (1969). *Mechanics of Continua*, Wiley, New York.

[5] R. A. Howard, (1971). *Dynamic Probabilistic systems*, Wiley, Chichester.

[6] D. L. Isaacson, and R. W. Madsen, (1976). *Markov chains theory and applications*, Wiley, New York.

[7] J. Janssen, *Semi-Markov models: Theory and Applications*, (1986). ed. J. Janssen, Plenum Press, New York.

[8] S. I. McClean, (1976). The two stage model for personnel behaviour, J.R.S.S. A 139; 205-217.

[9] S. I. McClean, (1978). Continuous time stochastic models for a multigrade population, *J. Appl. Prob.*, 15, 26-37.

[10] S. I. McClean, (1980). A semi-Markovian manpower model in-continuous time, *J. Appl. Prob.*, 17, 846-852.

[11] S. I. McClean, (1986). Semi-Markov models for manpower planning, in : J. Janssen ed. *Semi-Markov Models - Theory and Applications*, Plenum, New York, pp.238-300.

[12] A. Mehlmann, (1979). Semi-Markovian manpower models in continuous time, *J. Appl. Prob.*, 16, 416-422.

[13] G. Tsaklidis, (1996). The evolution of the attainable structures of a homogeneous Markov system with fixed size, *J. Appl. Prob.*, 33, 34-47.

[14] G. Tsaklidis, (1997). The continuous time homogeneous Markov system with fixed size as a Newtonian fluid, *Applied Stochastic Models and Data Analysis*, 13, 177-182.

[15] G. Tsaklidis, (1999). The stress tensor and the energy of a continuous time homogeneous Markov system with fixed size, *J. Appl. Prob.*, (to appear)

[16] P.-C.G. Vassiliou, (1986). Asymptotic variability ofnon-homogeneous Markov systems under cyclic behaviour, *Eur. J. of Oper. Res.*, 27, 215-228.

[17] P.-C.G. Vassiliou, and A. A. Papadopoulou, (1992). Non-homogeneous semi-Markov systems and maintainability of the state sizes, *J. Appl.Prob.*, 29, 519-534.

[18] A. A. Papadopoulou and P.-C.G. Vassiliou, (1994). Asymptotic behaviour of non-homogeneous semi-Markov systems, *Linear Alg. and Appl.*, 210, 153-198.

Index